Evolution and
Human Sexual Behavior

Evolution AND
Human Sexual Behavior

PETER B. GRAY

JUSTIN R. GARCIA

HARVARD UNIVERSITY PRESS

Cambridge, Massachusetts, and London, England

Copyright © 2013 by the President and Fellows of Harvard College
All rights reserved
Printed in the United States of America

First Harvard University Press paperback edition, 2016
First Printing

Library of Congress Cataloging-in-Publication Data

Gray, Peter B., 1972–
    Evolution and human sexual behavior / Peter B. Gray and
    Justin R. Garcia.
        p. cm.
    Includes bibliographical references and index.
    ISBN 978-0-674-07273-2 (hbk. : alk. paper)
    ISBN 978-0-674-66000-7 (pbk.)
    1. Sex (Psychology) 2. Sex (Biology) 3. Sex. 4. Human evolution.
    I. Garcia, Justin R., 1985– II. Title
    BF692.G687 2013
    306.7—dc23   2012037630

Contents

Preface

> For may we not infer as probably . . . that marriage between near relations is likewise in some way injurious,—that some unknown great good is derived from-the union of individuals which have been kept distinct for many generations?
>
> —Charles Darwin, *On the Various Contrivances by Which British and Foreign Orchids are Fertilized by Insects, and on the Good Effects of Intercrossing*

While Darwin's journals and books inform us of his ideas, his experiments, and his theories, they are largely silent on the most intimate of details. We do know that he was concerned over potentially negative effects of fathering children with his wife and cousin, Emma; the epigraph above reflects this concern, though Darwin's worries were later assuaged through research conducted by his son George. We do know that his marriage was a fruitful union, for he and Emma had ten children. We also know that, despite any fears of inbreeding, he felt a deep love for his wife and enjoyed a long, committed, marriage to her.

We can make some level-headed guesses about the sexual world Darwin inhabited based on the time and place in which he lived. Rather than attempt to personalize, in a sordid and uneasy way, what actually happened in Charles and Emma Darwin's bedroom, let us instead imagine a typical couple of the late nineteenth-century English gentry. Doing so will help illustrate how any given individual's sexual behavior bears the sexual marks of his or her evolutionary ancestors in addition to the cultural environment in which he or she was raised.

Such a generalized person probably had wondered about sex during his or her upbringing, engaged in curious sex play with childhood acquaintances, masturbated during pubertal years, and likely married a long-term partner. Such a person probably had intercourse on a bed, engaged in sex during night hours, presumably in the most typical of sexual positions (the missionary position), and at some point produced offspring. This person likely attempted to shield the children from direct view (and earshot) of parental sex; engaged in less sex during times in which a woman was on her period or late in pregnancy, or in the spell after having given birth to one of her offspring; and this person probably experienced a decline in intercourse frequency with advancing age. These are characteristics of behavior that applied to men and women of the time; with the exception of having sex on a bed, these are also, as we will see, quite typical patterns among people the world over. Yet we will also see variation on these themes (for example, variable liberty granted for masturbation, and variation in potential marital partners).

None of this sexual patterning may strike you as unusual. Now step out of your bipedal shoes and suppose you were a chimpanzee in an African rainforest. From a chimpanzee's-eye view, what would you think of these kinds of human sexual practices? You might have a host of questions. Why have clothes that require removal? Why have sex on a bed? Why have sex away from one's offspring and other adults? As a chimpanzee, you know that sex is easier without clothing that interferes with the act; clothing also impinges on one's ability to readily view the signals that facilitate a sexual encounter, such as a female's sexual swelling or a male's erect penis. Sex also works just fine without any special linen surfaces or candlelight—really, any spot will do. There is nothing wrong with sex outdoors, on the ground, and under sunlight. You would also know that sometimes it is preferable to sequester yourself and a partner during sex (away from a dominant male, perhaps, who might be irked if he caught the two of you), but that most times sex in view of fellow chimpanzees is fine. Well, maybe not if you are a mother whose nearby offspring resent your displaced passions and attempt to separate you from your mate (now that is parent-offspring conflict!). You would also puzzle over the missionary position. As a chimpanzee, most of your sexual behavior entails males mating with females from behind. Yet those strange humans rarely resort to that position.

Our imaginary-chimpanzee thought experiment reminds us that we take many details of human sexual behavior for granted. We may not realize that some of these features of our sexuality beg for explanation. We also may not realize how an evolutionary perspective can help contextualize and explain the specifics of our sexual lives. Our aim in this book is to present a concise, accessible, current, and integrative account of human sexual behavior. We seek to cover a wide human sexual terrain, synthesized within an overarching evolutionary perspective. We seek to combine substance with an enjoyable narrative. We hope the readers who come along on our tour will include academics, students, and the curious coffee-shop customer (several sat near Peter in a Las Vegas Starbucks as he wrote many of the words in this book; we wonder how many may have speculated on what he was actually writing about). And we hope that Darwin himself would have enjoyed this evolutionary perspective on human sexual behavior if he were alive today.

We also recognize our own biases in this exploration—how our own worldviews, family life, and personal and professional experiences shape our views of human sexuality. Peter, the lead contributor to this work, is a biological anthropologist at the University of Nevada, Las Vegas. Justin is an evolutionary biologist at The Kinsey Institute for Research in Sex, Gender, and Reproduction, at Indiana University, Bloomington. This project is our honest attempt to synthesize a large body of divergent literature in an accessible, integrative, and scholarly way. We are both active researchers, with diverse experiences, and genuinely intrigued by studies of human nature. We asked ourselves how our own views might color our interpretations, not to mention how this book may shade others' views of us. With this in mind, we have tried to make the book evenhanded—to avoid advocating any particular cause. Popular science writer Mary Roach put it best in "Foreplay" at the beginning of her book *Bonk: The Curious Coupling of Science and Sex:* "This book is a tribute to the men and women who dared. Who, to this day, endure ignorance, closed minds, righteousness, and prudery. Their lives are not easy. But their cocktail parties are the best."

QUESTIONS FOR THE SEXUALLY CURIOUS

Good science begins with a question. A compelling book should too. We have already raised some questions concerning human sexual behavior, examples of questions that we will address throughout the book. There are many more good questions that readers might ask. Why do sex differences in behavior exist? How do patterns of human same-sex sexuality compare with those of other animals? How does an evolutionary perspective make sense of a world of Internet pornography, breast enhancement, and sex tourism? We could keep going, but these examples reveal the flavor of the kinds of questions we hope to address in the course of this book.

To illustrate the thought process of evolution-based human sexuality research, let's take a Las Vegas example. Many are familiar with the successful advertising campaign that exclaims, "What happens in Vegas, stays in Vegas." The city cultivates a sexual luster, carries the banner of Sin City proudly, and permits unusual natural experiments in human sexuality.

Through a study led by our colleague Michelle Escasa-Dorne, we sought to address several questions concerning men's physiological responses to sexual stimuli while visiting a Las Vegas sex club (Escasa, Casey, and Gray 2011). More specifically, we sought to test whether men's testosterone levels increased during a visit to this sex club, whether men actually engaging in sex experienced greater increases in testosterone compared with those who simply viewed sexual behavior, and whether the magnitude of men's testosterone responses declined with age. To address these questions we asked the men participating in the study to provide saliva samples, from which their testosterone levels were measured. They also answered a series of interview questions. And then they went about their business in a naturalistic setting. (Lest you think we are male-centric, we also conducted a follow-up study on women in the sex club [Garcia, Escasa-Dorne, and Gray, in review], and we'll discuss those findings later in this book.)

Men's testosterone levels increased during the visit to the sex club, and those increases were more pronounced among men engaging in sexual behavior compared with those men only observing. No age-related differences in male testosterone responses were observed. This study might strike one as bizarre, distasteful, or worse. Yet, to our knowledge, it represents the largest human naturalistic study of its kind. The earliest of human studies investigating men's testosterone responses to sexual stimuli, conducted in the 1970s, entailed small sample sizes (those samples of 1 individual, authored by a medical doctor, leave little to be imagined). Further, most lab-based studies (think of Masters and Johnson's sex-in-a-lab studies) have found more robust increases in men's testosterone responses to audiovisual sexual stimuli—pornography—compared with actual sexual behavior. (A European research team may have been the first to study how men's brains respond during manual stimulation by a partner while inside an fMRI machine.) These results tell us that social context matters to human sexual response, including its physiological measurement.

The fact that men respond to sexual stimuli with testosterone increases fits within patterns widespread among vertebrates. Indeed, the so-called challenge hypothesis (Wingfield et al. 1990) suggests that male testosterone levels are elevated in response to male-male competition as well as to sexual behavior, and this framework has garnered significant support from studies of species from fish to dwarf mongooses, rhesus monkeys to chimpanzees. In other words, the findings of this human study place us squarely within the fabric of life, including evolved, adaptive physiological responses to sexual stimuli.

Yet many features of the sex club study are surprising. People have sex in public view. That surprises us, even if it would not surprise a chimpanzee. People sometimes have sex outside of dyadic (couple) relationships. And some people engage in group sex, with their primary relationship partner possibly also "playing" (sex club lingo for sexual activity) or perhaps just sitting nearby and watching. How does this social patterning fit within the evolutionary and cross-cultural scope of human sexual behavior? People have sex in the wee hours of the night. That's strange, considering that Old World monkeys, apes and we humans are largely diurnal (active during day hours), with sensory equipment (eyes and such) designed to assess mate attractiveness under regular sunlight (rather than artificial light). And Old World monkey and ape sexual behavior typically occurs during daylight hours too. Although chimpanzees and bonobos, our closest living relatives, sleep in nests in trees at night, primatologists do not think they're having sex in them (their beds are for sleeping, after all).

So by starting with one example—investigating men's testosterone responses to sexual stimuli—we open doors to all kinds of additional considerations. We place human sexual behavior in wider theoretical and empirical contexts, and we wonder about the wider variation in human sexual expression. These are some of the foundations of this book as a whole.

ANOTHER SEX MANUAL?

There are plenty of books on human sexuality. There are standards, from Darwin's *Sexual Selection and the Descent of Man* (1871) to Masters and

Johnson's *Human Sexual Response* (1966), to the mid-twentieth-century twin hits by Kinsey's team at Indiana University, *Sexual Behavior in the Human Male* (1948) and *Sexual Behavior in the Human Female* (1953). There are more recent wonderful reads, such as Potts and Short's *Ever Since Adam and Eve* (1999), Levay and Baldwin's *Human Sexuality* textbook (2009), and Dixson's *Primate Sexuality* (1998) and *Sexual Selection and the Origins of Human Mating Systems* (2009). In one nineteenth-century classic, *The Sexual Relations of Mankind*, the Italian polymath Mantegazza proclaimed, perhaps ahead of his time, "I am not engaged in writing a book on morality; my book is a page from the natural history of mankind."

A casual glance at contemporary sex books online or in a bookstore reveals that many are how-to manuals, with the *Kama Sutra* available in more versions than the sex positions illustrated in it. Many readers find these books helpful in navigating their own self and sexual exploration, and in finding mutual sexual satisfaction with a partner. However, in the pages to come, we do not provide advice—we are squarely in the business of "how come," not "how to." Of the available contemporary scholarly books on human sexuality, virtually all are health-based textbooks or are focused on a limited aspect of human sexuality at the expense of a wider, integrative framework. Here we seek to fill the gap and provide an eye-opening account of human sexual behavior from an evolutionary perspective. In some cases we may raise more questions than we have been able to answer, but our sincere hope is that in these pages readers will be informed and entertained by the rich patterns and variations that exist in human sexual behavior, and encouraged to ask more questions.

As an overview of the scientifically inspired approach to human sexuality, many books address the evolution and expression of sex differences in mating from an evolutionary psychological or behavioral ecological perspective. These works include Symons's *Evolution of Human Sexuality* (1979), Fisher's *The Sex Contract* (1982), Townsend's *What Women Want, What Men Want* (1999), Low's *Why Sex Matters* (2000), and Buss's *Evolution of Desire* (2003). Others highlight the cross-cultural record of human sexual diversity, such as Gregerson's *The World of Human Sexuality* (1994), or feature one or a few rich ethnographic case studies of sexuality, such as Herdt's work on the Sambia (1981), Malinowski's *Sexual Lives of Savages* (1929), or Suggs's *Marquesan Sexual Behavior* (1966). Still others, like Diamond's 2008 *Sexual Fluidity*, tackle intriguing aspects of female sexuality, but within Western samples and history. Some recent books tackling the evolution of same-sex sexuality approach an important and misunderstood topic but are imbalanced in their emphasis on same-sex sexuality from an evolutionary perspective (Kauth's *Handbook of the Evolution of Human Sexuality* [2007]) or uneasily mix politics with science (Roughgarden's *Evolution's Rainbow* [2004]). One of our preferred human sexuality textbooks, Bolin and Whelehan's *Human Sexuality* (2009), confounds a straightforward evolutionary approach with redundancy and unnecessary postmodernist jargon. Many other excellent works exist that primarily feature one or even several focused aspects of human sexuality, such as love (Fisher's 2004 *Why We Love*), attractiveness (Swami and Furnham's 2008 *The Psychology of Physical Attraction*), or the debates surrounding the function of female orgasm (Lloyd's 2005 *The Case of the Female Orgasm*). Other empirically rich books are clinical in their orientation (Bancroft's 2008 *Human Sexuality and Its Problems*). Some of the best-read books on human sexuality are informative but too thin in substance (Diamond's *Why Sex Is Fun* [1997]); some are immensely fun but too popular in their style and examples, such as Moalem's *How Sex Works* (2009), and Roach's *Bonk* (2008). And others are both popular and terribly misleading, in pursuit of a cause the evidence doesn't support (such as Ryan and Jetha's 2010 *Sex at Dawn*).

We have read all of these books, and many more, and we will cite most of them, where appropriate. But we believe that the state of sexual science still has a space for this book. We have attempted to make *Evolution and Human Sexual Behavior* integrative, accessible, concise, and current. We do not say radical things about human sexual behavior. We do not emphasize new theoretical or empirical research that suddenly illuminates human sexuality in an altogether new way. While we sprinkle some of our own sexuality-related research throughout this book, most of what we say is represented in the many scientific books and articles floating among us, just not in the form that we present: all together under a single cover.

The lack of a singular book like this one was painfully apparent when one of us began plotting out which books to assign for a class on the evolution of human sexuality (we have both taught classes on human sexuality, Peter in anthropology, Justin in health sciences). We found that we liked all kinds of elements in the works available, but none suited our taste perfectly for this classroom purpose. Two works were most aligned with our aims: Ford and Beach's *Patterns of Human Sexual Behavior* (1951), and Frayser's

Varieties of Sexual Experience (1985). Both of these books cover the breadth of evolutionary theory, comparative animal research, cross-cultural sexuality, sexual behavior across the life course, and mechanisms of sexual behavior, all within a relatively concise format. However, both are quite dated. Because it covers a similar breadth of theory and research and in a current, accessible mode, we think *Evolution and Human Sexual Behavior* will find readers among the curious, students, and teachers of human sexuality, especially when an evolutionary perspective is the backbone to a course, whether in anthropology, psychology, biology, or the health sciences. A wider audience interested in an evolutionary and integrative account of human sexuality will also, we hope, enjoy this book.

AN EVOLUTIONARY PERSPECTIVE ON HUMAN SEXUAL BEHAVIOR

There are important intellectual reasons for adopting an evolutionary perspective on human sexuality. One is that it puts humans directly in the fabric of nature rather than apart from it. We are subject to the same evolutionary processes as any other species, although we also play an inordinately powerful role in creating our own selective environment (what biologists call niche construction). We do not have to invoke any special force or magic to account for who we are and how we behave.

Through the lens of evolutionary science, we also have an account of human sexual behavior that must be consistent with the known laws of physics, chemistry, and the biomedical and social sciences. Any claims made about unique human sexual practices or beliefs must be aligned with what is known from these wider scientific realms. This will prove especially important when we put human sexual behavior in comparative perspective; if theory in nonhuman animals indicates that sexual dimorphisms (characteristics that differ between the sexes) are most parsimoniously explained by sexual selection, then that is where we need to start in understanding human sexual dimorphisms (this will be relevant to making sense of why, generally speaking, females have more fat than males, and why males but not females can grow beards). Further, when we unpack the processes by which females and males are created and differentiated during development, we build on the mass of findings from the biomedical sciences in specifying the genetic and hormonal mechanisms by which females and males arise. We do not get to just argue about socialization, not without weaving those socializing influences from an individual's environment into the developmental process.

An integrated and evolutionary perspective on human sexual behavior helps accomplish what E. O. Wilson called, in his 1998 book of the same title, "consilience." This is an approach that seeks to unify rather than divide the nature of human scientific understanding. No single field (whether anthropology or sociology or biology or psychology or medicine) has a monopoly on our understanding of human sexuality. In this approach, all pieces of the scientific puzzle must ultimately be compatible (of course, there will always be disagreements about exactly how the pieces fit together, but at least in outline the enterprise makes sense). Mate-choice surveys in different populations around the world tell us something about sexual preference; neuroendocrine mechanisms of human and nonhuman animal sexual response tell us what happens in the body during sexual arousal and behavior; in-depth ethnographic studies of same-sex sexual behavior give us insight into the cultural context in which they are embedded. An overarching evolutionary perspective helps put all of these pieces together in a coherent, sensible way. The terms (or jargon) used to describe this effort, and our attempted explanations, also ideally facilitate rather than impinge on interdisciplinary scholarship.

There are many good reasons for trying to understand human sexuality as best and realistically as we can. New findings, for example in genomics and epigenetics, require interpretation within an evolutionary scope. Predicting the future—say, of outbreaks of sexually transmitted diseases— demands the best of educated guesses. Health considerations (such as the impacts of aging or obesity on sexual function), reproductive reasons (understanding the biology of infertility to aid reproduction and fertility efforts, to develop effective contraceptives), legal bases (informing the politics of birth control, same-sex marriage, childhood sexuality), economic rationales (why cosmetics or pornography yield huge financial stakes), and simply personal interest all can inform why we might care about a scientific, evolution-grounded, and integrative view of human sexuality.

THEMES OF THE BOOK

A sketch of this book begins, like this preface, with Charles Darwin. An evolutionary framework, and sexual selection theory in particular, guides our discussion of the roots of human sexual behavior. An understanding of sex differences traces to the fact that females typically serve as the reproductively limiting sex; accordingly, female reproductive success tends to be ultimately constrained by access to resources, while male reproductive success tends to be ultimately constrained by reproductive access to females. From those humble beginnings, general trends of male-male competition and female choice arise and shape other features of anatomy, physiology, and behavior. Furthermore, the nature of competition and even sexual unity can take more subtle forms, including what are referred to as sperm competition and cryptic female choice. We survey some of the comparative evidence illustrating the ways in which reproductive anatomy and physiology, secondary sexual characteristics (such as male musculature), and behavior (from sexual coercion to parental care) play out in other species between females and males, with an emphasis on nonhuman primate examples.

While humans carry the sexual baggage of a hominoid, or great ape, our lineage has undergone shifts in the past few million years in the expression of its sexuality. Postulating ancestral mating behavior in multi-male, multi-female groups, shifts toward long-term sociosexual bonds likely emerged within the past two million years of human evolution. In other words, our sexual behavior has undergone some remarkable and recent evolutionary turns. Among hunter-gatherers and, indeed, most of humanity today, most sexual behavior occurs within these long-term pair-bonds, placing an emphasis in mate choice on partner characteristics compatible with long-term relationship viability and fertility. Anatomical, physiological, demographic, and genetic data (for example, DNA paternity testing) all point toward relatively low sperm competition pressures in our recent ancestors.

Against this evolutionary backdrop, expressions of human sexuality vary across and within societies over time. The specifics of sociosexual bonds can range from the long-term monogamous to polygynous, from frequent patterns of mate shifting to lifelong unions. Features of subsistence ecology and demography help account for some of that variation. The frequency and specific forms of sexual behavior vary across societies; also varying are tolerance of extrapair partnering, prostitution practices, and attitudes toward masturbation. An evolutionary framework can help explain specific sexual activities within local social contexts. For instance, matrilocal societies foster more open female sexual expression, during adolescence but also in adulthood. As another example, male same-sex sexual behavior is more common than female same-sex sexual behavior, both cross-culturally and among nonhuman animals.

The developmental origins of sex differences arise during sex differentiation. Sex differentiation also initiates processes that give rise to some of the sociosexual variation within the sexes, including developmental influences in which elevated perinatal androgen levels may predispose some females toward more masculine behavior and same-sex sexual behavior. From there, we might leap forward to the juvenile years. In humans as in other primates, juvenile years are times of sexual interest and exploration, although cross-cultural differences place more or fewer restrictions on that sexual exploration. Then puberty transitions individuals into the potential reproductive realm, upping the stakes for competition, mate choice, and child rearing. Across these eventual reproductive years, pregnancy and postpartum periods are associated with reductions in the frequency and variety of sexual expression, although the specifics of these patterns also vary (for example, we see shorter postpartum sex taboos among more polygynous societies). The age-specific effects of fertility and mortality that operated on our ancestors resulted in declines in reproductive function in both females and males with advancing age. Across the life course, neuroendocrine mechanisms—studied through various methods, from functional imaging of males exposed to erotic videos to assays of saliva samples from females for assessment of steroid hormone levels—both orchestrate and respond to the ongoing processes of sexual expressiveness.

This is the body of human sexuality work we will cover. While some of it may seem more ripe with jargon than passion, we try to present all of the material in a comprehensible and defined fashion. Still, we do not avoid the substance essential to a worthwhile read. To begin, we scan the evolutionary landscape for the roots of human sexuality as we turn to Chapter 1.

Evolution and
Human Sexual Behavior

CHAPTER ONE

The Evolution of Sex, Sex Differences, and Human Sexuality

> The human being, like the immortals, naturally places sexual intercourse far and away above all other joys—yet he has left it out of his heaven! The very thought of it excites him; opportunity sets him wild; in this state he will risk life, reputation, everything—even his queer heaven itself—to make good that opportunity and ride it to the overwhelming climax.
>
> —Mark Twain, *Letters from Earth*

WITH DUE RESPECT to Mark Twain, humans are hardly the only creatures whose passions are stoked by the prospect of sexual intercourse. Indeed, not far from Las Vegas, where one of us lives, some unusual vertebrates scamper around the deserts of the U.S. Southwest and northern Mexico. Among the many species of whiptail lizard are some that have abandoned sexual reproduction. These species reproduce clones of themselves—that is, they are parthenogenic (Nelson 2011).

Among these parthenogenic whiptails, females live without males, and females produce more females. However, there are some interesting twists in how they do this. Females must go through the mating motions with other females in order to reproduce. They engage in mating postures with other females, who play the role of males—one female mounts the other and they pseudocopulate in order to facilitate the cloning process and egg development.

Why have some species of whiptail lizard evolved without sex? Genetic research suggests such parthenogenic species represent the amalgamation of sexually reproducing species. These parthenogenic species have double the number of chromosomes. This suggests a mechanism by which sexual reproduction is hindered, but it does not tell us whether parthenogenesis has been favored by selection. Yet consideration of the lizards' desert ecology could suggest some adaptive reasons for just such a selection process: imagine how difficult it is to find a potential whiptail mate, let alone to survive. The search costs of mate seeking, especially in novel terrain, could act against sexual reproduction. Indeed, among some other vertebrates that fluctuate between sexual and asexual reproduction, such as the island-dwelling Komodo dragon, the very real challenges of finding a mate might favor an ability to reproduce asexually.

THE EVOLUTION OF SEXUAL REPRODUCTION

The asexual reproduction we see in parthenogenic lizards is rare among vertebrates (Roughgarden 2004). Instead, sexual reproduction is the norm and has evolved repeatedly on Earth, in many and various lineages of plants and animals. For this reason it must be the case that sex offers selective advantages that outweigh its costs. Notwithstanding recent biomedical success with cloning (actually only partial success, since cloned mammals live shorter lives than their noncloned counterparts), which makes sexual reproduction unnecessary, human sexual reproduction is part of a legacy extending back at least a billion years.

So just what are the adaptive benefits of sexual reproduction? For that matter, what are the costs?

The costs spill out whenever you see excess male mortality (deaths related to mating competition, as among adolescent males when they do things that reduce their survival odds) or, less dramatically, the routine costs of contemporary human courtship—money spent on things like flowers, chocolates, jewelry, and romantic nights out, or the other expenses of seeking sexual partners. Were it not for the need to reproduce sexually, organisms, including people, would have no need to devote time and effort to seeking, courting, and maintaining mates. Were it not for the need to reproduce sexually, organisms, including people, would be less likely to engage in same-sex competition for mates, competition that can range from hurtful jealous bickering to physical aggression. All of this comes under the cost of one of life's most primitive, complex, and essential pleasures—sex.

Another major cost of sexual reproduction is the loss of half your genetic legacy. Sexual reproduction requires that your gametes mix with those of another organism to produce an offspring who carries both your genetic heritage and your mate's. Thus, sexual reproduction effectively serves as a 50 percent estate tax on one's genetic legacy, a significant cost to you as an organism. For such costs, organisms surely must receive significant benefits, but what are they?

In fact, sexual reproduction offers several major potential benefits (Geary 2010). One is that harmful mutations (changes in the genetic code) can be weeded out. In other words, sexual reproduction can take deleterious mutations, shuffle them, and yield offspring that may or may not have them. Without this shuffling, an asexually reproducing organism would be stuck with passing the harmful mutation to its offspring, perhaps to the demise of its own lineage. Another benefit is the generation of novelty (Rice 2002). Sexual reproduction enables a shuffling of the genetic deck, creating novel combinations that may prove beneficial in the face of shifting selective pressures. To be sure, the shuffling also breaks up winning genetic hands, but that cost can be offset by the benefits of novelty.

Imagine you are an animal who is encountering a new landscape that exposes you to differences in temperature, precipitation, predators, disease, and other environmental factors. Sexual reproduction may enable adaptation in the face of these new environmental pressures. An organism's genome does not know which genetic contributions would be best suited to a new environment, so the production of a new variety of combinations may yield blends that make one organism better adapted than another. As evidence that this hypothetical scenario describes biological reality, we have the example of some species, such as aphids, that vacillate between sexual and asexual reproduction, reserving sexual reproduction for times when they must cope with greater environmental uncertainty (Roughgarden 2004).

Multicellular organisms can also face eternal back-and-forth challenges wrought by pathogens (Hamilton and Zuk 1982; Zuk 2007). As pathogens reproduce themselves in huge numbers and with rapid generation times, organisms may survive by employing an array of defenses. One major defense takes the form of the adaptive immune system. Sexual reproduction enables the production of novel immune markers that, by chance, may be beneficial when challenged by an unpredictable pathogen. This is why the immune systems in humans and other organisms are among the most rapidly evolving components of our physiology, and why in genomics research we repeatedly find that the immune system changes and adapts rapidly: it must do so to fend off the unending variety of pathogens that use humans and other organisms as resources or way stations on their own route to reproductive success.

ORIGINS OF SEX DIFFERENCES

If adaptive cost-benefit ratios favor sexual reproduction, does that translate into sex differences in the phenotypes, or observable characteristics, of an organism? In other words, how does sexual reproduction lead to distinctly different sexes, such as the males and females around us (and variations on these general two sexes), and to the sex differences in anatomy, physiology, and behavior that we also see around us? To be sure, there are species in which sexual reproduction does not require or produce two distinct sexes (for example, clownfish, like the stars of the animated movie *Finding Nemo*, are sequential hermaphrodites and can change from male to female during their lives, although the movie ignored this biological reality). There are also species in which sexual reproduction requires and produces more than two sexes (e.g., some slime molds), but these examples are far removed from our human-centered world (Low 2000).

Through classic laboratory experiments conducted with fruit flies in the mid-twentieth century, Angus Bateman (1948) helped lay the theoretical and empirical foundation for an understanding of sex differences. Bateman documented that reproductive success in female fruit flies provided with adequate food differed little whether the flies had one, two, or three male mates. The male who mated with her first apparently provided enough sperm to fertilize the female's eggs, such that additional sperm from additional males had little reproductive benefit to her (there was no difference in the number of fly hatchlings). In that lab setting, the female's offspring may not have needed adaptive immune differences or other variations to ward off environmental challenges. While these findings (that a female's reproductive success did not differ by the number of mates she had) may not seem remarkable, they differ from those observed in male fruit flies. Male fruit fly reproductive success increased when males were allowed access to an increasing number of female mates. Males with three mates had a greater number of offspring than those with two mates, who in turn had more off-spring than those with one mate. Perhaps eventually males could run out of sperm—or steam—and thus the marginal fitness benefits to having a greater number of mates would gradually decrease to zero, but that was not the case over the range of mates considered.

These experiments showed that female reproductive success was associated with sufficient resources, such as food, whereas male reproductive success depended on access to females. This result became the basis of what biologists call Bateman's principle. Research with many organisms, not just fruit flies, suggests that Bateman's principle applies quite widely in nature, even extending to humans (Andersson 1994; Geary 2010).

Bateman's (1948) original report suggested that the sex difference in gamete investment is the root of the sex difference in reproductive constraint. Females typically invest more in their gametes (their eggs), than males do in theirs (their sperm), leading to these asymmetric relationships in reproductive constraint (the fact that females tend to be ultimately constrained by access to resources, such as food, and males by reproductive access to females). In the years following Bateman's initial observation scholars have refined and elaborated on this logic.

Robert Trivers (1972), in a book chapter that has since become a classic in the literature, suggested that reproductive asymmetry traces to relative parental investment (rather than gamete size). The sex that invests more in parental care tends to be the reproductively limiting one, while the other sex tends to exhibit more competition for reproductive access. This insight helped move the basis for sex differences closer to humans' phylogenetic home, among mammals—where parental care is common—but it also had its limitations. One of these limitations was that it did not easily account for exceptions, such as mouth-brooding frogs, in which males might provide the bulk of parental care but also be more apt to compete among themselves for reproductive opportunities with females. Accordingly, Clutton-Brock and Vincent (1991) advanced the concept that "potential reproductive rate" was the better foundation for reproductive asymmetries; the sex with the lower potential reproductive rate is the one over which competition will occur. Even more recently, Hanna Kokko and colleagues have hearkened to demographic factors such as adult sex ratios that can change the gradient of sexual selection (Kokko and Jennions 2008). In contexts with more females than males, females may compete more among themselves over males, while a sex ratio with more males means more male-male competition over females.

Amid the layers of theory, we find the empirical foundation of sex differences. The sex whose potential reproductive rate is lower tends to be the one whose reproductive success is most closely tied to resources such as food, and the sex whose potential reproductive rate is higher tends to be the one whose reproductive success is most closely tied to competition for reproductive access. Applying this to mammalian sexual behavior, imagine a female investing in gestation, lactation, and other forms of parental care. Compare that with a male investing in sperm that fertilize her egg. The female's intensive forms of investment will typically mean that female mammals, within a given species, are reproductively limited by access to resources such as food. Males within a given species of mammal are reproductively limited by access to females, placing greater emphasis in their lives on competition with other males in courtship and in access to females.

SEXUAL SELECTION

While Charles Darwin had discussed sexual selection in his classic 1871 book, *The Descent of Man, and Selection in Relation to Sex,* all of the observations of fruit fly sexual behavior, and the theorizing about potential reproductive rates, came long after his death. Yet these later observations and theories only formalized Darwin's original theory.

Sexual selection theory arises in part from the observation that many traits seem detrimental to reproductive success. Why does a peacock display a magnificent train (sometimes mistakenly called a peacock "tail")? Wouldn't this same train help a predator find it, rather than a less flamboyant peacock or a peahen? Wouldn't this same train be a curb to efficient walking or flying? Wouldn't this train be a waste of the energy required to make those beautiful feathers shine? How, in the name of natural selection, could such a trait evolve if it seemed to have such downsides?

As Darwin puzzled over traits such as the peacock's train, the bird-of-paradise's coloration and mating dances, female chimpanzees' sexual swellings, and male deer antlers, he surmised that they were produced by competition for access to mating opportunities. As members of one sex competed with each other for access to the other sex, they might evolve traits that would improve their chances of success. Thus, the male stag beetle's weapons (large mandibles for fighting other males) are the product of sexual selection; his weapons are not designed to help him find more food or to deter predators, but to improve his chances of reproductive success.

Darwin identified two types of sexual selection: intrasexual selection and intersexual selection. Intrasexual selection refers to selection within members of the same sex, whereas intersexual selection refers to selection between the sexes. Darwin emphasized male-male competition as the main form of intrasexual selection, and female choice as the main form of intersexual selection. He saw many traits, like those horns, antlers, and other armaments in mammalian males, as the product of male-male competition. He suggested that the peacock's train could have been favored by female choice. If females preferred mating with males possessing more colorful feathers or elaborate trains, then their mating choices could also drive these traits to excess, to the point where they might even be counterproductive to survival. Later scholars built on this idea to demonstrate that many traits, such as a peacock's train, are "honest indicators" of an individual's overall health. Such signals indicate presence of disease or unfavorable genetic mutation, they might reflect quality of diet and thus foraging ability, and they might demonstrate the individual's vigor, its ability to put on a flashy show and still avoid depredation (Zahavi and Zahavi 1997).

Sexual selection can also lead to other patterns of sexual behavior besides male-male competition and female choice. In many species we see female-female competition (Rosvall 2011). The female sexual swellings that Darwin observed can be viewed as one product of female-female competition, with females displaying such vivid mating advertisements in order to benefit from enhanced mating opportunities. Male choice clearly operates too, especially when males provide indivisible, hard-to-obtain resources such as food: in these cases, males may seek to optimize investment of their limited resources in the female or females who will provide the greatest return. And as recent work by scholars such as Hanna Kokko reminds us, the demographic specifics of a population may modulate the specifics of intrasexual and intersexual selection (Kokko and Jennions 2008). Change the sex ratio dramatically, for example, and you can also change the ways in which competition plays out.

Sexual selection can account for many differences between the sexes, or sexual dimorphisms (Andersson 1994; Geary 2010). Although some sexual dimorphisms may be related to differences in behavior that relate to survival (for example, when females and males feed on different resources), the kinds of dramatic ones we have highlighted so far in this chapter—those large male antlers, for example—trace instead to sexual selection. Differences in body size dimorphism in association with mating systems are thus thought to represent, in part, the workings of sexual selection. One-male polygynous species (in which a male has several female mates) tend to have extremes of body-size sexual dimorphism. Multi-male, multi-female species (in which multiple males and females live and mate together) tend to be characterized by lower body-size sexual dimorphism. And monogamous species (with mated pairs of one male and one female), are relatively monomorphic, or lacking in sex differences in body size (Dixson 2009). While other factors may be at play (for example, larger female body sizes may be favored in order to enhance fertility [Ralls 1976]), larger males within a species generally seem to use that extra weight during male-male competition. So for many overt sex differences in anatomy, physiology, and behavior, we wonder whether those differences owe their existence to sexual selection.

As is true of many scientific theories, exceptions can help prove the rule. The exceptions, in cases of sexual selection, are sex-role reversals (Eens and Pinxten 2000). These are scenarios in which typical patterns are reversed, accompanied by changes in the kind of sex differences we would expect to see. Take jacana birds or Wilson's phalarope as examples. In most species of birds, males and females are similar in their size and coloration; in fewer, males (like male peacocks, or the turkeys served on many Thanksgiving tables) are more colorful and larger than females. But among jacana birds and phalaropes, the females tend to deposit their eggs with males, who subsequently provide the bulk of parental care. Females in these species also tend to be larger and more colorful than males. The interpretation is that males in these sex-role-reversed species are the reproductively limiting ones. Accordingly, females compete more among themselves than do males, and males practice more mate choice, resulting in females evolving the kinds of traits—such as colorful feathers—that enhance their chances of reproductive success.

Are there any sex-role reversals among mammals? No. The more extreme sex differences in potential reproductive rates in mammals seem to have prevented sex-role reversals from evolving. There is no ready mammalian counterpart to jacanas or, for that matter, other well-known sex-role reversers, such as pipefish and sea horses (male sea horses go so far as to become pregnant and give birth). Among mammals, the closest thing to a sex-role reversal is seen in the spotted hyena.

Possessing enlarged clitorises (about the size of hyena penises), which led Aristotle to speculate that they were hermaphrodites, female spotted hyenas are also slightly larger than their male counterparts, and they engage in extreme female-female competition, including over territory and the carcasses of their favorite hunted or scavenged prey (Glickman et al. 2006). Female spotted hyenas give birth through their clitoris. Many females die while giving birth, and offspring mortality is even higher, especially for a female's first litter. The functional reasons for their heightened female-female aggressiveness and body size are controversial, as are the developmental and endocrine bases of hyena sex differences (more on this in Chapter 4). And while spotted male hyenas slink around the females, males do not provide parental care, affirming that this is not a true sex-role reversal.

SPERM COMPETITION AND CRYPTIC FEMALE CHOICE

One of the few biological processes that Darwin did not say anything about is the reproductive competition waged during and after intercourse. This includes competition between the ejaculates and genitalia of males, as well as the processes by which a female's anatomy and physiology "choose" a potential victor to fertilize one of her eggs. Much of the evidence for this competition lies under the covers of science—in observable features of reproductive anatomy and physiology, in areas of science now referred to as sperm competition and cryptic female choice.

Females may evolve traits that enhance their ability to "choose," but males in turn evolve traits that give their own sperm the best chance of reproductive success. In some species we see enhancement of the male seminal fluid, including the various components of the ejaculate (water, proteins, lubricants) in addition to sperm (which is the only part of the ejaculate that carries genetic information). In wolves (and even some large domestic dogs, as many a worried pet owner has learned), the male penis continues to swell during mating so that the male becomes stuck inside the female following ejaculation. Although this may appear to be painful for the male wolf—and may make some men cringe—it ensures that no other males can attempt to mate with that female until the stuck male's sperm has had some time to travel through the female reproductive system. Indeed, all kinds of interesting traits in males would otherwise be puzzling without the evolutionary lens provided by studying the effects of sperm competition.

Among the phenotypic highlights of sperm competition are the largest testicles in the world (belonging to a right whale), which weigh nearly as much as a small car (Birkhead 2000), and superlong sperm cells produced by male flies that, when stretched out, are many times longer than the male's own body (the equivalent of a human sperm stretching the length of a football field). On the less well studied female side—we unfortunately have far fewer examples—we know that in some species females have developed extended oviducts (tubes connecting ovaries, where eggs are released, to the uterus, where fertilized eggs implant), with bends and twists that present a challenging obstacle course (and a good way to "choose" a strong candidate) on the way to the egg (Eberhard 1985).

Research on sperm competition began in the early 1970s with studies of insects, but it is findings on relative testicle size in mammals that have attracted the most attention. All else being equal, relative size of male testicles should be related to body size; larger animals, within a given lineage, should tend to have larger testicles. But what sperm competition theorists predicted and have found support for—in cetaceans, Australian marsupial mammals, and primates too—is that species in which females mate with multiple males around the time they are ovulating are also species in which males have relatively large testicles (Smith 1984). The idea is that in order for their sperm to compete within a female's reproductive tract, males must produce a larger volume of ejaculate. For that, having larger testicles helps. We will come back to this line of evidence in several places in the book, but suffice it to say that humans and our closest primate kin exhibit notable differences in relative testis size that argue for species differences in the magnitude of sperm competition pressures.

As for cryptic female choice, this concept first took shape in Eberhard's 1985 book, *Sexual Selection and Animal Genitalia*. There he highlighted research on insects but also described traits in other groups that seemed to trace to cryptic female choice. He described an obvious point: in nearly all animals with internal fertilization, males have the intromittent organ (sea horses being the striking exception, with females having the intromittent organ used to deposit eggs in males' pouches), whereas females have an anatomy and physiology that screens male sexual behavior and gametes to try to do what is best for their eggs. Across various species of animals, some anatomical and physiological manifestations of cryptic female choice include the abilities to store sperm until needed (like many species of cats can do), to alter the rate of sperm transport within the reproductive tract depending on the female's physiological response to mating, and to control the duration of copulation in order to alter the composition of a partner's ejaculate and, in turn, the probability of fertilization. As a mammalian example of cryptic female choice, in species where females mate with multiple males around the time of ovulation, these same females tend to have longer oviducts than females of species that do not mate with multiple males (Dixson 2009).

LIFE HISTORIES, MATING STRATEGIES, AND MATING SYSTEMS

Any organism faces budget constraints. Challenged by limited time and energy, an organism channels those limited resources into what life-history theorists refer to as competing expenditures on growth, maintenance, and reproduction (Hawkes and Paine 2006; Muehlenbein 2010; Stearns 1992). For a mammal like ourselves, growth tends to end around adolescence, but there are clearly species and population differences in growth trajectories. Maintenance refers to investments, for instance in the immune system and in behavior, designed to keep an organism alive. Reproduction can be divided into mating effort (consisting of competition, courtship, and mate guarding) and parenting effort (including its variable forms, such as lactation, carrying, and protecting young). Because of limitations, trade-offs may occur among these competing expenditures. As one example, reproductive effort may come at expense to growth; adolescent women who become pregnant and give birth illustrate this precise allocation challenge of limited energetic resources: should the mother divert resources to a gestating fetus or her own growth?

Females and males face different life history allocation challenges based on the reproductive asymmetries we have specified, and this can vary among lineages. The life history expenditures that maximize a female's reproductive success (measured by the health and number of offspring she bears who reach reproductive age) may differ from those that maximize a male's reproductive success. Where extra male bulk pays in successful male-male aggression, males may invest more in growth and less in maintenance than females of the same species, a pattern consistent with the larger male body sizes of most terrestrial mammals (Andersson 1994). Among many spiders, like black widows, females are larger than males; the extra female investment in growth probably allows her to lay more eggs that survive to become the next generation of widow-makers. Where offspring survival depends on maternal care, as is the case among mammals, females may have a greater payoff to staying alive to help care for them, and accordingly female mammals have generally more sensitive and active immune systems that serve interests of maintenance and survival.

Female mammals place a considerable investment in reproduction. Specifically, they invest heavily in parenting effort, with cyclic efforts (such as fluctuations across the ovulatory cycle) devoted to mating effort. Females may grow for as long as is necessary to be ready for the energetic and social challenges of reproduction. When contrasted with their counterparts, male mammals spend their reproductive effort differently, investing more in mating effort than parenting. Males may devote more effort to growing in order to add traits such as size and muscle that will be useful in mating contexts. Males may ratchet down their investment in maintenance at the same time they escalate effort for high-stakes mating competition.

How an organism plays out its life history allocation can be referred to as a strategy. Among mammals, female life history strategies emphasize parenting effort, whereas male life history strategies feature mating effort. These coarse generalizations mask much more species and population variation, including the mixed mating and parenting effort in specific contexts and life history stages (that is, at different ages). These strategies are also not necessarily conscious ones. Indeed, humans may be unique in occasionally reflecting consciously on issues surrounding these kinds of strategies ("Should we wait for my promotion before trying for our first baby?"), whereas for animals generally the term *strategy* is simply meant to characterize a pattern of investment that seems designed for some outcome. While a woman may consciously decide to engage in mating effort with someone she cares for, she may simultaneously employ contraception to avoid an unwanted expenditure of resources on parenting.

Female mating strategies, as Darwin pointed out, often entail considerable mate choice. Given that female reproductive success tends ultimately to be constrained by access to resources, it makes sense that females should scrutinize their mate or mates to maximize their own reproductive success (Buss 2003; Clutton-Brock and McAuliffe 2009). In practice, this means that females are expected to favor mates who provide benefits such as protection (against predators, against other males, to benefit herself and perhaps her offspring), resources (food, territories containing food), and genetic variation among offspring (especially in the face of pathogen pressures). Depending on the species and context, this may entail females choosing a single mate at a time or it may favor extreme randiness, such as a female chimpanzee who may mate with virtually every male in her social group across her cycle, even if she preferentially mates with the dominant male closest to her ovulation time. Among muriqui monkeys of South America, where females tend to have many mates, one female was observed to mate with fifteen males over several days, some of whom waited in line for their turn to copulate with her (Strier 2011).

One scheme that captures further nuance in female mating strategies considers distinct patterns of proceptivity (females actively seeking a mate or mates), attractivity (a female being a sexual stimulus to a potential mate), and receptivity (a female accepting a potential mate's mating intentions) (Beach 1976). When the time is right, such as around the time of ovulation and when a desirable mate is available, females can exhibit a lascivious proceptivity, all the best to ensure that a precious egg does not go to waste, or unfertilized. Females may exhibit proceptivity at other times when benefits favor it (for maintenance of a relationship with a mate, for example, something that we will argue is decisive in humans). Strategic deployment of receptivity can similarly enhance female reproductive success: accepting the advances of a desirable, socially dominant male who might provide benefits to her offspring, while she shuns the advances of other males offering fewer such benefits. This scheme assumes that females are as strategic as males—so much for the "coy" female of a past literature in natural history (see Hrdy 1981).

Male mating strategies are expected to be less discriminating than those of females (Buss 2003; Geary 2010). The reproductive costs and benefits may favor males seeking to mate with many females, especially if the costs of doing so are low. However, males may also face limitations that result in channeling their efforts toward a specific mate or mates to maximize their reproductive success. Matings with a fertile female may be worth greater effort. Mating with a long-term partner may be pursued instead of a wider array of mates if this would jeopardize that ongoing partnership. We will later discuss why for humans a long-term partnership is so critical to survival, and why this puts us on a unique evolutionary and sexual trajectory.

Put it all—or at least female and male mating strategies—together, and the result is a mating system, the interaction between female and male interests (and perhaps others, if kin or competitors have their say). In some contexts, female interests may get the upper hand; in others, male interests may have greater expression. In many primates, males attempt to sexually coerce females, narrowing the breadth of female choice (Muller and Wrangham 2009). Altogether, this is sexual politics. Females and males enter the arena with their own agendas and they are not the same, even if certain arenas foster greater alignment than others.

Mating systems can be defined in several ways, depending on which stakeholder's view one takes (Geary 2010; Low 2000). In polygamous mating systems an individual male or female mates with multiple partners. Polygamous systems can be polygynous (a male with multiple females) or polyandrous (a female with multiple males). Monogamy refers to a single female and single male mating. A multi-male, multi-female system involves more than one female and more than one male mating.

Mating system terminology seems simple enough, but it has further complexities. Most reports of human polygamy really are cases of polygyny, such as we would see in a sampling of fundamentalist Mormon communities in the United States. More substantively, the meaning of these terms depends on perspective: whether the perspective is synchronic (looking at a slice of time) or diachronic (across time). A monogamous relationship followed by another monogamous relationship with a different mate followed by another monogamous relationship with yet another mate is really serial monogamy, or polygyny if seen from a diachronic perspective. A male may be mated polygynously, but if the several females he mates with mate only with him, from their vantage they are monogamously mated; in such cases, the term *polygyny* usually is used, just the way it is used in describing gorilla sexual behavior, as we shall see. Further, widespread availability of genetically based paternity testing in humans and in other animals has revealed that many mating systems that had been labeled on the basis of observed social behavior really should be labeled differently based on unobserved genetic realities. Research on many species of birds and some mammals—for instance, gibbons (more on them also to come)—revealed that socially monogamous individuals were raising offspring sired by extrapair mates (Birkhead and Moller 1998).

Phylogenetic realities of nonhuman mating systems have implications for our understanding of our own mating systems. Since about 90 percent of bird species are socially monogamous, research on their mating and parenting behavior has stimulated theoretical and empirical work in humans. For example, in systems of cooperative breeding (where many family members help raise offspring), available opportunities for reproduction may determine whether an adolescent bird stays to help its family or flees the nest to start its own family. Does the same apply to people (such as those who stay at home longer during an economic recession)? Parallels at this level build on analogy or convergence. Among mammals, however, more than 90 percent of species are socially polygynous (Clutton-Brock 1991). The contrast between mammals and birds likely traces to the fact that both bird males and females can tend to an egg (that is, incubate it in a nest), whereas mammalian young begin their lives in mother's pouch or uterus. At any rate, marked sex differences in reproductive cost and benefit yield more extreme sexual selection in mammals, shaping anatomy, physiology, and behavior to meet sex-specific life history challenges. This all sets an evolutionary framework in which homology, or shared ancestry, factors into some of our reproductive realities today, such as why women lactate but men rarely (not never) do.

PRIMATE SEXUALITY

The first primates evolved around 70 million years ago. These small, bush-dwelling creatures were among the primary beneficiaries (as were mammals generally) of the mass extinction of their largest competitors, the dinosaurs, some 65 million years ago in the aftermath, apparently, of a meteor's hitting Earth in the area of what is today the Yucatán Peninsula. Fossil remains of some of these early primates, along with research on a variety of living primates, enable us to characterize broad features of their sexuality.

As studied today, strepsirrhines, which include the lemurs of Madagascar and their cousins, the haplorrhine monkeys of the Americas, exhibit a range of mating systems, from polygyny to monogamy and even, in the case of some small, arboreal South American monkeys such as marmosets, occasional polyandry, a mating system quite rare among mammals generally (Campbell et al. 2011; Strier 2011). In many species of lemur, females dominate males and have a similar body size, and, as is true of many other strepsirrhines and haplorrhines, most of their sexual behavior occurs around the time that females are ovulating. Indeed, in some haplorrhines, such as squirrel monkeys, the anatomy and physiology of the female reproductive tract are in condition for intercourse only during the time around ovulation. Strepsirrhines and haplorrhines are also more reliant than other primates on smell for the detection of potential competitors and mates in their forested world, scent marking (leaving their scent on a tree or other place to advertise their presence) and rubbing urine on themselves to aid this form of identity recognition.

The catarrhines, the group that includes both Old World monkeys—vervet or rhesus monkeys being the most common examples—and apes, evolved approximately 30 million years ago (Campbell et al. 2011). While their sexual behavior occurs most frequently around the time that females are ovulating, they also can deviate from this pattern, with greater flexibility in mating behavior being determined by a mate's dominance rank or other aspects of social context. All wild catarrhines are diurnal, meaning they are more active—including sexually active—during daytime hours. Some catarrhines display one-male polygyny, leaving lots of bachelor males on the social fringes desperately searching for mating opportunities, but many others, such as rhesus monkeys, live in multi-male, multi-female groups.

In all species of catarrhines, females spend the bulk of their reproductive years pregnant or lactating, and thus it is relatively rare for a female to experience ovulatory cycles (Campbell et al. 2011). Among Old World monkeys, females tend to live in matrilineal groups, in which daughters inherit their rank from their mothers, and males immigrate to multigenerational female systems. As is the case with mammals generally, in most matings males mount females from behind (in what can be called ventral-dorsal sex). Juveniles, and even adults, may engage in a considerable amount of same-sex sexual activity, depending on the demographic specifics of an animal's social group (for example, the availability of different potential partners). Mountings frequently are for establishing dominance rather than for reproduction, indicating that once sexual behavior evolves it can also be co-opted for other purposes, including alleviating boredom or simply pleasure (Poiani 2010).

The first apes evolved approximately 20 million years ago. While technically we humans also can be called "apes," since we have traits in common—for instance, we lack tails and we have a mobile shoulder joint—scientists over time have tended to place us in our own group: humans. The apes flourished during the Miocene period, but climate change and the loss of forests resulted in the extinction of many species, including one of our personal favorites, Gigantopithecus. In nature, however, one organism's loss often is another's gain, and extinction of some apes meant expanded ranges, numbers, and diversity for Old World monkeys such as rhesus.

The so-called lesser apes, including about a dozen species of gibbon and siamang (another personal favorite, as will become evident at the start of Chapter 3), are the most distantly related to us of the living apes (Bartlett 2011; Strier 2011). These lesser apes live in mainland and insular rainforests in southeast Asia. They live in territorial family groups characterized by mutual male and female defense; in other words, neither females nor males take kindly to potential intruders they perceive as threats to their ongoing social bonds. Females provide most of the parental care, though siamang fathers can be seen carrying offspring and occasionally providing food. Their remarkably long arms, which dangle near their ankles, help make them the acrobats of their forests.

The mating system of gibbons and siamangs is monogamy. Couples tend to spend all day and night in fairly close proximity and may produce multiple offspring together. It would be difficult to differentiate males from females in a shadowy forest: they have similar body sizes (monomorphism) which, as we will discuss further, is evidence consistent with their monogamous social behavior. Males also have relatively small testicles, consistent with low sperm competition pressures and meaning that only rarely do females mate with multiple males around the time they are ovulating. That said, the advent of genetically based paternity testing has revealed that even some of these primate models of monogamous behavior cheat on each other, for a fraction of offspring are fathered by nonresident males. It is important, too, that among these apes actual sexual behavior occurs relatively infrequently. They are sexually active when females are cycling and during pregnancy, but their relatively low rates of copulation belie any simple argument that maintenance of monogamy requires frequent mating.

We are next most closely related to orangutans (Strier 2011). Orangutans live on two islands, Sumatra and Borneo, and these two populations diverged long enough ago that biologists have begun to think of them as separate species. When we consider human genetic diversity, it is worth noting that, despite their deep divergence, orangutans from the two islands would be difficult for nonexperts to tell apart. In contrast, different humans are easy to distinguish despite having far less genetic variation among populations, a consequence of our recent human origins. Orangutan females are dispersed throughout their forests in order to be efficient in acquiring ripe fruits, their preferred food (fragrant durians are relished as much by orangutans as by humans, even if it takes a stick or another ingenious method to open some of the thorny fruits). Orangutan females thus live apart from each other and have minimal social contact because of the patchy availability of their food. Female isolation may be relieved by time spent with dependent offspring. Adult males, on the other hand, are true loners. For males, then, sex is a good reason to have social contact, and it is almost equally so for females. Put another way, during the relatively rare occasions when an adult female is cycling, she is keen to find a preferred mate, and males are keen to find her. Since female orangutans have the longest interbirth interval (time between successive births) of any mammal—around eight years—these are marvelous cyclical times indeed.

The mating system of orangutans can be described as dispersed polygyny (Dixson 1998; Knott and Kahlenberg 2011). A female may mate with one or several males, typically when she is cycling, maybe when she is pregnant, but hardly ever when she is lactating and not cycling. If she hears the loud forest call of a dominant, large adult male when she is cycling, she may preferentially seek him out as a mate, whether to enjoy benefits of his capacity to protect her and her offspring or, as some suggest, to obtain his "good genes." Females are considerably smaller than these full-fledged males. In fact, orangutans exhibit some of the most pronounced sexual dimorphism found in primates. Orangutan males come in one of two developmentally and socially influenced types. The typically younger, less dominant male is about the same size as a typical adult female, and he lacks such traits as the large fleshy cheek pads that make it possible for a fully developed adult male to announce his presence in a thick forest (the cheek pads visually mark adults but also help male vocalizations resonate). The larger male morph is generally preferred by females and has a presence that frequently scares away smaller male competitors. Yet orangutan males, often including those who are relatively smaller, also engage with varying frequency in forced copulation (Muller and Wrangham 2009). In other words, males of any size sometimes force themselves on females, whose signs of distress indicate that they do not want to participate. Yet, the fact that orangutans have relatively small testicles leads to the conclusion that sperm competition is not a significant factor in orangutan sexual behavior.

Like orangutans, populations of gorillas—who live only in Africa—exhibit a remarkable amount of genetic variation considering how similar they are in appearance (Strier 2011). Gorilla mating systems also exhibit some variation across the forests where they live, but these are variations on a pattern of polygyny. Some groups, for example the mountain gorillas made famous by primatologist Dian Fossey, typically live in one-male polygynous groups, in which a single adult male mates with multiple females. These females generally are relatively aloof from one another rather than engaged in any sort of female bonding. In other gorilla populations, sometimes two adult males share matings with multiple females, or at least a more dominant male appears to allow the presence of another male, perhaps because otherwise his tenure would be threatened by outsider gorilla males. The relatively small testicles of gorillas indicate that a female rarely mates with anyone but the alpha male in her social group.

Gorillas exhibit extreme sexual dimorphism in body size, with males growing to the largest size of any living primate (apart from some "supersized" humans) alive today (Dixson 1998). The males' extralarge size, abundance of muscles, and impressive canine teeth are thought to be important adaptations to male-male competition, for instance when a resident alpha male in a group fights would-be successor males. In this polygynous system, there are plenty of bachelor gorilla males seeking the dominant position in a group, biding their time by practicing dominance displays and same-sex sexual behavior until they might succeed in gaining reproductive access to the group's females.

When an outsider does succeed in overthrowing a resident gorilla male, a frequent consequence is infanticide, as the new alpha male may kill the offspring of the former alpha (Wrangham and Peterson 1996). The new alpha needs the group's females to return to a state of ovulatory cycling, so that he can mate with them and father his own offspring. To expedite the process he might kill those females' young nurslings. In some groups, infanticide accounts for an estimated 40 percent of gorilla mortality, and may lead to a female mating with the very male who killed her previous offspring. From the female's perspective, this new male may be a better protector than the last one. Infanticidal gorillas are not unique in the animal kingdom. Infanticide under similar circumstances has been documented in a wide variety of animals, from langurs to prairie dogs, lions to jacana birds.

Of all creatures on Earth, chimpanzees and bonobos are the most closely related to humans (Campbell et al. 2011; Cohen 2010; Wrangham and Peterson 1996). Genetic and fossil evidence suggests we last shared a common ancestor with chimps and bonobos around 6 million years ago in Africa. In terms of sexual behavior, chimpanzees and bonobos prompt all kinds of comparisons, and some of the differences between their sexual behavior and ours are really quite striking. Those differences, in fact, are so striking that they inspire a major theme of this book: we humans and our immediate hominin ancestors have undergone some profound shifts in sexuality compared with our closest kin and with other apes as well. Some of the most remarkable of human evolutionary changes have occurred in the evolution of our sexual behavior.

Genetic data indicate that bonobos (once called pigmy chimps) and "common" chimpanzees diverged around a million years ago. While technically we are equally related to both species, several lines of evidence suggest that bonobos are more likely derived from a chimpanzee-like creature: bonobos are found only south of the Congo River (in contrast to the chimpanzee's wider geographic range); bonobos are less genetically diverse than chimpanzees; the only putative chimpanzee or bonobo fossil ancestor material has been obtained in East Africa, closer to chimpanzee ranges; and the geography of the fossil record of hominins (humans and our close extinct kin on the branch since last sharing a common ancestor with chimpanzees and bonobos) suggests a branching more in line with chimpanzees than bonobos. All of this, then, provides some support for viewing bonobos as recently derived apes and chimpanzees as more representative models of an early hominin ancestor.

Chimpanzees live in multi-male, multi-female groups. The nature of food distributions may result in females (and males) fissioning and fusing throughout the daytime hours in search of clumped resources, like trees that provide preferred fruits. Females may forage alone or in small groups, sometimes including their offspring. As a core ecological principle of social evolution, females distribute themselves in relation to resources, such as food, and then males map on to where the preferred female mates are. Male chimpanzees also frequently employ sexual coercion (Muller and Wrangham 2009). Harassing and physically attacking a female, even when she is simultaneously sought as a desirable mate, may deter her from mating with other males. Yet females also mate strategically in order to garner the best outcomes within their social context.

What factors guide the mating strategies of chimpanzee females? Females seek to mate with nearly all males in a group, possibly to confuse paternity (Stumpf 2011). By mating with so many males, females make it difficult for any given male to know whether he is the father of any given offspring, and he thus may be kinder toward her and her offspring than he otherwise would be. In this way, females combat male harassment and coercion of themselves and their offspring by fostering uncertainty about paternity. The male chimpanzee's testicles (like those of bonobos) are relatively immense, a product of this high degree of female multi-male mating. And selection has equipped females with the tools to incite this mating riot: a sexual swelling on her rear creates a dramatic visual enticement to potential mates. At the same time, however, as cycling females reach their peak fertility around the time of ovulation they preferentially seek, by displaying the proceptivity discussed earlier, a dominant male or males with which to mate.

One of the most noteworthy aspects of chimpanzee sexual behavior is that, by several measures, males appear to prefer mating with older females who are still capable of reproducing (Muller, Thompson, and Wrangham 2006). These can be females who already have offspring and characteristics such as elongated nipples from frequent suckling of their existing progeny. Since the tenure of any relatively dominant male chimpanzee may be short, these males seek to optimize their mating efforts by choosing females of proven fertility and care-giving capabilities, rather than the younger, less proven females. Indeed, many species besides chimpanzees recognize the value of an older female's experience. Trinidadian guppies (much like the kind of guppies you can find in a local pet shop) are a prime example. They engage in "mate copying," whereby younger females watch and copy the mating choices of nearby older females. The male chimpanzee's preference for older females is probably a fairly widespread male mating preference in species with multi-male, multi-female primate groups, such as rhesus monkeys (Maestripieri 2007). A preference for an older female mate is at odds with typical human mating preferences (notwithstanding media attention to the "cougar" phenomenon of young men dating much older women), and it is one significant example of the major differences in sexual behavior between chimps and us.

Bonobos, in contrast, have earned a reputation as a "make love, not war" creature. Some of that tendency may trace to the fact that they have been studied much less than chimpanzees, due in part to the diseases, deforestation, and civil unrest that plague people seeking to study bonobos in the Democratic Republic of Congo, the only country where they live in the wild. The challenges of studying wild bonobos have meant that findings about captive bonobos have received disproportionate attention. Craig Stanford (1998) has pointed out that captivity might encourage greater expression of sexual behavior than is the norm in the wild. (If you're not looking for food all day, why not have sex?) Yet others, like Frans de Waal (2006), note that even in captivity, bonobos behave differently than chimpanzees—that there are real species differences in behavior, apparently including their sexual behavior too.

In parallel with chimpanzees, bonobos also live in multi-male, multi-female groups, and bonobo females mate throughout their ovulatory cycles with most or all group males. Female bonobos seem more capable of living together than chimps, perhaps because of differences in the feeding ecology where they live (plant food distribution may allow more consistent social grouping and less fissioning and fusing among bonobos), which in turn enables the formation of female coalitions that can blunt some of the sexual coercion that is seen so much among chimpanzees. Bonobo females also appear to strategize matings in order to confuse paternity, and when at peak fertility they seek out matings with more dominant males. Like chimpanzees, bonobo females display sexual swellings to provoke male sexual excitement.

Female-female coalitions in bonobos appear to be cemented by same-sex sexual activity that includes females vigorously rubbing their genitals together in face-to-face embraces, and males valiantly engaging in "penis fencing." Accounts of these behaviors suggest they frequently end in orgasm. They also seem to dispel tension, such as that which arises over available desirable food resources. Other forms of bonobo sexual behavior that garner scholarly attention include open-mouth kissing, fellatio, and short copulations.

The discussion thus far has barely scratched the surface of what we know about primate sexual behavior. Yet variation in mating systems, in same-sex sexual behavior, in frequency of sexual behavior, in mating positions, in traits (like sexual swellings) that have been shaped by sexual selection to enhance female polyandrous mating—all these things help us contextualize our own sexual practices. At a minimum, these observations counter simplistic views of same-sex sexual behavior as unnatural (it is plenty practiced in nature), of sex as being solely about reproduction (it is and is not, as female nonfertile proceptive sex illustrates), and of the missionary position as being desired by all animals (it is the most common human sexual position, as we will see, but is regularly employed only by bonobos among other primates). This brief survey of the varieties of primate sexual behavior also has introduced general socioecological principles that account for species variation in sexual behavior and that enable some inferences, as we will see, about the evolution of human sexuality. As one example, the associations between relative testis size and sperm competition observed across primates can incorporate humans into the analysis; humans have relatively small testicles, indicative of low (though not necessarily nonzero) sperm competition pressures in our recent ancestors.

THE EVOLUTION OF HOMININ SEXUALITY

Here are some fantasies we have for discoveries and evidence that would give us a better understanding of the evolution of hominin sexuality. Archaeologists discover 5-million-year-old rock art depicting hairy hominin-like creatures in a multi-male, multi-female group and a male having sex with a female from behind. Paleontologists discover a hominin social scene, preserved by a volcanic explosion (think Pompeii, just a lot longer ago), that includes a female mating with a male while her offspring—of an estimated five years of age—seems determined to separate them. Ancient DNA uncovered from a 1.5-million-year-old *Homo erectus* specimen carries genetic markers that reveal signatures of long-term pair bonds and reduced sperm competition pressures.

Setting aside these fantasies, we confront the real evidence available to anyone who would try to reconstruct the evolution of hominin sexuality.

Some evidence comes in the form of inferences from what is known about the sexual behavior of living nonhuman primates. Other evidence comes from applying general principles of sexual selection (such as mating system and body size dimorphism) to the scant hominin fossil record as we try to discern patterns (such as monogamy or extreme polygyny). And still more evidence can be drawn from analogies based on recently studied human hunter-gatherers; a small and largely recent archaeological record; studies of a wider array of human sexual behavior in more complex societies and a variety of niches; and inferences drawn from human genetics and physiology (indicators of sperm competition pressures and cryptic female choice).

The resulting picture is illuminating, even if it is not definitive. As we hope to show, there is evidence that our hominin ancestors around 6 million years ago may have lived and mated within multi-male, multi-female groups similar to those of living chimpanzees, and subsequently shifted toward longer-term sociosexual bonds, especially over the past 2 million years within the genus *Homo*. Our recent *Homo* ancestors may have exhibited slight polygyny and experienced low, nonzero sperm competition pressures; females preferentially mating with males of higher social status, occasionally while mated to another male; and most males mating monogamously, except for a few, higher-status males who perhaps had a second long-term partner. Now let us look in detail at the evidence and rationales for these kinds of inferences.

The comparative primate evidence, and especially the patterns among chimpanzees and bonobos, provides a rationale for the reconstruction of a multi-male, multi-female mating system in early hominin groups. Furthermore, the earliest fossil material that may represent the first hominins is found in East and Central Africa, dates to around 5 to 7 million years ago, and includes creatures similar in body size to today's chimpanzees. The genetic evidence also indicates that this is about when humans and chimpanzees last shared a common ancestor.

The hallmark of a hominin is habitual upright bipedalism. This trait can be seen in the earliest putative hominin fossils, for example *Sahelanthropothus tchadensis*, a find dated to around 7 million years ago in Chad. This individual had a skull about the size of a chimpanzee's, and yet the hole and orientation of the foramen magnum, where the spinal cord would have inserted, indicates (at least to some scholars) that this creature may have been upright (Klein 2009). Another example is a remarkable fragmented yet complete early hominin named *Ardipithecus ramidus,* dated to about 4 million years ago in Ethiopia, who appears bipedal but with some anomalies (White et al. 2009). Those anomalies, like other finds in the fossil and archaeological record, suggest that hominin bipedalism may have evolved in two waves.

The first wave, at the dawn of hominins, was a form of bipedalism in which these individuals still spent a considerable part of their lives in trees (perhaps including sleeping in nests in trees, like chimpanzees and orangutans do). These early hominins had relatively longer arms (useful for climbing and dangling from tree branches) and curved fingers and toes (also useful for grabbing onto branches). That same *Ardipithecus* specimen even had divergent toes, which would have helped it grasp branches with its feet. Interestingly, these upright bipeds also apparently evolved in wooded environments, so may have moved bipedally on the ground between forest patches or, as some paleoanthropologists suggest, in the trees themselves. The second wave of hominin bipedalism began around 2 million years ago with the genus *Homo* (a Latin reference to human, since it's also our own genus), when fossils indicate that arms became shorter, legs relatively longer, finger bones were no longer curved, and these individuals began living in a wider array of environments, including more open landscapes. All of this suggests *Homo* fully committed itself to bipedalism; there were no more retreats to the trees. Regardless of these fascinating stages in hominin bipedal evolution, best guesses would be that hominins engaged in sex in daylight, in the nude, but may have begun experimenting with the social expression of sexuality as well as the positions in which sex occurred. Shifts in locomotion likely affected variations in sexual behavior. How, for example, would a pelvis shaped for bipedalism influence the orientation of the vaginal barrel and thus intercourse?

The early hominins gave way to australopithecines (Klein 2009; Wood and Lonergan 2008). The so-called gracile forms of *Australopithecus* evolved by around 4 million years ago, from one of the earlier hominins. Gracile australopithecines lived in both eastern and southern Africa, and they gave rise to so-called robust australopithecines (or what some called *Paranthropus*), with the "robust" referring to its larger teeth and jaw-muscle attachments, in turn presumably related to derived dietary differences. The robust australopithecines went extinct around 1.3 million years ago. However, our hominin ancestry is thought to include at least some species of gracile australopithecines.

Perhaps the best-known of gracile australopithecines is the *Australopithecus afarensis* fossil of a female who was given the name Lucy, and who lived in Ethiopia around 3.4 million years ago. Lucy is a wonderfully preserved specimen, clearly bipedal, and with a body size and cranium size akin to a modern chimpanzee's, as well as many other resemblances. To put it crudely, Lucy can be thought of as a bipedal chimpanzee. And yet, we have reason to think that she did not mate like a chimpanzee.

There are at least two indicators of mating system that have been derived from comparative animal research and that can be applied to the fossil record (Dixson 2009; Strier 2011). The most prominent and commonly used of these is body-size sexual dimorphism. In more monogamous species, the two sexes tend to have similar body sizes (monomorphism: think of those gibbons), whereas polygynous species tend to exhibit greater body size dimorphism (think of those gorillas). Although a few scholars disagree, most estimations of body-size dimorphism among Lucy and her kind tend to be quite high (see Plavcan 2012). In fact, the typical estimates of body-size dimorphism among gracile australopithecines suggest that they were polygynous, with size differences maybe even as extreme as those seen in gorillas today. This has led some researchers to suggest that gracile australopithecines may have mated in gorilla-like systems (Geary 2010), whereas others suggest instead a continuation of a multi-male, multi-female system. Either way, these ancestors do not resemble modern humans in body size, in cranium size, or in body-size dimorphism, suggesting that their sexual ways were quite different from ours.

The other major indicator relating morphology to mating system is canine-tooth dimorphism (Dixson 1998; Strier 2011). Canine teeth can be used to catch prey, as seen today in male and female dogs alike (and in other canids). These teeth can also be used as weapons, particularly in same-sex competition, and sometimes for show, including as a signal (sometimes false) of the ability to inflict harm if necessary. For this reason, more polygynous species tend to exhibit greater canine dimorphism. Consider the highly dimorphic canines of gorillas: males use their larger canines in male-male combat (rather than in the capture of prey items). Compare gibbons, in which both females and males have pronounced canines that are similar in size, which

Body size dimorphism in the genus *Homo* clearly indicates a shift in sociosexual behavior, but the timing of this shift is difficult to pinpoint (Dixson 2009; Gray and Anderson 2010). Some fossil samples suggest that shifts toward a slightly polygynous system occurred by around 1.8 million years ago with *Homo erectus*. Others suggest that the shift occurred more recently—around 500,000 years ago with *Homo heidelbergensis*. And still other researchers might contend that it was only with the origins of modern humans in Africa, around 200,000 years ago, that the shift fully occurred. Like many changes we can discern in the fossil record, seeing this change as an abrupt shift masks some more gradual trends.

We imagine a scenario unfolding in several steps during the social evolution of *Homo* to yield modern human sexuality (see Foley and Gamble 2009). Consistent with general principles of socioecology, we start with a view that females distribute themselves in relation to food resources and predation pressures, while males distribute themselves in relation to sexually available females (Wrangham 1979; Strier 2011). If early *Homo* females began obtaining different foodstuffs (perhaps including tubers), which they also began processing more with tools before consumption, this dietary change would enable them to cluster in larger female groups. The tools used for such endeavors would likely be lost in the archaeological record, given preservation biases against plant-based materials, but such a proposition would be consistent with the fact that most primate tool use is by juveniles and adult females, with females using tools for acquiring resources such as insects (Strier 2011). Additionally, with early *Homo* leaving the trees for good, they would have encountered greater terrestrial predation pressures, which could serve as another selective factor favoring larger female groups. Larger groups of females would have attracted more males, who would have found it difficult for a single male to monopolize matings, reducing reproductive skew (or variation in reproductive success among males), and favoring more egalitarian male-male relationships. Long-term sociosexual bonds may have originated through mate guarding, later giving rise to more affiliative pair bonds and paternal care, and also freeing males to engage in more risky foraging strategies (for example, hunting). The result of this sketch is long-term, mostly sociosexually monogamous units, along with the occasional polygynous one, within an unusual social system that is multi-male, multi-female.

Does that all seem far-fetched?

they then can use in same-sex competition. With these kinds of ass
guiding our thought process, what does the hominin fossil recor
about canine dimorphism?

This record indicates that, going as far back in time as we can det
both male and female hominins had relatively small canine teeth. T
really no evidence of pronounced male canine teeth anywhere among
inins. Some see this as evidence of reduced male-male competition
shift toward long-term sociosexual bonds early in hominin evolution; (
Lovejoy's (2009) view, for example, is that the similarities in canine t
size indicate that *Ardipithecus ramidus* formed long-term, largely soc
monogamous bonds. Others, however, say that this morphological ind
tor, while useful for other groups of animals, is useless for the homin
because the very nature of bipedalism means that the males fight differen
when they do fight (Dixson 2009). Rather than using angled frontal
tacks in which canines are at the fore, physical aggression associated wi
hominin bipedalism benefits more from grappling and punching with tl
arms. It can also benefit from the use of weapons, when they are used. S
others, like Alan Dixson (2009), altogether dismiss the use of canine dimor
phism as a means of deriving inferences about hominin sexual behavior.

Other morphological signatures of hominin sexual behavior may ultimately be found. Measures of craniofacial dimorphism, such as brow ridge
size and jaw robusticity, exhibit some relationships with mating system
among apes and humans, but the implications for the evolution of human
sexual behavior are not yet clear (Schaefer et al. 2004). One new suggestion is
that digit ratios may be useful cues to the mating system (Nelson et al. 2011).
Here, a digit ratio refers to the ratio between an individual's second and
fourth digits, with lower ratios more commonly found in males, but lower
ratios also more often found in more polygynous species (more on digit ratios in Chapter 4). Extrapolated to the hominin fossil record, scholars suggest that early hominins may have been polygynous—that gracile australopithecines may have shifted toward a more monogamous system, and then
throughout the genus *Homo* this digit ratio indicator points to polygyny,
although it declines in prevalence from *Homo erectus* to *Homo heidelbergensis* and Neandertals (these creatures are all members of the genus *Homo*,
with *erectus* preceding *heidelbergensis* and Neandertals) and ultimately to
modern humans.

HUNTER-GATHERERS ON THE SCENE

The above scenario is consistent with most of the recent hunter-gatherer ethnographic data. Across hunter-gatherers studied by anthropologists over the past century or so, and combined with insights from archaeology, we know that foragers tend to be slightly polygynous (Ford and Beach 1951; Marlowe 2010). Indeed, in a phylogenetic study of hunter-gatherer marital practices, Robert Walker and colleagues identified "a deep evolutionary history of limited polygyny and brideprice/service that stems back to early modern humans and, in the case of arranged marriage, to at least the early migrations of modern humans out of Africa" (Walker et al. 2011: 2). Most males within a hunter-gatherer society have one partner (wife), whereas a few men with higher social status related to hunting prowess or skills as a shaman may have more than one wife. The main exceptions to these patterns among foragers, exhibiting a higher degree of polygyny, tend to be found in Australia, but aspects of demography, ecology, and cultural transmission in the past few thousand years from New Guinea may help explain their unusual patterning (see O'Connell and Allen 2007).

A woman's mating strategy appears adapted to maximize reproductive success by mating with a man who provides resources such as food and makes it possible for her to have more babies (shorter interbirth intervals) during her reproductive years. She generally seeks to assure him of paternity rather than confuse his concept of it, as we have seen in chimpanzees or bonobos. She grows jealous and angry if a partner cheats on her. For his part, the male also seeks long-term sociosexual partners with whom he can have children and provide extensive parental care. A male might seek an additional mate, but the rarity of fertile females, challenges of male competitors, and protests of an existing partner make that difficult. Males also tend to be jealous and angry if a partner cheats. As we will discuss in Chapter 3, infidelity by either sex rarely occurs without consequence. The sociosexual behavior of foragers is also regulated by other adults (for example, parents frequently arrange first marriages) in ways that have no obvious nonhuman primate parallels: no one has seen a gorilla mother signaling a prospective suitor of her daughter to get lost, or chimpanzee males negotiating daughter exchange.

If the hunter-gatherer ethnographic picture fits the preceding hominin evolutionary narrative, what other lessons about human sexuality (Ford and Beach 1951; Gregersen 1994) might we glean from forager sexuality?

Most hunter-gatherer females spend their reproductive years pregnant or lactating, and thus it is quite rare that women are cycling (Ellison 2001). This has highly significant ramifications—such as partnered fluctuations in sexual behavior across these reproductive spells—that, surprisingly, have been little explored, or for that matter, hardly taken into account in recent research on evolved mating psychology; we will explicitly tackle this issue in more detail in Chapter 10. Another finding on hunter-gatherer sexual behavior is that sex is typically covert. Whereas primatologists can readily observe chimpanzee or baboon sexual behavior taking place out in the open, for obvious reasons few researchers have dared to try such observations with humans (studies by Alfred William Masters and Virginia Johnson in the 1950s and 1960s are among the notable exceptions). Among foragers, covert sex may occur outside of a camp during daylight, or it may take place at night (unusual for an Old World monkey or ape) near a fire or within proximity of children. Also, modesty typically governs female genital displays. Even in hunter-gatherer societies where women are altogether naked, they typically adhere to standards that might, for example, prohibit them from opening their legs while sitting. As far as same-sex sexual behavior, this has been documented in some forager societies (such as the Aranda of Australia), but certainly not all (the Bofi of the Central African Republic). There is little evidence of lifelong expression of same-sex sexual orientation, however, with same-sex sexuality typically age-structured (such as among young men in some more polygynous Australian aboriginal communities) or intermixed with reproduction (the same individuals who engage in same-sex practices also have children).

Apart from these forager data (and the comparative primate and fossil evidence), there are several other lines of evidence that can be tapped to reconstruct the evolution of hominin sexuality. Besides human testis size, there are additional physiological and genetic indicators of sperm competition pressures. We will unpack that body of evidence in more detail later in the book, but for now it suffices to say that, like relative testis size, other measures similarly point to relatively low recent hominin sperm competition pressures—just as we would have expected from the fossil and hunter-gatherer data.

The archaeological record offers a few hints of past human sexuality, but most of these are quite recent. A 5,000-year-old European rock art depiction of a man with an erect penis standing behind a domesticated animal suggests bestiality was not born yesterday (Gregersen 1994). An intricately carved "Venus" figurine from around 35,000 years ago in Europe in which the vulva is clearly shown makes one wonder about pubic hair removal practices, and the fact that some of these figurines show elaborate hair dress raises questions about female ornamentation practices that might elicit interest from a mate (Adovasio, Soffer, and Page 2007). We will occasionally touch on examples from the archaeological record, but these few examples alone suggest some of the limitations of its use.

The wider cross-cultural and historic record of human sexuality illuminates patterns of evolved human sexual proclivities. As an example, when despots are able to mate with whomever they might like, they often have a preferred mate, but they also commonly have additional mates, with some extreme examples of men arranging harems to service their sexual and procreative fantasies (Potts and Short 1999). In those same harems you might also find women sexually dissatisfied, turning to alternatives of same-sex behavior with or without technological assistance (such as dildos, which, by the way, are also found in the archaeological record), an indicator that female proceptivity has its cyclical peaks that demand attention.

The record of human cultural evolutionary change also attests to population-specific trajectories in sexuality. Modern humans evolved in Africa around 200,000 years ago, largely spreading around the globe only in the past 60,000 years (Stanford, Allen, and Anton 2011; Tattersall 2012), expanding the scope of sexual practices along the way. Genetic data indicate a small degree of interbreeding between modern humans and Neandertals, and between modern humans and so-called Denisonovans (cousins of Neandertals, known almost exclusively from their ancient DNA), during this expansion out of Africa (see Reich et al. 2010). The dynamics of cultural transmission (Richerson and Boyd 2005) have given rise to population-specific beliefs and behaviors such as conception beliefs, marital norms (such as daughter betrothal), specified allowances for extramarital sex (during specific ceremonies, for example), and sexual behaviors (such as preferred positions during intercourse), an indication that human sexuality must be understood within its cultural context.

All said, in our attempt to put together all of the potentially relevant and available pieces of evidence, in this chapter we have sketched a scenario of the evolution of hominin sexuality. The limitations to this exercise are certainly profound. But, to underscore a theme recurring throughout this book, we seek an integrative view of human sexuality that does not privilege any single line of evidence but rather incorporates the best available and relevant evidence within a common evolutionary framework. In the next chapter we explore the landscape of human sexuality cross-culturally, observing the array of sexual practices that have arisen during the expansion of modern humans globally, and also noting underlying patterns that speak to the power of evolutionary forces in shaping beliefs and behavior.

CHAPTER TWO

The Garden of Variety

Cross-Cultural Variation in Human Sexuality

> No one supposes that all the individuals of the same species are cast in the very same mould. These individual differences are highly important for us, as they afford materials for natural selection to accumulate, in the same manner as man can accumulate in any given direction individual differences in his domesticated productions.
>
> —Charles Darwin, *The Origin of Species*

WHILE WAITING FOR a bus near Montego Bay, Jamaica, Peter stepped inside a fast-food restaurant. Standing in front of him in line were a man and woman who acted and looked like they were romantically involved. The woman, speaking with an American accent, was white, somewhat overweight, and not especially attractive. The man, speaking with a local accent, was black, appeared younger than she, was more attractive, and allowed her to pay for his meal. After ordering and receiving their food, they continued their romantic touches and glances. Their interaction served up many interesting questions about human sexuality.

It turns out that similar male-female dynamics occur in a number of the tourist-oriented beach settings along the Jamaican coast (Kempadoo 2004). What some characterize as "romance tourism" also shades toward prostitution, as women from the United States, Europe, and elsewhere travel to hot-blooded exotic places seeking a few days of attention and sex with local men. Often these women have "low mate value," meaning that they are unlikely to receive similar attention from attractive men in their regular lives at home. But for the cost of travel and supporting a lover, they can obtain a kind of romance-novel courtship and excitement normally absent from their life. They also can suspend reality in anonymity, without friends and family questioning their behavior. Indeed, female romance tourism flourishes elsewhere in the world too, with Thailand now a destination for Japanese women and Bali the choice of Australian women.

For Jamaican men, these foreign women offer otherwise unavailable economic and social opportunities. In exchange for providing attention and sex, local men may receive meals and entertainment, and gain access to other resources (maybe a visa to another country) that can be difficult to obtain any other way. The men may strive to fashion a desirable profile, exercising to stay toned or grooming themselves to fit an attractive "rent-a-rasta" model. They may preferentially seek out older, less sexually desirable women who have more resources to offer. And the men often engage in sexual behavior—for instance, oral sex—that is uncommon in their more immediate cultural background.

Cultural variation in human sexuality can lead to encounters such as the one Peter observed in the Caribbean. Cultural variation can also give rise to a growing demand in other realms of sexuality, from fetish markets in pornography, to mail-order brides. Cultural variation can also be carefully studied by researchers and become the subject of wide-ranging scholarship. The aim of this chapter is to cover this cross-cultural garden of sexual variety.

MEASURING CROSS-CULTURAL SEXUALITY

How do you collect data on human sexual behavior? Should Peter have conducted an in-depth, open-ended interview with the romantic couple over burgers and fries, asking them about their sexual attitudes and behavior? Should he have asked them to log in to a password-coded website to answer a predetermined set of questions concerning their sexual behavior? Should he have followed them back to their bedroom to see with his own eyes what they were up to?

Measurement challenges such as these make human sexuality one of the most daunting subjects for scholarly research. Observations employed in studying chimpanzee or baboon sexual behavior are rarely feasible in people. Researchers are almost always restricted to questionnaire- and interview-based information. But it is not easy to know whether people provide accurate answers to questions about their sexual behavior: some respondents may downplay their sexual experiences, others may exaggerate. Still others may balk at sharing information (such as an extramarital affair) they feel is not other people's business or that might be harmful to themselves or others. In cross-cultural context, additional issues arise: actions, attitudes, and meanings may vary, making it a special challenge to identify comparable variables in different cultural contexts.

Despite these challenges, researchers have employed a variety of methods in cross-cultural studies of human sexual behavior that have yielded a variety of data sets (see Davis and Whitten 1987). The most common methods—interviews and questionnaires translated into local languages—have been supplemented by participant observation and talking with locals as they go about their daily activities, although there always are limitations to what can be learned this way about sexual behavior. Much cross-cultural information on sexual behavior has been collected by anthropologists working in specific communities, or by governmental or health researchers using standardized instruments. More recent efforts, led by psychologists, have relied on questionnaires, mostly administered to college students in many parts of the world. Cross-checks between respondents (for example, between partners) and validation from other sources of information, such as clinical findings, help ensure accuracy. Much of the available information highlights behavior; less has been done on the attitudes toward and meanings given to sexuality.

Quantitative cross-cultural studies of human sexual behavior in small-scale societies of hunter-gatherers, horticulturalists, agriculturalists, and pastoralists tend to emphasize marital patterns. This makes sense, because much of human sexual behavior occurs in the context of long-term relationships such as marriage. Studies of marital patterns employ the same terminology (monogamy, polygyny, and so on) that we used in the discussion of animal mating systems in Chapter 1. One should also be aware of the limitations of marital-systems data: these data can result in normative descriptions of whole societies (characterizing a society as polygynous rather than looking at individual mating patterns within a society), marital behavior can be a misleading indicator of sexual behavior (ignoring, for example, the prevalence of prostitution and extramarital affairs), and some of the marital data we still use today are quite dated, meaning that they may describe a world of the nineteenth or twentieth century. So the availability of older data is a mixed blessing. On one hand, the data provide a fascinating and important record of variation; on the other hand, so many practices are changing at such a rapid rate that we cannot assume that older data remain valuable for understanding sexual behavior today. Rapid change in acceptability of masturbation and the consumption of pornography in China (into the 1980's, many Chinese adolescents were warned masturbation was bad for their health) is but one example of a case in which older data do not tell us much about behavior today (Parish Laumann, and Mojola 2007).

With these limitations in mind, let us look at some of the quantitative compilations of data about sexual behavior, beginning with studies that rely on large, often representative probabilistic samples. These compilations have been collected or financed by governments, services (such as Demographic and Health Surveys conducted in various developing countries), international organizations (for example, the World Health Organization), and private industry (such as pharmaceutical companies). These studies are remarkable in their scope and sample sizes, though they also lose much in context (in explaining the meaning of the data). An example is the 1992 U.S. National Health and Social Life Survey. This survey reported on 3,432 interviews with American men and women age eighteen to fifty-nine years old (Laumann et al. 1994). A more recent study, from the 2010 National Survey of Sexual Health and Behavior, reports the latest U.S. adult probability study of sexual behavior, based on approximately 6,000 participants, men and women, age fourteen to ninety-four years old. In this latter study we learn, among other things, that 63 percent of men age twenty to twenty-four and 44 percent of women in the same age group said they had masturbated in the past month (Herbenick et al. 2010a).

An interdisciplinary body of more qualitative, cross-cultural scholarship exists, which we will draw upon here too. Some of these studies are classic ethnographic sources grounded in anthropologists' participant observations and interviews; Malinowski's (1929) *Sexual Lives of Savages,* based on his research on the Trobriand Islanders, is one example. Collections of thematically organized ethnographic reports are available on romantic passion (Jankowiak 1995) and other aspects of sexuality (Suggs and Miracle 1993). Also available are compilations of qualitative, cross-cultural sexuality data from small-scale societies of hunter-gatherers, horticulturalists, pastoralists, and others (for example, Ford and Beach 1951; Gregersen 1994). These compilations draw on several resources, such as the Human Relations Area Files (HRAF) or one of several samples spun off from it, such as the Standard Cross-Cultural Sample, or SCCS. The studies in the SCCS were chosen in order to include 186 small-scale societies from different regions and linguistic groups (Murdock and White 1969). A compilation of qualitative data on sex and gender from countries around the world is available free of charge on The Kinsey Institute website or in published form (see, for example, Francouer and Noonan 2006). The archaeological record can also be a resource. Material remains of human sexuality in times past may be excavated, but they often require additional levels of interpretation (how were those 30,000 year old "Venus" figurines viewed by their makers?); still, we occasionally draw upon international examples from the archaeological record too (Voss 2008).

All these sources, and the methods underlying them, together serve as the foundation for understanding the cross-cultural record in human sexuality we will present in this chapter. By taking a cross-cultural approach to human sexual behavior, we avoid the mistake of assuming that our own lives represent all of humanity. We can document and explain the variations in sexuality that may be of interest, perhaps if only because they may be encountered through travel, among colleagues in the workplace, or on the Internet in our globalized world.

CROSS-CULTURAL VARIATION IN MARRIAGE PATTERNS

Suppose you were chosen for the emperor's harem, to spend your days with other women and the harem's eunuchs until those rare times when you were rotated into service. Maybe you and your family would feel proud of you for being selected; maybe your children would fare better than they otherwise would have, even if you were forced from service by the time you reached thirty years of age. Or suppose you were married to three brothers in a highland Nepalese agrarian community. The first brother may be your preferred spouse, and you really enjoy sex and an emotional connection with him more than with his brothers, your other husbands. But you have to play the appropriate role and occasionally indulge the less-practiced, younger brothers, one of whom has barely begun puberty. You do this to facilitate a sense of economic and familial harmony in this challenging set of circumstances. Or to step into another set of shoes, imagine that you sit at a wedding altar, awaiting your first sight of your spouse. Your parents have selected your bride through careful thought and negotiation with her parents, and have determined your future sexual and reproductive life through their decision. You will, years later, talk about how happy you are with your spouse, and how your intimacy with her grew over the years.

Exactly these sorts of scenarios have played out around the world (Low 2000; Murdock 1949). While the days of despotic emperors, harems, and eunuchs have faded into history books, occasional polyandrous marriages remain, and certainly arranged marriages do too. These various arrangements offer a glimpse of the cross-cultural scope of human marital patterns, and the crux of mating, in which they are intertwined. Yet, how do they fit within the wider quantitative and qualitative cross-cultural sources of data?

In one cross-cultural survey of 849 societies of hunter-gatherers, horticulturalists, pastoralists, and others, findings suggested the vast majority allowed polygyny (Murdock 1967). In fact, approximately 30 percent of these societies had generalized polygyny (at least 20 percent of men having multiple wives) and another 50 percent had more modest polygyny (fewer than 20 percent of men having multiple wives). Almost all of the remaining societies exhibited social monogamy; only 4 were characterized by polyandry. In a related cross-cultural sample of 186 societies, similar patterns were seen. In other words, most small-scale societies of horticulturalists, agriculturalists, and pastoralists, have allowed polygynous marriages.

These patterns tell us several important things. One is that most marital patterns reflect slight polygyny rather than strict monogamy. This is consistent with other lines of anatomical, physiological, and behavioral evidence indicating that slight polygyny has been stamped into our human form. Recent genetic data also point toward slight polygyny in our natural history (though with culturally variable frequency, such as higher rates of polygyny among the Yoruba of West Africa compared with an East Asian sample; Labuda et al. 2010). However, overwhelmingly, for most people in most societies, marriages are socially monogamous (one wife with one husband). Further, these unions entail long-term emotional and sexual dynamics extending over years, decades, and even lifetimes. This has bearing on the psychological and physiological substrates that facilitate these long-term sociosexual unions. In fact, this is such an important pattern—that most human sexual behavior occurs within the context of long-term bonds—that we will obsess over this very topic throughout the next chapter. Another lesson from these patterns is that polyandrous marriage is rare. The highland Himalayas represent one of those rare niches where it occurs, where the scarcity of land and basis of inheritance has favored multiple brothers marrying a single wife in order to keep a viable farm intact. The rarity of human polyandrous long-term unions is consistent with general mammalian and primate patterns.

A recurrent pattern in cross-cultural marriage arrangements is that men with higher status and resources tend to be the ones with a greater number of wives (or often other sexual partners, including concubines) (Betzig 1986; Low 2000). Some religious and cultural systems quite explicitly tie the logic of this economic reality to marriage. Islamic codes, for example, specify that a man should marry a second, third, or fourth wife only if he can economically support those additional women and the additional children born from such unions; moreover, he is supposed to divide up his resources equally among co-wives rather than favor a single wife, such as a first wife. Through many centuries, then, it was the wealthier men in an Islamic society who could afford to marry multiple wives. Today the prevalence of polygynous marriage has fallen dramatically in most Islamic countries, sometimes leaving the more rural and poorer communities still practicing polygyny while wealthier families (who have become more incorporated into formal educational and international circles) shun it.

Apart from Islamic marital guidance, plenty of cross-cultural and historical evidence underscores the pattern by which men of higher status and resources have a greater number of sexual partners. Taken to extremes, for example, Incan emperors could command access to hundreds of potential sexual partners. A Moroccan emperor, Moulay Ismael the Bloodthirsty, has been said to have fathered an estimated 888 children through his access to a large harem over many years. The ancient Egyptian pharaoh Ramses II cavorted with secondary wives and mistresses in addition to a supreme wife. Many Chinese emperors had access to concubines, who were rotated through to the emperor based on careful tracking of their menstrual cycles. During the Middle Ages in Europe, powerful men, including churchmen, mated polygynously but married monogamously, for only sons born to legitimate wives could be legitimate heirs; in contrast, illegitimate, or bastard, children were denied inheritance of paternal resources. And even in the nineteenth-century, the king of Siam (now Thailand) wound down his day by choosing a sexual partner from the royal harem (Betzig 1986; Potts and Short 1999). Less dramatically, Polynesian chiefs had multiple wives while other men might have one, and, prior to the intrusion of monogamous proscriptions, European tribal elite males also sometimes married polygynously. Even the Bible, especially the Old Testament, is ripe with descriptions of polygyny, eunuchs, and marital considerations benefiting higher-status males.

It is not just the sexual side of marital life that varies cross-culturally: the emotional side does too. In one formulation, there is extreme variation ranging from "intimate" to "aloof," with intimate relationships marked by couples eating, sleeping, and spending leisure time together, unlike aloof couples, who do not do such things together (Whiting and Whiting 1975). This variation can be partly explained by factors such as the degree of polygyny (more polygynous societies tend to have less marital emotional intimacy) and intersocietal violence. As an example of an aloof marital relationship, imagine you live in a highland New Guinea society characterized by high degrees of intersocietal violence; your shield and spears are essential tools of male survival. In the face of violent threats from neighboring peoples, you form strong bonds with other men to foster successful coalitions for defense and attack. These male-male bonds may be cemented through interactions in men-only houses. The same strong male-male bonds, however, come at the cost of intimate family life. Moreover, a wife in your community may have been stolen from a neighboring enemy group, providing more grounds for conjugal tension, even if the new couple might now want to have children together (Gilmore 1990).

Contrast that situation with a small-scale hunter-gatherer society in which men and women spend a considerable amount of relaxed time in each other's company and with other group members. Suppose there is little threat from other hunter-gatherers (though neighboring pastoralists or intruding urbanites might yet be concerns), meaning that male-male bonds matter less. Imagine that most marriages are monogamous, and fathers are also highly invested in their children. These are the cornerstones of a more intimate marital relationship.

Along with emotional intimacy, female sexual behavior varies among cross-cultural marital systems (Broude and Greene 1976; Frayser 1985). In other words, in some societies females are given more or less freedom to express their sexual desires. As an example of greater female sexual liberation, across a swatch of Amazonian societies a concept of "partible paternity" is upheld, whereby a child is held to have one or even several fathers who had sex with the child's mother when that child was gestating (Walker, Flinn, and Hill 2010). Females typically have a main partner (spouse), but they also commonly have other sexual partners who provide resources to them such as fish and game meat. While men do not relish their wives' affairs (we will discuss sexual jealousy later), and women may be beaten or abandoned as a consequence, women nonetheless regularly exert their sexual choices. A key factor about these societies is that most are matrilocal. Females live with their more immediate biological relatives, such as their own mothers, who provide some degree of support against a cruel spouse and a familial social nest for the wife to fall back on if a marriage fails.

Now contrast that relative sexual freedom with a social world in which female sexuality is shrouded, clipped, or bound. Such cases are more likely to occur among patrilocal (men stay in their natal, or birth, groups), patrilineal (inheritance and descent are traced through the male line) societies than the matrilocal ones noted above. Imagine the narrow streets of Lamu, Kenya, with people and donkeys amid an architecture that was imported from the southern Arabian peninsula—an architecture with high walls and courtyards and, important to our present point, an architecture that enables keeping women out of sight of potential male sexual threats. To help narrow women's sexual options, women could be cloistered inside these high-walled homes. When walking outside, such as to the market or to visit family and friends, they wear black veils, shielding all but parts of their face, hands, and feet from public view. These are but a few ways in which female sexuality can be partially hidden.

European chastity belts used during the Renaissance and various forms of genital cutting represent additional attempts to constrain female sexual freedom (Wilson and Daly 1992; Potts and Short 1999). If a husband is concerned about his wife's sexual fidelity while he is away at work, a lock-and-key chastity belt could help satisfy his concern while literally constraining her sexual options. In the case of female genital cutting, which we'll return to later in this book, this can run the gamut of practices from removal of the inner labia (labia minora), or the inner labia and clitoris, to an extreme of clitoris removal along with infibulation, when the labia are sewn together except for a small opening that allows for urination and the flow of menstrual blood but prohibits intercourse. In the case of infibulation, a woman who is married and seeks to become pregnant may be unsewn to allow intercourse, and sometimes she may be resewn if, say, her husband is to be away for an extended time. These practices function to reduce female sexual pleasure and to reduce the chance of sexual intercourse disapproved by family and husband. Interestingly, a woman's female relatives often reinforce the tradition and conduct the actual genital cutting, believing that a young woman's mating prospects would be harmed if she avoided the practice. Female genital cutting has also occasionally occurred in the United States, to "treat" masturbation, under the false presumption by the medical community of the time that women were not and should not be driven by sexual desire. This went on during the era of Victorian prudishness. One case, reported in 1894, involved a New Orleans physician who performed a clitoridectomy to halt the masturbation practices of a fourteen-year-old schoolgirl (Duffy 1963). Female genital cutting, often done "for cultural, religious, or other non-therapeutic reasons," has been documented, to lesser or greater extent, in societies around the globe (Gregersen 1994). Today the World Health Organization is making an effort to end these practices.

CROSS-CULTURAL VARIATION IN SOCIOSEXUALITY

Apart from cross-cultural variation in marital systems, in recent years psychologists have investigated international patterns in various features of men's and women's mating psychology and behavior. Often the participants in these studies have been drawn from universities in developing countries, but in some cases community-based samples have also been included. These efforts help reveal similarities and differences in people's sexuality, both between the sexes and across contexts.

One of the first major efforts to reveal cross-cultural differences in sexuality was led by evolutionary psychologist David Buss (1989; 2003). Marshaling collaborators from various corners of the world, he investigated mate preferences among some 10,000 participants in thirty-seven societies. Among his findings were that love, emotional stability or maturity, and a dependable character were ranked the highest among traits preferred in a mate by both men and women around the globe. These preferred traits accord with what one would expect for maintenance of a successful long-term sexual and reproductive relationship. Buss also found sex differences in mate choice. Men valued female attractiveness, whereas women valued male status and resources. These patterns fit with evolutionary foundations of human sex differences: men, constrained by reproductive access to females, place an emphasis on attractiveness in a mate, whereas women, constrained by access to resources like those provided by high-quality mates, view the ability to provide them as an important quality in a mate (Symons 1979). But what about cross-cultural variation?

There was considerable variation in men's expression of concern over female premarital sexual behavior. In India, Indonesia, and China, men highly valued chastity in a mate. Contrast that with the Netherlands, Sweden, and France, where chastity was not highly valued by men. There was also variation in the degree to which women rated male traits like industriousness and ambition. These qualities mattered more to women in Nigeria and Iran, for instance, than in other countries, such as Belgium and Australia.

Age preferences in a mate also varied. Across the entire sample, women tended to prefer older mates and men tended to prefer younger mates. This pattern reflects several considerations: on average, human females pass through puberty around a year or two before males, therefore the younger a women is, the higher her reproductive value (or her future reproductive potential). Older men, on the other hand, are more likely to have acquired status and resources. In most European and North American countries, women preferred to mate with men about two or three years older than themselves—that is, not much older—but the preferred age discrepancy tended to be slightly higher in countries in other regions (for example, Bulgaria, Colombia, Iran, and Nigeria).

Evolutionary psychologist David Schmitt carried forward Buss's studies with a similar international, questionnaire-based investigation of human sociosexuality through the International Sexuality Description Project (Schmitt 2003; Schmitt 2005). Drawing on approximately 16,000 respondents from fifty-two countries, he found consistent sex differences in sociosexuality. Across various time frames (from in the next month to in the next thirty years), men expressed more interest in having multiple sexual partners than women (Schmitt 2003). This pattern accords with an evolutionary basis of human sex differences, according to which female reproductive success benefits less from a greater number of sexual partners than does male reproductive success. Still, there was also international variation in the number of future desired sexual partners. In East Asian samples, 17.9 percent of men and 2.6 percent of women said they wanted more than one sexual partner over the next month, whereas the comparable numbers from South American samples were 35.0 percent of men and 6.1 percent of women.

In a related analysis, drawing on a slightly smaller sample, Schmitt (2005) also investigated sex differences and international variation in scores on the sociosexual orientation inventory (SOI), a measure of how sexually open versus restricted a person is. Across the entire sample, men reported less restriction on their sociosexuality compared with women. However, other patterns also emerged. The sociosexuality scores of men and women within particular countries were highly correlated, meaning that if women in a society were relatively restricted so too were the men of that same society. Sociosexuality scores also varied considerably across societies. The most restricted societies included Taiwan, Bangladesh, South Korea, Zimbabwe, and Hong Kong; the most open included Finland, New Zealand, Lithuania, and Slovenia.

While documenting variation is interesting, what factors help account for it? One relevant factor is demographic: the operational sex ratio (OSR) (Schmitt 2005). The OSR is the ratio of sexually competing males to sexually competing females; in other words, it is a population measure of sexual competition. Societies with a relatively higher proportion of women tended to have less restricted sociosexuality, a pattern consistent with a political tilt toward the preferred sociosexuality of the relatively rarer men. Fertility rates were also negatively associated with sociosexuality, meaning that the higher the average fertility within a society, the lower the average sociosexuality; that makes sense since the presence of children can intrude on sexual latitude, a topic on which we will focus in Chapter 10. A number of variables, including the Human Development Index (a measure of societal quality of life incorporating economic and social factors), were associated with more open sociosexuality, especially among women. This pattern is likely picking up a signal: women in long-lived, highly educated, economically prosperous societies are spending more of their time educating themselves and working than having children. As a result, women in these circumstances exercise a more open sociosexuality, enjoying the sexual present, and prospecting more openly for their Mr. Right.

These well-replicated findings of sex differences in mate choice can often be difficult to accept, especially for those of us who identify as feminists. While there are many versions of feminism, academic/scholarly feminisms, to the surprise of many students, are not necessarily the same as political/ activist feminism. In fact, academic feminism has been applied to evolutionary science, and takes an interest in considering the active role of both women and men in human evolution (Fisher, Garcia, and Chang, 2013). The data show that women more generally desire an older mate with resources, and men more generally desire a younger, attractive, and presumably fertile mate. Moreover, this is what evolutionary theory leads us to expect. The data tell us what *is* and provide little support for some feminist agendas that would prescribe what *ought* to be. Politics and contemporary variations aside (for now), it is important to note that the mating model we can construct from these data supports an evolutionary account of human mating psychology, similar to those proposed years ago by anthropologists such as Donald Symons (Symons 1979) and Helen Fisher (H. E. Fisher 1982).

SEXUAL BEHAVIOR GOES GLOBAL

If you have Internet access within reach, try an image search for "Moche sex pots." What do you see? You find ceramic vessels that were buried with elite Moche individuals during the first millennium A.D. in Peru (Weismantel 2004). But they are not just any ceramic vessels—this pottery is decorated with figures in all sorts of sexual positions, including men having anal sex with women (although earlier scholars mistakenly assumed these were men having anal sex with other men), women holding babies while having sex, and couples engaged in sex in the "missionary position" (of course, centuries before missionaries ever reached that region). The interpretations of these depictions vary. One view of baby-toting women having anal sex is that this could be a way to avoid becoming pregnant with another child while still breast-feeding an existing one (more on lactational amenorrhea later). Maybe these sexual acts illuminate a symbolic basis by which the elite propagate their reign and lineage. A sure answer is elusive. However, the illustrations of sex on these pots does get us thinking more about actual sexual behavior in cross-cultural context.

Researchers have recently compiled the most comprehensive survey of international human sex data in history, using nationally representative sex studies and demographic and health surveys (Wellings et al. 2006). The report by Kaye Wellings and colleagues is a marvel, compiling data, in some cases longitudinal data, from fifty-nine countries, from Ghana to Kenya, Turkey to the Philippines, Australia to Bolivia, all from studies published between 1995 and 2005. Like other international efforts, this compilation documents similarities across studies but also illuminates variation across countries and regions.

Around the world, the age of first intercourse varies for both women and men. In Bangladesh, more than half of all women have had intercourse by age fifteen, whereas only a few males have had intercourse by that age. This reflects an early age of marriage in Bangladesh among females but not males. By contrast, around 40 percent of males in Brazil have had intercourse by age fifteen, sometimes by relying on prostitutes for their sexual initiation, whereas only around 10 percent of females in Brazil have had intercourse by fifteen. In the United States, by age sixteen nearly 40 percent of young men and young women have had their first intercourse (with slightly more young men than young women), and by age twenty-one nearly 85 percent have had sex. There is also considerable variation across countries in the percentage of men and women who have had sex before marriage. In countries like the United States, more than 90 percent of men and women have had sex before marriage, as has been the case for more than a half century (A later study [Finer 2007] put the percentage even higher, arguing that some 95 percent of Americans experience sex before marriage; again this finding is consistent over the past several decades.). In Mali, Nepal, and Indonesia, by contrast, fewer than 25 percent of women have had intercourse before marriage.

Men are generally more likely to report having had multiple sex partners in the past year than are women. In some cases this reflects premarital sexual experimentation; in other cases, polygynous marriages; and in others a combination of marriage and affairs. Still, the countries in which women report having had multiple sex partners in the past year are mainly in North America and Europe, and the data largely reflect female premarital sexual behavior. These behavioral patterns are consistent with international attitudinal measures in which men in these same countries report little concern about a woman's premarital chastity. Additionally, the spousal age discrepancies, which are greatest in African countries like Burkina Faso (14.7 years) and Mali (13.2 years), and lowest in countries like the United States (2.2 years) and Australia (1.9 years), generally fit with spousal age preference data previously reported from international surveys.

One of the most noteworthy patterns in this international sexual behavior compilation is that, in all countries, married men and women report a higher likelihood of sexual behavior in the past four weeks than do their unmarried counterparts. This pattern reinforces the reality that most human sexual behavior occurs within long-term relationships such as marriage. This pattern is particularly pronounced among women in some countries, like Rwanda, Indonesia, and Nepal, where very few single women report sex in the past four weeks but the vast majority of married women do. The pattern is less pronounced among women in Western countries, like the United Kingdom; Eastern European countries, such as Romania; and in some African countries, like Togo. In the European and North American contexts, the pattern reflects more common premarital sexual behavior, whereas in African countries the pattern in good part reflects lower rates of marital intercourse in polygynous marriages.

Leave it to a condom maker—Durex—to finance a large, twenty-six-country study of sexual behavior, conducted in countries such as India, Russia, and Japan. While they may develop better marketing tactics based on their well-designed survey (which was based on a questionnaire administered to more than 25,000 people in 2006), they have also created a rich data set concerning human sexual behavior that is freely accessible on the Internet. One of the most interesting variables measured in this survey was frequency of sexual behavior. Findings from this survey indicate that the frequency of sexual behavior (and sexual satisfaction) differs across countries. At one extreme, in Greece, 87 percent of respondents reported having sex weekly, and 51 percent report being satisfied with their sex life. Contrast that with Japan, where 34 percent report weekly sex, and 15 percent report being sexually satisfied. Patterns in other countries, from China to Germany, India to Mexico, were generally somewhere in between these poles. The frequency of such sexual activities as anal sex, oral sex, and looking at erotic materials are all reported, though unfortunately these data are not presented according to country.

SEXUAL CONTEXTS

In his classic nineteenth-century tome on human sexuality, *The Sexual Relations of Mankind* (originally published in 1885), Paolo Mantegazza had much to say about masturbation. He wrote,

> Masturbation is a thing so natural and spontaneous in the man who is without a woman, and the woman without a man, that it tends to spring up at all times and in all countries. . . . Colleges, monasteries, schools, all institutions which bring young men together, are the breeding places and seminaries of masturbation. . . . With the possible exception of those countries where polygamy prevails, masturbation is everywhere a good deal more common in the man than in the woman. . . . It is only in polygamous countries that the woman, in the course of her long periods of idleness and the prolonged abstinences imposed upon her in the harem or the zenana, at once learns to masturbate. (Mantegazza 1935: 79)

It is fascinating to compare Mantegazza's writings on masturbation to the 1848 annual report of the Massachusetts Lunatic Asylum in the United States, where nearly 32 percent of admissions were attributed to "self-pollution," a euphemism then used for masturbation (Duffy 1963). Mantegazza's reflections also stimulate discussion of the contexts of sexual behavior, particularly behavior outside of the long-term relationships and affairs that we have emphasized thus far in this chapter.

Another central feature of the context of human sexuality is *where* people have sex. Allusions to "the bedroom" during discussions of sex suggest that is where many of today's youth were spawned. But bedrooms are recent architectural inventions in only some societies. In his recent research on sexual hook-up behavior among emerging adults on college campuses (which we will discuss further in the Chapter 6), Justin found that sexual activity occurred in a variety of places, ranging from clubs to cars to a twin bed in a dormitory, perhaps even with a roommate sleeping a few feet away (or waiting in the hall after seeing the sock on the doorknob, signaling that he or she should stay out). But looking beyond college campuses and across cultures, where *do* people have sex?

According to Ford and Beach's (1951) survey, the preferred location for intercourse was outdoors in twelve societies, and indoors for thirteen societies. An important factor determining this preference was privacy. Of the fifteen societies in which people lived in their own home or in a partitioned room, twelve of the societies were reported as preferring to have sex indoors. Of the ten societies where such living situations did not hold, in only one of them did people still prefer having sex indoors. To illustrate this pattern, take the Siriono, horticulturalists and hunters in Bolivia. There, Holmberg (1946) notes, "Much more intercourse takes place in the bush than in the house. The principal reason for this is that privacy is almost impossible to obtain within the hut where as many as fifty hammocks may be hung in the confined space of five hundred square feet."

When do people have sex? In the United States, sex is more common on weekends, an unsurprising finding in the context of a fast-paced work schedule during the week. In Alfred Kinsey's original mid-twentieth-century studies, he and colleagues reported that Sunday mornings were a particularly common time for marital sex. In a wider cross-cultural scope, the timing for sex clearly varies in relation to issues such as privacy, work, and religious considerations. During the holy month of Ramadan, for example, Muslims are not supposed to engage in sex during daylight hours, when they are fasting (and the lack of food and water throughout the day could impinge on this desire in the first place). In their study of the Aka hunter-gatherers' sexual behavior, Hewlett and Hewlett (2010) note that one factor perhaps facilitating frequent night-time sex is the lack of electricity. Sex provides one form of intimate recreational activity when few others are available. In the wider sample studied by Ford and Beach (1951), there is no clear pattern of preferring sex during day or night, with both occurring.

What are the most common positions for sexual intercourse, and do these vary cross-culturally? The most common position is the missionary, or ventro-ventral position, with the couple facing each other, the woman on her back and the man above (Gregersen 1994). This often results in substantial body-body contact during coitus. This is a common position adopted across age groups in the United States, and among such diverse groups as the Crow of North America, the Chagga of Tanzania, and the Miao of China, as well as other societies. Some have suggested that the "missionary" aspect refers to its association with missionaries' encouraging peoples to give up alternative sexual practices. However, as Priest (2001) has detailed, this is a misnomer, and while the church did advocate procreative sex occurring in a face-to-face man-on-top "missionary" position, this name is derived from historical inaccuracy. While there is certainly variation, this is a commonly recorded sexual position historically and across cultures, with stylistic specifics of the position (such as placement of the woman's legs) often varying by common cultural practice.

Another regionally common sexual position is the "Oceanic" position, referring to the islands in Melanesia, Micronesia, and Polynesia (Ford and Beach 1951; Gregersen 1994). In this position, a man squats or kneels in front of a woman lying down. They can minimize bodily contact, or they can pull themselves toward each other to enable more intimacy and deeper penetration. This position can offer increased stimulation of the clitoris and surrounding vulvar areas. In Polynesia, sexual positions with women on top are commonly reported, consistent with a cultural concern for women's sexual pleasures; as an example, among the Marquesans, positions include the woman astride the man, sitting on the man's lap and erect penis while facing him (Suggs 1966).

Positions in which women are on top offer the greatest potential for female sexual enjoyment. Women can control the rhythm and movements more, and may obtain more vigorous clitoral stimulation, which in turn provides enhanced possibilities for pleasure and orgasm. Ford and Beach (1951), found relatively few societies where this position (woman on top) is the favored one, though it is often used along with other positions: "The variant with woman squatting upon the man is a secondary position for a number of peoples, and it is significant that they describe this method of copulation as the one which brings the greatest satisfaction to the woman" (34–35). However, there are some indications that female-on-top positions are associated with higher sperm loss after intercourse and lower sperm exposure to the cervix, which leads one to wonder whether conception is slightly lower in female sex-on-top positions. Surprisingly, however, we are unaware of any data demonstrating whether conception is more or less likely with sex in an Oceanic position versus the missionary position versus a ventral-dorsal position (male from behind, colloquially called "doggy style"). We hypothesize that a ventral-dorsal position may maximize fertilization success given the location of the cervix and pooling of semen after ejaculation, but we leave it to future scientists to settle this question.

Other positions enter the equation in some societies and in some contexts (Ford and Beach 1951; Gregersen 1994). Positions in which a man and a woman both lie down, facing each other, are preferred in some societies, such as the Masai and Kwakiutl. Positions with both partners lying down and with the man from behind with the woman's back pushed to his chest, are often late-pregnancy resorts, when concerns over the pregnancy and female comfort lead to temporary adjustments in sexual position. Positions in which a couple is standing, with the man from behind, are rare, despite being the most common sexual position among other mammals. In some cases positional variations may be related to quick sexual encounters (including extramarital affairs) when time and privacy are important factors. Various other positions for intercourse require situating them in context. The sexual variation detailed in the *Kama Sutra* (an ancient Hindu text that contains, among other things, a sex manual) reflects a capacity among an elite in India to read and indulge in acts few others would have had time for at the time this book was illustrated (potentially during the third century, although some historians have debated the exact time period). The same could be said of paintings of Chinese urbanites having sex on a swing, or the erotic vignettes found on an ancient Egyptian papyrus scroll-painting (sometimes called the Turin Erotic Papyrus), showing a burly man having sex with an attractive woman in various positions. And sexually graphic, twisty, naughty, stone facades displayed on historic Indian temples can in part be seen as advertisements to patrons seeking to pay for the flexible services of a temple prostitute.

HOMOSEXUALITY

Among the Sambia, horticulturalists of highland New Guinea, preadolescent and adolescent males engage in same-sex sexual behavior (Herdt 1981). More specifically, they fellate (perform oral sex on) older males. These practices may continue for years, and they occur within a social context in which young males are supposed to shun contact with females. The semen they ingest is, according to Sambia ritual, imbued with important properties that help turn boys into men. Yet it does not typically turn them into men with a lifelong same-sex sexual preference. As adults, only about 2 to 3 percent of Sambia men have reported a same-sex sexual orientation (a sexual identify not exclusively tied to sexual behavior) in adulthood, a figure comparable to the reported 6 to 8 percent of men who identify as gay in the United States. Among the Sambia, these same-sex sexual activities are age-graded practices that eventually give way to the fellators becoming the fellated, and eventually becoming the next wave of married men and fathers. These activities are part of socially expected development. Furthermore, related male same-sex practices occur among various other New Guinean societies, indicating that these are part of a wider cultural milieu.

The cultural context of these same-sex sexual practices also includes polygyny, a degree of sex segregation, and a considerable degree of male-female animosity. One interpretation of these practices is that they help apprentice younger males with older ones, keeping potential male competitors sexually active and yet removed from the wives of older married men. If adolescent and young adult males have unalienable sexual desires, this is one way to channel them away from sexual competition with more politically established older males. The bonds forged between younger adult males can also be quite important. In a context of high intergroup violence, these relationships may facilitate mustering the manpower to both ward off and take resources from enemies. So same-sex sexual activities may foster enhanced male-male bonds. Still another view of these practices is that they are relatively common aspects of sexual development in humans, and indeed in many social mammals. Through sexual experimentation and involvement in sex, including with members of the same sex, males learn about their place in the social hierarchy and about sexual pleasure. Showing, too, that these same practices are far from essential, they have fallen by the wayside in more recent years, in the face of rapid social change in New Guinea (Herdt 2006). If the social pressures favoring these practices disappear, so too do the practices.

The sexual experiences in Sambia open up a wider discussion of homosexuality. What do we know about same-sex sexual activities from a cross-cultural perspective? How often, and in what circumstances, are they associated with lifelong (rather than age-graded) same-sex sexual orientations or alternative sex roles? While gay districts may be flourishing in cities around the world today, we want to place discussions of homosexuality in a wider cultural context. In Ford and Beach's 1951 compilation of seventy-six societies, among twenty-eight were reports that same-sex sexual practices "on the part of adults are said to be totally absent, rare, or carried out only in secrecy" (136). In the remainder of the societies, they were socially accepted. The vast majority of these data points refer to male activities, with anal intercourse the most commonly practiced same-sex sexual behavior. Cross-cultural analysis indicates that male-biased sex ratios and a higher prevalence of polygyny are associated with more favorable attitudes toward male same-sex sexual behavior and higher rates engaging in it (Lehr et al. 2012). Same-sex female activities have been both less well studied in cross-cultural contexts (Blackwood and Wieringa 1999) and, where they have been studied, appear to be less common than male same-sex activities, despite attracting less moral or legal attention (fewer laws proscribe female-female sexual behavior compared with same-sex male behavior, such as "sodomy" laws). As an example, Aranda females were reported to engage in mutual clitoral stimulation and simulated intercourse, but few other details have been provided on this Australian aborigine society. The term *lesbian* traces to the Greek island of Lesbos, where the poet Sappho was born more than 2,700 years ago; her words describe passions for both women and men, but whether this entailed consummation in same-sex sexual behavior is not clear.

Cross-cultural patterns indicate that male same-sex sexual activities in societies other than the Sambia have age-graded or context-specific elements to them (Kirkpatrick 2000; Poiani 2010). In other words, same-sex sexual activities tend to be practiced by the same males who at some point will marry and father children, and are distinct from a lifelong same-sex orientation. Examples include other Melanesian societies related to the Sambia, and also the Azande of Central Africa, Siwah of North Africa, Japanese samurai (who might be apprenticed to males with whom they maintained emotionally and sexually close relationships over years), and classical Greek society. Yet, this same-sex sexual behavior does not necessarily characterize an individual's sexual orientation; that is, there is a distinction between same-sex sexual behavior and identifying as gay or lesbian. In parts of Latin America and North Africa men who adopt a penetrative role (as opposed to a recipient role) in anal sex are less likely to view themselves as homosexual, for instance. We could add other contexts in which men might be unable to access available female sexual partners, so they resort to engaging in same-sex sexual practices with other males; prisons serve as one example, and so do some all-male boarding schools. Same-sex sexual behavior in prisons is particularly illuminating, as inmates may serve terms of many years within a unique, often aggressive, same-sex environment. With some reports suggesting that nearly one-fourth of male inmates have been rape victims, the sexual politics of prison depict an image of sexual behavior not predicated on sexual orientation, but rather on physical pleasure and power. Male inmates who willingly form same-sex sexual relationships with other males may consider themselves heterosexual, but in the absence of available females or in an effort to avoid rape, they will engage in same-sex sexual activities, including forming romantic relationships. This type of facultative or situational sexual expression (occasionally termed situational sexuality) serves as an example of how sexual orientation, an internal sexual preference and identity, can operate independently from sexual behavior.

That said, the cross-cultural record also indicates some cases of lifelong and distinct nonheterosexual sexual orientations or gender roles (Aldrich 2006; Roughgarden 2004). Among the Mojave who lived in the region of the Nevada-California border, for example, both heterosexual and two-spirit men (*alyha*) were recognized, as well as heterosexual and two-spirit women (*hwami*). And the Mojave were hardly alone in the Americas for these patterns. The sometimes pejorative term for a similar alternative male role, "berdache," could be found among a number of Native American societies, including the Zuni and Navajo of the Southwest. Archaeologists have attempted to find skeletal and burial evidence of occupational specialization as well as the ritual spaces held by two spirits, helping to extend an appreciation for the depth of these roles in the history of various Native American societies (Voss 2008).

In ancient Greece, same-sex sexual behavior took a particularly interesting form. The practice of pederasty ("boy love") was widespread, in which an older man would mentor a younger man, providing education, resources, and emotional commitment. Yet a form of gender roles persisted, with the active role of penetrator being associated with masculinity and status (typically the older man), and the passive role of penetrated partner being associated with femininity and lower-status youth (typically the younger man). It was a military asset for members of a pederastic relationship to fight together, to promote bravery and loyalty to each other on the battlefield; in fact, the Sacred Band of Thebes was a special troop of men and their boy lovers. The age range for the "boy" was generally between twelve and seventeen years of age, as pubertal development in the later teen years, such as body hair, was seen as indicative that the younger penetrated partner was ready to depart as a man able to marry and copulate with women. Yet sexual and romantic relationships between two adult men were less acceptable, due to the importance of being a masculine penetrator. Thus, individuals who engaged in passive homosexuality past an acceptable age of pederasty would have been considered to have "made a woman" out of themselves, and be subject to substantial social prejudice and derogation.

Elsewhere in the world, the *xanith* of Oman are males who dress like women and serve as prostitutes to men. In Thailand, kathoey, commonly known as ladyboys, are males who dress like women and in modern times have often received hormonal replacement therapy, cosmetic surgery, or complete gender reassignment surgery (surgical breast implants, and removal of the penis and testes); these individuals frequently work in entertainment and as sex workers. In a number of Polynesian societies, a third gender category exists, known as *mahu* in Tahiti and *fa'afafine* in Samoa. These are typically roles held by genetic males who display feminine characteristics and gender roles, and who may provide sexual services such as fellatio to other males. To a lesser degree, masculine females may also occupy a role as *mahu*. As another case, *Hijra* in India are low-status males who present themselves as transvestites or who have been castrated. The explorer and ethnographer Richard Francis Burton's recordings of his nineteenth-century travels in North Africa gave the impression that young gay males were plentiful there, but they seemed to be so partly because they were paid for providing services to outsider men, effectively serving in an early example of sex tourism.

OF SEX ALONE, WITH PROSTITUTES, AND WITH BEASTS

Las Vegas is graced with the Erotic Heritage Museum. Situated next to an exotic dance club not far from the Strip, the museum offers one of the largest assemblages on planet Earth of sexual images, films, and paraphernalia, generally. In wandering its halls, we notice that a fair number of the glass cases harbor phallic-shaped objects serving one of several functions: as deflowering devices (for females, before making their sexual debut) or as dildos. The exhibits prompt the question (well, really, lots of questions): How long have dildos been around and are they widely used? This is another of those archaeological conundrums. The shape of these objects suggests several potential uses, and thus definitively determining what is a dildo or not is tricky. That said, a 2010 finding of bone in Sweden looked the part, and at more than 3,000 years old, it is but one of other early examples, with others including known dildos from ancient Egypt. One religious practice in ancient Rome involved the woman squatting over a wooden phallus of a deity after marrying, so that if she were a virgin she would harbor no resentment toward her husband for the pain of first penetration. The use of wooden dildos and even cassavas and bananas among the Azande in Africa shows that various materials have been used for this purpose.

Sex objects like these also facilitate self-pleasuring. But is masturbation (with or without an assistance device) found everywhere? Some biblical invectives against males spilling their seed suggest that desire may be present but variably encouraged (in some cases, folk legends about health risks of masturbation have been used to frighten adolescents and young adults from practicing it). An earlier reference to Aka sexuality indicated an absence of understanding of either male or female masturbation. The Aka apparently have enough premarital and marital sex to preclude such an outlet. Among the Kagaba of Colombia, male masturbation is viewed as an awful waste of semen. It is also frowned upon by the Mehinaku of Brazil. Across societies, male masturbation is more commonly reported than is female masturbation, especially during adolescence, and female masturbation more often tends to entail use of sexual implements (such as dildos).

In an unusually detailed account of female masturbation and same-sex sexual behavior, Isaac Schapera (1941) describes practices among the Kgatla of southern Africa that suggest females use other sexual outlets when male offerings are inadequate. He writes,

> Women who do not take lovers almost invariably practice masturbation. This generally takes the form of rubbing the finger about in the vagina, or of playing with the labia minora. Some women have small sticks covered with rags, which they keep carefully hidden when the husband is at home, and use as a substitute penis whenever he is away. Occasionally two very close friends sleep together and indulge in homosexual play by fondling each other's labia, or, more usually, one slips her fingers into the other's vagina and rubs them about, while the latter, 'feeling very nice', responds by moving her body about as if copulating. But there do not seem to be any genuine homosexual affairs among women, involving the total exclusion of men. All those with whom I discussed the matter said that the practices just described were employed merely to obtain gratification in the absence of their lovers or husbands. The vast majority of women, if not all, prefer normal coitus to all other forms of sexual activity. (Schapera 1941: 183)

Billed as the oldest profession, prostitution has been around quite awhile in complex states with a division of labor. It was recognized in the archaeological remains of Pompeii (Voss 2008); it has been quite common historically throughout much of Europe; it has a deep history in China; and by some accounts it had a presence in precontact America. In Thailand, the history of prostitution extends back at least to the fourteenth century, with women earning money that way and often men achieving their sexual initiation in the arms of a prostitute (Bolin and Whelehan 2009). However, prostitution is lacking in small-scale egalitarian societies of hunter-gatherers, meaning that it does not exist across all societies. Still, any sort of exchange of sexual favors for resources can shade into such a dynamic. Among the Bari of Amazonia, for example, the practice of offering fish or meat to a pregnant woman for sex would meet some definitions of prostitution.

Prostitution is overwhelmingly a service provided by women to men (Bolin and Whelehan 2009; Symons 1979). Most male prostitutes service men rather than women. This is true both of typically local prostitutes' services and of the widespread international industry in sex tourism, where clients travel to destinations for sex. The prices for women's sexual services also run a wider gamut than for men's sexual services (see, for example, Pisani [2008] on prostitution in Indonesia); attractive, socially engaging women can command high-end prices that do not have ready parallels in male prostitution. In the Jamaican anecdote at the start of this chapter we find one of the examples closest to female-oriented prostitution, with females providing resources in exchange for sexual favors. Note, though, that the intertwining in that same example of romance and exoticism suggests that this prostitution-like case takes a different form from male-oriented prostitution, which features sex more explicitly. It is also worth noting that in interviews that Justin and his colleagues have conducted with female prostitutes in the United States, they have found that not all are single; in a handful of interviews the women describe having loving relationships with their boyfriends or husbands outside of work, and several describe scenarios in which men are paying for sexual services but also looking for intimate exchanges, such as talking and cuddling (in some cases spending the whole of a customer's purchased time on the latter and never getting around to actual sex).

Attitudes and behaviors related to prostitution vary around the world. Governments in many countries have carved out a niche for legal prostitution, from parts of Mexico to Canada to Australia to even some counties within the State of Nevada. When large groups of males are clustered together without the availability of long-term partners, this often gives rise to some sort of sexual outlet, whether with nonhuman animals (more on this momentarily), with other males (think prisons), or with prostitutes. The rampant spread of syphilis and other sexually transmitted diseases that accompanied male troop presence in World War II Europe is just one manifestation of the ready use of prostitutes' services in France and elsewhere, and may help account for a view still held today by many Americans, that the French are great lovers.

Returning to an international compilation of sex data discussed earlier, Wellings and colleagues (2006) report variation across countries in the percentage of men who say they paid for sex in the past year. In parts of central and southern Africa, a slightly different definition of providing money, gifts, or favors for sex suggests that approximately 11 to 14 percent of men have engaged in such exchanges. Across various regions of China and in Hong Kong, the percentage of men who pay for sex is about 11 percent; for East and West Africa, around 9 to 10 percent; for the Caribbean, about 6 to 7 percent; for South Asia, around 3 to 5 percent; for Southeast Asia, around 3 to 10 percent; and for Latin America, North America, Europe, central Asia, North Africa, and southwest Asia, it is less than 3 percent. In a detailed account of the sex industry in Indonesia, Elizabeth Pisani (2008) notes that single men are more likely to frequent prostitutes, a pattern that likely holds true widely throughout the world.

Attitudes toward prostitutes themselves vary within and between societies. Some view prostitutes as victims of coercion and male power, but given the variation in the experience of prostitutes, it is difficult to generalize so broadly (Bolin and Whelehan 2009). In cases where women have been forcefully pushed into prostitution or misled about the true nature of a job they sought elsewhere, this view certainly has support. A number of women traveling from Eastern Europe and central Asian countries to jobs in Western Europe have encountered these sorts of coercive realities. In other places where prostitution is legal and women can choose to move in and out of the occupation by choice, they may find it more lucrative than other forms of work, and safer than non-institutionalized prostitution in terms of potential sexual/physical violence and condom use (Weitzer 2010). Indeed, in some cases, adding more fuel to the political dimensions of prostitution, parents may facilitate their children's entry into such work because it pays better than alternatives (see Taylor 2005 for a related discussion involving Thailand). In a discussion of sacred prostitution in India, for example,

Nagendra Singh comments that girls are often pushed into the lifestyle: "Their parents pressurise them, tempt them and plead with them to go and help their family monetarily. Sometimes, they are asked to sacrifice their comforts and dreams for the good of their younger sisters, and brothers who are to be married" (Singh 1997: 196). In these latter cases, moral, religious, and political arguments (from wives who loathe such competitors, or perhaps parents who shun these realities when raising their daughters) help account for the ambivalence over prostitutes and the wider occupation.

Beyond masturbation and prostitution as sources of nonrelationship sexual expression, human sexual glances occasionally cross the species boundary, opening up discussion of bestiality (Beetz and Podberscek 2005; Gregersen 1994). The topic raised eyebrows when, in Kinsey's classic mid-twentieth-century studies, it was noted that a considerable number of men and women reported sexual experiences with nonhuman animals—approximately 8 percent of men and 3.6 percent of women. For many males, these reports were related to living on a farm and having sex with animals like cows. And these Americans were hardly alone. Although largely symbolic, many of the exterior temple sculptures at Khajuraho, India, include people having sex with animals, such as a man having intercourse with a horse. In a nearly 10,000-year-old cave painting in northern Italy, a man about to have sex with an animal is depicted. In ancient Greek mythology Zeus took the form of a male swan and seduced Leda, queen of Sparta.

Bestiality (also called zoophilia) has been reported in many societies. One of the patterns associated with it is that more males than females tend to engage in bestiality. It is especially common among young, unpartnered males who have close access to potential nonhuman animal consorts. These situations can range from males having sex with llamas among the Aymara to males mounting cows, camels, sheep, horses, or other herd animals in various Eurasian pastoralist societies. Among the more surprising of partners are some birds, such as geese; by attempting to penetrate these animals' cloacae and then killing them, some males apparently relish the sensation offered by the birds' last muscle contractions. Interestingly, when females engage in bestiality it tends to take a different turn. Dogs are some of the more commonly used partners, providing cunnilingus or, less commonly, engaging in intercourse with women.

Among the most gut-wrenching images of bestiality we have seen was an image of a shaved, restrained female orangutan, shown in a presentation by world-renowned primatologist Birute Galdikas, who conducted the first long-term field studies of these close kin of ours. This orangutan was used to provide sexual services to human males in Southeast Asia, where apparently a small interest developed in paying for sex with the animals. Orangutans, however, have almost returned the assault. A number of human field researchers have been grabbed and nearly sexually molested by orangutans themselves, with only the researchers' clothes perhaps halting some otherwise forced copulations initiated by orangutans.

Stretching the scientific imagination about bestiality, one scholar recently asked whether a sexual orientation toward nonhuman animals exists (Miletski 2005). Suggesting that "zoosexuality" meets the core elements of a sexual orientation—namely, affectional orientation, sexual fantasy orientation, and erotic orientation—Miletski finds that a few people in a sample engaging in bestiality would meet this definition. We do not advocate one way or another for recognizing such a concept, but we do marvel at the remarkable variation in human sexuality highlighted by this single example, not to mention this entire chapter. With that observation, we now leap into deeper reflections on a theme of this book: the importance of long-term bonds to human sexuality, including the role of romantic passion, the psychological experiences of pair-bonding, its neurobiological substrates, and more.

CHAPTER THREE

Love and Maybe Marriage

Patterns of Pair-Bonding and Romantic Love

> The human craving for romance, our drive to make a sexual attachment, our restlessness during long relationships, our perennial optimism about our new sweetheart—these passions drag us like a kite upon the wind as we soar and plunge unpredictably from one feeling to another. These emotions must come from our ancestry. So I shall propose that they evolved with genesis, to drive our ancestors in and out of relationships.
>
> —Helen Fisher, *Anatomy of Love*

WHILE DARWIN CONTEMPLATED the joy of marriage, some of his predecessors pondered love and possibly marriage with perhaps more artistic flair. In *The Symposium,* Plato describes a dialogue on the nature of romantic love. Several clever men assert their views, including that love's origins are old among the gods of Olympus and love's presence can bring men honor, that it is the source of good health and happiness, keeps us youthful but brings us wisdom, and turns us all into philosophers. During Aristophanes's speech, he provides an origin myth for why we mortal humans seem to crave romantic love so much. He contends that in the beginning there were spherical creatures inhabiting the earth, they were all male, all female, or "androgynous" (both male and female). In an act of retaliation, Zeus, king of the gods, chopped each in half. And so, the story goes, humans were born from the halving of that more complete self, but now we spend our lives seeking another person to love in order to feel whole. We live in an attempt to reach our original unhalved form, to be complete, to be in love. Those are some tall poetic charges for one feeling or emotion—it is fitting then that Plato is also credited with the famous quote, "at the touch of love, everyone becomes a poet."

But, what does science have to say when it comes to "love"? In Chapter 2 we reviewed several marital patterns, but marriage is, in many ways, a social and legal construct that may or may not be formed around a love relationship. In this chapter we will continue considering cross-cultural patterns of human sexuality, with greater emphasis on the context in which sex most often takes place: in long-term pair-bonds. Attachment bonds, including child-caregiver relationships and friendships, are extremely important to the survival and health of individuals, from Old World monkeys to apes and humans, and from hunter-gatherer groups to modern industrialized societies (Carter et al. 2005; Hewlett and Lamb 2005). Attachment bonds are a motivational desire for humans, and the romantic pair-bond is an especially important one. Because we are highly social animals, the protection and nurturing of human offspring is a collaborative effort. This raises the question of what role romantic love, a seemingly inordinately profound experience, plays in human evolution and human sexuality. This also advances some very basic questions about the nature of love, including whether love is a feeling, an emotion, or something deeper (perhaps a motivation). Is romantic love just a Western notion? Is there a biological basis to the power of passion?

In this chapter we explore the highs and lows of romantic attraction and attachment, and attempt to show that despite what some theorists have argued, there is plentiful evidence that romantic love is much more than a modern-day theme of poetry and art—it is a human universal, a motivational drive engrained in our evolved biology. We contend that love is an adaptation to promote survival and reproduction, and that it coincides with cooperative breeding models of human evolution. The nature of these relationships is integral to appreciating the context of most human sexuality. Many sex researchers, for a long time concerned with individuals, are increasingly becoming interested in couples and the dyads in which love and sex exist, that is, the sociosexual context. Around the world, love relationships have driven humans to do extraordinarily dramatic things, from building shrines to waging war; the efforts lovers put into potential long-term relationships characterize the most intense of all sexual encounters—compare these with the relatively minimal efforts individuals put into pleasing a prostitute or courting a short-term sexual partner.

HUMAN ROMANTIC LOVE AND BONDING

Across history and cultures, humans have often attempted to reconcile a desire for sex with a desire for romantic love. Within the United States and elsewhere, emerging adults are often told not to confuse the two, the presumption being that the power of passionate love is stronger, deeper, and more meaningful than sexual activity in the absence of love. Humans produce material culture that highlights this conflict: works of art, music, and literature from many different societies are devoted to the struggles of love. In Indian literature, for example, the Moghul prince Saleem and Hindu slave Anarkali fall fervently in love with each other, causing Saleem to wage war against his disapproving father, the great Moghul emperor; no match for his father, Saleem watches Anarkali sacrifice her life in order to save his and have one last night of passion together before she is entombed alive. According to ancient Greek mythology, Eros's arrow caused Helen of Troy to leave her husband, Menelaus, and his effort to bring her back started the epic Trojan War. Set during the Eastern Jin Dynasty, the Chinese legend of Zhu Yingtai and Liang Shanbo's love depicts a deep emotional intimacy that grew over many years until Zhu's father arranged for her marriage to another man, which caused Liang to fall irreversibly ill and die; on her wedding day Zhu hurls herself into Liang's grave, from which their spirits then emerge together as two beautiful butterflies dancing in the wind. In William Shakespeare's late-sixteenth-century classic *Romeo and Juliet,* both main characters take their own life at the sight of their lost beloved. And even today, when we lecture on love in our respective human sexuality courses at different universities time zones apart, the unmistakable smiles of absolute joy and glossy eyes of broken hearts are an inevitability.

While various literary examples may highlight cross-cultural commonalities, this certainly isn't enough to demonstrate a human universal. However, a landmark cross-cultural analysis using standardized behavioral acts and emotional expressions might tell us more. The study we refer to found that romantic passion was observed in the ethnographic data of every society that commented on the topic, which included 88.5 percent of the 166 societies sampled (Jankowiak and Fischer 1992). The findings for the remaining societies were all inconclusive, as data did not exist on the appropriate questions—meaning there were no negative findings. Over the years, William Jankowiak has updated these data to include more societies that were originally classified as inconclusive, so that most recently, 151 out of the 166 societies have now been documented as expressing some form of romantic love (Jankowiak and Paladino 2008). However, like all aspects of mating and reproduction, cultural rules vary for appropriate expressions of love (Hatfield and Rapson 2005; Jankowiak 2008). An interesting relationship exists between sexual desire and love, which in Euro-American culture are often assumed to occur together. This is more than just a perception in the United States, where sexual dissatisfaction is a known risk factor for relationship dissolution and divorce (Karney and Bradbury 1995). On the other hand, Hewlett and Hewlett (2008, 2010), in their ethnographic work on two central African groups, the Aka foragers and Ngandu farmers, detail how both groups have a sense of love that is distinct from sexual desire. Sexual intercourse among the Aka and Ngandu is for the explicit purpose of procreation, not for pleasure. Aka couples have sexual intercourse an average of three times per night, with two days of postcoital rest; Ngandu couples average two times per night, with three days of postcoital rest. On average, U.S. couples in the same age category are reported to have sex 86 times per year, compared with Aka couples' 439 times per year, and Ngandu couples' 228 times per year. It is important to note that these numbers do not take into account different reproductive states, and as we will discuss later, rates of sexual behavior are likely much lower during pregnancy and lactation. At first it might appear as if the Aka and Ngandu are more passionate, more sexual, or more loving than couples in the United States—but such a comparison would be rash, and baseless using the frequency-of-sex data alone. As it turns out, both Aka and Ngandu people believe sex is "work," albeit more pleasurable than "day work," but nonetheless work for the purpose of creating children. Both groups identify love as a separate but important attachment bond with a partner that facilitates survival and the raising of offspring (Hewlett and Hewlett 2008, 2010). As one Ngandu man is quoted as saying: "I love my first wife the most, she is closest to my heart. She helps me and gives me food and respects me. We did not have children together; she was not able to. Now she does not menstruate, and we no longer have sex. I have sex with my second wife, to take care of the desire, but it is the first wife I love the most" (Hewlett and Hewlett 2008).

Researchers have identified a variety of traits that characterize what "being in love" actually means, including the ideas that the person one loves takes on "special meaning" and that we experience a host of responses to the one we desire: intrusive thinking, focused attention, labile psychophysiological responses, longing for emotional reciprocity, emotional dependency on the dyad, empathy and responsibility, impulse to make a good impression, sexual desire often including sexual exclusivity, craving for emotional union, and a sense of involuntary passion (Fisher 1998; Harris 1995). Although on average the sexes will differ in their expressions, the desire for romantic love, much like the desire for sex, appears to be largely sexually monomorphic—that is, despite the sex differences characterized in Chapter 1, pair-bonds are generally a prerogative for both men and women. While men and women may initiate courtship with different expectations, and desire different traits in a potential mate, both sexes typically seek and find pair-bonds. A common misperception is that relative sex differences in potential reproductive rate and parental investment will cause men to solely prize sexual desire and women to prize emotional gratification. In a charming series of studies on expression of love, Ackerman, Griskevicius, and Li (2011) actually showed quite the opposite. Despite what most people believe, that women say "I love you" first and desire it sooner (they even demonstrated this false perception among study participants), their studies showed that it is actually men who tend to confess their love first and to report feeling happier than women when hearing their partner say it back. With women as the choosier sex, it likely behooves men to be focused on their partners and to demonstrate affection and commitment. But what are the implications of a simple expression like "I love you," or *Jeg elsker dig* (Danish), *Aloha Au Ia 'Oe* (Hawaiian), *Bi shimbe hairambi* (Manchu), *Ngiyakutsandza* (SiSwati), or even *Dw i'n dy garu di* (Welsh)?

To operationalize the term, Hatfield and Rapson provided the following definition for passionate love: "A state of intense longing for union with another. Passionate love is a complex functional whole including appraisals or appreciation, subjective feelings, expressions, patterned physiological processes, action tendencies, and instrumental behaviors. Reciprocated love (union with the other) is associated with fulfillment and ecstasy, unrequited love (separation) with emptiness, anxiety, or despair" (Hatfield and Rapson 1993: 5).

The desire for love relationships, and the outcome of raising offspring with both maternal and paternal investment, strongly exists for both sexes. A romantic relationship can even have profound influence on one's health. But this bond goes beyond social graces. Couples have been shown to coregulate physiological responses, and relationship satisfaction and positive marital communication are associated with less stress and lower cortisol levels, reduced inflammation, and improved wound healing (Kiecolt-Glaser, Gouin, and Hantsoo 2010). Married men and women are more likely to survive a heart attack (Chandra et al. 1983), and epidemiologists have long suggested that people who are unmarried have a higher risk of death and a lower life expectancy (although that's not to say everyone who is married or cohabiting is in love). Love can even heighten sexual pleasure. In the 2010 National Survey of Sexual Health and Behavior, men who had their most recent sexual encounter in the context of a romantic relationship reported greater arousal, pleasure, orgasm, and erectile function than men with an uncommitted sexual partner. And in another national study of 7,463 adults, across all age categories, couples with the highest marital satisfaction also reported the highest frequencies of sex (Call, Sprecher, and Schwartz 1995).

Romance has been studied by a wide variety of scholars, from psychologists to classicists, anthropologists to historians, and thus there exist a multitude of ideas and theories on the topic, some more scientific than others. One common theme, however, is that love is a constellation of feelings and emotions, idiosyncratic to each individual and expressed variably in different contexts and cultures. Elaine Hatfield and Richard Rapson apply a widely used model for differentiating "passionate love" and sexual desire—a sense of infatuation—from a less intense form of "companionate love," which includes deep attachment and commitment (Hatfield and Rapson 2005). The triangular theory of love, proposed by Robert Sternberg, suggests that the type of love one experiences is based on the relative combined intensity of three components: intimacy (liking), passion (infatuation), and commitment (Sternberg 1988). William Jankowiak details a cross-cultural tripartite conundrum the conflict being an internal one among the sexual imperative, the romantic, and the companionate (Jankowiak and Paladino 2008). Helen Fisher has argued that humans have evolved three distinct yet connected behavioral-brain systems to direct mating and reproduction: lust (sex drive and libido), romantic attraction (passionate love), and attachment (bonding) (Fisher 2004). All of these particular classifications are intellectually warranted and have biological, psychological, and cultural evidence on their side.

Fisher contends that humans have evolved three primary brain systems to ultimately direct reproduction, and passionate love is but one aspect. She has argued that mammals and birds engage in a variety of preferential mate choices, and romantic attraction is a primary motivation. The complex process of human romantic love is thus a developed form of the general brain systems for mating attraction seen in other species. Perhaps Darwin was not so far off when describing mate choice in courting birds as "love" (Darwin 1871). In this view, attraction evolved to stimulate individuals to desire and pursue specific partners. Lust or libido (sex drive) is generally a precursor, motivating individuals to crave sexual engagement with conspecifics, while attraction allows one to then focus in on a particular partner in the crowd.

A generalized sex drive is a common feature of mammalian and avian reproduction, and females of many species experience fluctuations in sexual receptivity across the estrous cycle. As we will discuss later in the book, human females show subtle cyclical variations in sexual initiation but remain sexually receptive (physiologically) throughout the menstrual cycle. Compare this with the giant pandas of China: female pandas are receptive only during estrous, which occurs once per year and lasts only two to three days. Sexual motivation is fundamental to direct copulation and reproduction, but its form varies substantially depending on the species. Libido is influenced by a variety of psychosocial, developmental, and biological factors; in primates, it is centrally influenced by androgens (such as testosterone), and it appears to be processed in parts of the brain distinct from those involved in romantic attraction (Fisher, Aron, and Brown 2006)—we will return to details of the sexual response in Chapter 8.

PAIR-BONDS

Romantic love is characterized by focused attention and is reminiscent of a more general courtship attraction. It tends to be much stronger and more intense than sexual desire, in that individuals will go to much greater lengths for passion than for sex alone. This is why many societies that practice arranged marriages have strict rules to prevent young men and women from falling in love on their own accord. Take the Makassar of Indonesia, who practice traditionally arranged marriages. While young boys and girls may play together, young men and young women of marrying age are forbidden from doing so in an attempt to minimize the risks of their developing liaisons beyond the arranged marriage (Röttger-Rössler 2008). This policy recognizes the ability of individuals to fall in love naturally, and aims to promote the growth of a love relationship between those whose marriages have been arranged (by others). In mainstream Euro-American culture, marital practices only recently became focused on love and romance between partners (Coontz 2005). This is a likely return to a type of mating and pair-bond formation deeper in our evolutionary history, before sociocultural practices gave us the legalization of pair-bonds by way of marital practices. However, among lower classes such as the European peasantry, where property, political, and economic exchanges were not such an important issue, more individuals are presumed to have married for love.

The copulatory investment for human reproduction is, surprisingly, greater than a short-term sexual liaison would generally warrant. That is, one reason human reproduction almost always takes place in the context of a long-term bond is that it can take couples many months of "trying" (a polite way of saying "having lots of unprotected sex") before conception, and then the birth of offspring, occurs. Taken together with the importance of pair-bond relationships as a human universal, short-term sexual behavior appears to have exerted relatively minimal evolutionary pressure in terms of actually resulting in pregnancies. A vast majority of human reproduction occurs within pair-bonds or extended extrapair partnerships.

Romantic attachment, or a deep feeling of commitment, encourages individuals to pair-bond with a partner through the duration of parental duties for the raising of successful offspring. Pair bonding simply refers to the bonded relationship that develops between individuals of a species, typically in a breeding pair. Some individuals pair-bond for life, remaining socially monogamous from the time of union until death. Take, for example, two high-school sweethearts who stay faithfully in love, through triumphs and adversities, until their deaths some three-quarters of a century later. The Laysan albatross, like nearly all seabirds, is also monogamous; the male incubates eggs while the female hunts for food, and in the event one of the pair dies, it will often take the other several years to find a new mate and begin breeding again (H. I. Fisher 1976). Consider the Djungarian hamsters of Mongolia: the males assist in the birthing process, and one study shows a 95 percent success rate (still alive at time of weaning) for pups raised by pairs, compared with 47 percent for pups raised by the mom alone (Wynne-Edwards 1987). Or take the beautiful white mute swans Justin often sees proudly sailing the waters during the summer in Cold Spring Harbor, New York. Swans, a popular attraction at wedding locations along the water, often pair-bond for life—although "divorce" can happen (yes, this is the term used in the animal behavior literature), as can infidelity. Because swans, as it turns out, prioritize attachment, males have been known to pair-bond with other males in the absence of a female mate.

Among some species, pair-bonding may occur for a breeding season, with potential for pairing again the following season. Recurring pair-bonds are seen in some species of penguins during nesting season, and they are presumed to recur among gray foxes, which are solitary much of the year except when breeding and raising young. Many species will remain pair-bonded through gestation and the rearing of offspring and then form new pairs. In fact, in a study of fifty-eight different human societies, H. E. Fisher (1989) showed that divorce rates tend to peak approximately four years after the birth of a child, suggesting that once primary childrearing responsibilities are over, many unions dissolve. Fisher also demonstrated that divorce and remarriage are common cross-culturally, suggesting a pattern of "serial monogamy." Pair-bonds in other animals have all the highs (romantic love?), lows (divorce, infidelity), and even variation (life-long, short-term) that we see in human mating. Despite all the variation in mating systems, a consistent theme among humans is the desire to form romantic attachments, bonds with others to raise offspring, another cross-cultural and biological pattern. More than 90 percent of bird species will form pair-bonds, but less than 3 percent of mammals do (although among primates roughly 15 percent of species form socially monogamous pair-bonds). And while polygyny is permitted in some 83 percent of human societies, it is linked to an individual's rank and wealth; in most societies that permit polygyny, typically only 5 to 10 percent of men actually have multiple simultaneous wives (Frayser 1985; H. E. Fisher 1992). Moreover, polygyny may occur at substantial reproductive costs to females, such as reduced resource investment from males, reduced spousal attention, reduced fertility, and increased child mortality (see Strassmann 1997b). While males may attempt to maximize total number of offspring by having multiple wives, this generally does not occur at the exclusion of a primary relationship. Humans are unique in the long-term sociosexual-bond context of our sexual behavior. Fisher has argued for a model of evolved human mating strategies that involves long-term or serial monogamy and clandestine adultery (H. E. Fisher 1992, 2011). This model emphasizes both long-term sociosexual bonds and the occurrence of shorter-term sex within extrapair mateships. Importantly, and as we will discuss more in the final chapter, placing an emphasis on the long-term sociosexual bond-context of human sexual behavior suggests that contemporary uncommitted sexual behavior, say among two singles, may be a more recent historical phenomenon (Garica et al. 2012; Pedersen, Putcha-Bhagavatula, and Miller 2011).

MATE CHOICE

With so much at stake when it comes to pair-bonding, it is not surprising that research has found that humans make a number of trade-off decisions in choosing their long-term mates. Individuals try to find a balance that works—a partner that meets an acceptable combination of attractiveness across domains. But how exactly do we go about making those decisions? The classic story of mate choice takes place among peafowl, where the male peacock will display his glorious shimmering train of feathers in an attempt to attract the female peahen. This is an advertisement that he can avoid predators, can forage well (he eats foods high in carotenoids to maintain color), he avoids parasites that would diminish his plumage, and he does not lose battles with other peacocks who would attempt to pull his train feathers out and make him look less attractive (those peacocks fight dirty). A peacock is dressed to impress, and when on his game, the female peahens find him sexy. The attraction mechanism functions to allow one to focus in on one specific individual in the crowd. And, not so unlike peafowl's diet improving its feathers, some recent evidence suggests that a diet high in antioxidants might help our own human complexion and enhance the attractiveness of our skin. But in humans, the characteristics that we use to make actual mate choices are even more complex and varied.

One consistent finding across cultures is that men with status and resources are viewed as more desirable than other men, and women that appear attractive and young are more desirable than other women (Townsend 1998). This finding is relatively consistent in societies around the globe, but it often evokes feelings of discomfort in those rightfully concerned about the sexual double standard. The reason for this apparent sex-specific rule of attraction likely has to do with the relative costs of reproduction (Fisher 1982; Symons 1979). If women have to gestate and give birth to a child, those women who appear most fertile and able to produce young will be most valued. If men have greater reproductive potential but are expected to contribute external resources, those men who appear most able to abundantly provide resources will be most valued. But these patterns are undergoing dramatic change in the United States, Japan, and elsewhere: women are better educated today than men in many countries, and more and more women earn more than prospective or current partners (Mundy 2012). In such a context, men who are looking for a mate increasingly evaluate a woman's capacity to provide resources, and women pursue various mating strategies (by placing more value on a man's ability to provide direct childcare, for example, or by marrying a less educated man and encouraging him to further his education).

Beyond status, one of the first things humans look at, literally, in a potential mate is the face. And despite cultural variations, the characteristics of an attractive face are fairly consistent. Attractive faces tend to be characterized by smooth skin, bilateral facial symmetry, and several morphological characteristics (see Gallup and Frederick 2010). Both men and women say they prefer smooth and unwrinkled skin, with sores or discoloration evoking a negative reaction. Both men and women prefer symmetry, and although no one is perfectly symmetrical, greater overall facial and bodily symmetry has been related to overall health and is considered attractive. Both men and women dress and ornament the body and face to look more attractive; women more often use jewelry and objects to highlight aspects locally considered beautiful. Faces are perceived as attractive independent of factors such as gender, age, and ethnicity (Thornhill and Gangestad 1999).

An attractive face has been associated with other health factors as well. Men whose faces were rated most attractive were found to have higher-quality semen—that is, having more fertile sperm was associated with being rated more attractive by women who were shown men's pictures (Soler at al. 2003). In a study in which participants rated yearbook photos, researchers found that those with more attractive photos also had longer life expectancy (Hendersen and Angelin 2003) and a greater number of offspring (Jokela 2009). Even across cultures a facial photo alone is enough for others to fairly accurately rate the person's overall physical strength (Sell et al. 2009). Faces are often good indicators of other body-type proportions that men and women find attractive. The face contains a lot of information, and it can be a fast way to assess a person's overall physical quality and health.

Another mate-choice characteristic that has gained considerable attention is a woman's waist-to-hip ratio (WHR). Women with a low WHR are viewed as having a more attractive body type cross-culturally, and WHR of approximately 0.7 is often viewed as most attractive (D. Singh 1993). There is some historical and cultural variability (Swami et al. 2009), potentially a result of environmental conditions preferentially favoring the appearance of slightly higher WHRs. The logic underlying this measure is that the hourglass shape of a low WHR suggests appropriate fat stores and the ability of a woman to have a live birth. Women with lower WHRs appear to be more fecund (reproductively viable), and as such this may be an honest indicator that sexual selection has led men to find it attractive. Consumer products reflect this; in advertisements and other commercial images, nude models, film stars, and successful online escorts all have close to a 0.7 WHR (Saad 2011). In a study using photos of women who altered their WHR with cosmetic surgery, men in North America, Cameroon, Indonesia, and Samoa all found the postsurgical lower-WHR body types more attractive (Platek and Sing 2010; Sing et al. 2010). Assessing mannequins, blind men too, using tactile input rather than visual cues, found the 0.7 WHR more desirable than other body shapes (Karremans Frankenhuis, and Arons 2010).

On the other hand, men with a shoulder-to-hip ratio (SHR) that is more wedge shaped, with broad shoulders and narrow hips, are viewed as attractive. Men with a wedge shaped SHR are believed to have higher testosterone levels and to be physically stronger. And it is more than just being attractive—in one study both women with a lower WHR (hourglass) and men with a higher SHR (wedge) began having sex earlier and had more sexual partners than their peers (Hughes, Dispenza, and Gallup 2004).

In addition to physical characteristics such as facial symmetry, skin smoothness, and body shape, mate choice also happens on the bases of other senses, including sound and smell. The sound of someone's voice can determine how attractive they are perceived to be. As humans use much spoken language for vocal communication, particularly at night, the acoustic properties of a person's voice understandably contribute to mate selection. Much like birds such as male sparrows, that sing to attract a mate, or toads that call out to each other, humans use vocalizations to assess a mate. Or let's look at the example of the male red deer, which roars vigorously throughout the breeding season to ward off other males and to entice females into his harem. And entice the females he will—female red deer exposed to roaring come into heat sooner and are thus able to begin breeding sooner, the result of which is increased offspring survival (Judson 2002). Quality of voice can depend on pitch and clarity, and is associated with how attractive individual listeners rate the voice as sounding. The quality of a vocalization is more than enjoyable music, as it can have reproductive influence. In an Asian species of water striders, females have evolved a shield to prevent male forceful copulation, and will expose their genitalia for mating only after males have produced a vibration song they deem acceptable on the water's surface (Han and Jablonski 2010). Or take female canaries, which will produce larger eggs in their clutch if they are exposed to the sound of good-quality male canary song (Leitner et al. 2009). In other research, more attractive voices were associated with more attractive faces, and both men and women with attractive voices reported having more sexual partners, and were also more likely to have engaged in infidelity, than those with less attractive voices (Hughes, Dispenza, and Gallup 2004). Among the Hadza hunter-gatherers of Tanzania, men with a more masculine lower voice (that is, a more attractive voice) had sired more children than other men (Apicella et al. 2007). Voice can also be an indicator of fluctuating asymmetry (how symmetrical an individual's body appears, and a measure of phenotypic quality).

Body odor is another indicator of symmetry and overall health. Smell, or body odor, is important to the mating rituals of many organisms (Kohl and Francoeur 2002). For many insects, amphibians, reptiles, and mammals, chemical signaling allows an individual to mark a territory, signal sexual receptivity, or lure a mate in from far away. In some cases, pheromones (chemical signals) share important reproductive information with those of the opposite sex. First studied in silkworm moths, a small drop of female pheromone was found to attract numerous male moths from substantial distances. In a well-known pheromone study in mammals, a male hamster was put into a cage with another male that had been anesthetized, and the first hamster either ignored the anesthetized conspecific or simply pushed it around. But when the anesthetized male hamster was rubbed with female hamster vaginal secretions and the awake male was placed in the cage, he would attempt to mate with the passed-out male (Johnston 1986). In mammals, the vomeronasal organ (VNO) is a chemosensory organ located at the

base of the nasal septum that is believed to allow for reception of pheromones. It is disputed whether humans have a functioning VNO, but several
studies do suggest detecting the chemical signals of others may play a role
in human attraction and mating. A series of sweaty T-shirt studies have
shown that people find the smell of healthy and attractive individuals the
most appealing, and those with major histocompatibility complex (MHC)
genes dissimilar to themselves the most pleasant. If humans are able to detect pheromones, this would allow one to tell by smell if another person
were immunocompromised or too genetically similar to reproduce with
(and thus to copulate with). While the idea is greatly debated, some scientists, most notably James Kohl, have even suggested that pheromonal communication is the primary means by which human mating and social interaction occur (Kohl and Francoeur 2002).

We would be remiss to not remember how humans perceive the behavior
of others and learn from their surroundings. As mentioned earlier, fish such
as guppies and sailfin mollies are known to engage in "mate copying". Females watch the mating behavior of older and presumably experienced females and preferentially mate with the same males themselves. Humans
show mate-copying effects as well, with both men and women displaying
increased relationship interest in potential partners after perceiving them as
successful in the dating and relationship arena (Place et al. 2010). This reproductive priming effect was documented in a sample of U.S. university
students: the young men and women reported having increased dating opportunities with others shortly after they entered a new relationship (Platek,
Burch, and Gallup 2001). Our ability to assess the local environment and
learn from others, including in the domain of mating, has been instrumental in human evolution. This includes assessments of personality and behavior tendencies. Used with cross-cultural samples of thousands of adults
signed up for online dating, the Fisher-Rich-Island Neurochemical Questionnaire (FRI-NQ) was designed to assess participants for one of four
general personality dimensions and demonstrate preferential mate choice
for other personality types (Fisher 2009). Indeed, there are a variety of behavioral characteristics involved in mate choice, from humor to intelligence, and also a variety of physical traits we find attractive, from eyes to
teeth and smile, but the evolutionary significance and cross-cultural displays of these mate preferences become increasingly complex. Even within-
culture variations, as seen in local subcultures, value aesthetic differences in

dress and style (for example, goth, preppy, skater) that make an individual seem more or less attractive (Townsend 1998). There are tremendous individual and cultural differences in particular displays of beauty.

There exists a broad literature in evolutionary psychology on cognitive and psychosocial aspects of mate choice and relationship formation (see Buss 2003; G. Miller 2000). It is sometimes unclear whether these traits exist cross-culturally (that is, outside of Western samples) and what are the mechanisms underlying such putative mental adaptations. Yet many of the concepts raise interesting and important ideas about the nature of human mate choice. For instance, recent work on the construct of mating intelligence (Geher and Kaufman, 2013) suggests there are specific individual differences, across a range of mental adaptations, that assist in attracting and obtaining conspecific sexual and romantic interest. As one example, consider variation in cross-sex "mind reading," that is, the ability to assess mating interests of the opposite sex (Geher 2009). Work by Geoffrey Miller (2000) argues that intelligence and creativity are expressions of psychological adaptations that have evolved via sexual selection and mutual mate choice. Other work suggests that just as there is a wide range of mate preferences, individuals also have "deal breakers", or traits that cause instant mate repulsion, such as poor hygiene or laziness. The stakes for relationship formation are high, and so both men and women are expected to be choosy about whom they mate with and why.

SOCIAL AND SEXUAL MONOGAMY

Swinging swiftly through the rainforests of Southeast Asia, far above the forest floor, a pair of apes are singing a remarkable love song. Their chorus can be heard from great distances, and as their relationship grows stronger their songs become more synchronized. These dueting apes are gibbons— Hylobatidae, also known as the lesser ape. They roam tropical rainforests, and much like humans, they form intense, often lifelong pair-bonds with their mate. They were long the model of monogamy, engaging in a variety of behaviors we would associate with attachment. In some species of gibbons, if something happens to one of the partners, the other may find a new mate, but they will not sing the exact same duet—that is, the survivor will not sing the same love song to a new partner. Of course, gibbons are not the only creatures that sing to each other; a variety of species are known to perform duets during courting, from whales to birds. But gibbons continue their song throughout their intense pair-bond relationships. Perhaps reflecting our best effort to match the duetting of gibbons, roughly two-thirds of U.S. adults in a romantic relationship say they engage in romantic baby talk with each other—what some term "loverese" (Chang and Garcia 2010); talking to a lover in this type of high-pitched voice is inversely related to age, and is associated with passion and intimacy.

But just as gibbons swing through the trees, some have suggested their mating system is much like "swinging" too—not the dance style, but the lifestyle of being romantically attached to one partner but engaging in sexual behavior with others. In human "swinging" relationships, typically both partners agree to an open-relationship dynamic. This is different from polygyny, such as that practiced by the growing number of men in Tajikistan who marry multiple women, and this is different from polyandry, as is seen in the Kinnaur society in the Indian high Himalayas, where women often marry a set of brothers simultaneously (although this practice of fraternal polyandry is declining). While primatologists have characterized gibbons as monogamous, genetic testing in the 1990s started to reveal something peculiar—while socially monogamous, the gibbons are not always sexually monogamous. That is, while gibbons typically form an intense pair-bond, defending each other, sharing resources, and raising young together, they may have sex with one or more mates outside the pair. In fact, one reason for making the distinction of *social* monogamy is that with the advent of genetic paternity testing, we have found that there are nearly no species that are completely sexually monogamous. One notable exception is the owl monkeys of Central and South America, which long-term observations and genetic research suggest are truly sexually monogamous (Fernandez-Duque 2004). Their sexual monogamy also helps explain two striking characteristics of male owl monkeys: they exhibit incredibly low sperm counts and yet incredibly high paternal care behavior. Humans and many other species engage in a variety of mate-guarding behaviors to protect their monogamous bonds. But humans have found their own ways of handling infidelity and maintaining their attachments.

Earlier in this chapter, we presented a quote from a Ngandu man who described deeply loving his first wife but having sex with another wife for the purposes of procreation. In societies that promote sexual pleasure, sex beyond the pair-bond may occur for the purposes of sexual satiation and may operate independently from romantic love. One manifestation of social but not sexual monogamy can be seen in open relationships. Among gay males in the United States, for example, it has been estimated that nearly half of male-male romantic couples have open relationships, cherishing an emotional commitment to one partner while allowing for sexual behavior with others (Hoff and Beougher 2010). Interestingly, however, this pattern of open relationships appears to be nearly absent among lesbian romantic relationships. Consistent with the centrality of pair-bonds, many couples with open relationships report difficulty navigating issues of mutual comfort, jealousy, and trust. Among both sexes, regardless of sexual orientation, romantic attachments are an imperative, even in the face of one's potential interest in sexual variety.

Another example comes out of the fieldwork we have conducted with our colleague Michelle Escasa-Dorne at a renowned sex club in Las Vegas. Our primary projects there involve measuring the hormonal shifts that occur following different types of sexual activity in a non-laboratory setting. However, other enlightening things have also come of our (professional) experiences in the club. Well after midnight, while interviewing couples at the club for participation in our research, Justin and Michelle sat down with one particularly pleasant pair. The couple identified as heterosexual, had been together more than a decade, were married, had children, and had a vibrant swinging lifestyle. They frequently visited sex clubs and swinger parties, and the woman estimated that, other than her husband, she had had approximately seventy-five intercourse partners in the past year. But what was most remarkable about this particular couple was the degree of emotional intimacy that they exuded—laughing, holding each other, finishing each other's sentences. This couple was bonded. They were attached, comfortable with each other, and perfectly fine with engaging in this lifestyle together. To this couple, sexual behavior was a physical act, and they alone were the focus of each other's romantic, emotionally intimate attachments. To them, swinging was a way to preserve their loving marriage; it was not infidelity, and it was certainly not betrayal.

There are different types of love relationships with different partners, but the focused-attention characteristic of romantic love is typically directed toward one individual, with others being sexual relationships and/or other types of attachment relationships. There are cases in which people try to equally share the same type of love among multiple partners. In the United States, for instance, there are estimated to be nearly 500,000 polyamorous relationships in which the partners openly engage in romantic exploits outside their "primary" pair-bond. Many polyamorists contend that there is no primacy of one love relationship over another—unlike some polygamous marriages, where there may be a spousal hierarchy among co-wives and thus more friction. Author Jenny Block (2009) describes her initiation into what later become a polyamorous relationship, despite her being in love with her husband: "Three years into our marriage, I began to feel itchy. So I had an affair. She was beautiful. . . . I deliberately chose to have an affair with a woman, rationalizing that it wasn't as bad as sleeping with another man. (Simply by virtue of his gender, my husband could never be for me what she could be.) . . . And slowly I began to figure it out. For Christopher, sex with me was about loving me. And loving me was about caring for and respecting me. Although there are people who can manage that duality (or plurality), my husband simply couldn't. And I wasn't sure he should have to." She later writes that while she maintained her marriage and loved her husband, she also continued to have a female lover; her husband, on the other hand, chose to not take another partner.

Pair-bonds can take a variety of forms, running from "aloof" to "intimate" types of relationships, as discussed in Chapter 2 (Whiting and Whiting 1975). In many societies marital decisions are family decisions, often involving heavy input from parents and kin. There does not appear to be any other species that so heavily relies on the opinion of kin in establishing long-term pair-bond relationships. The role of family and friends can be overwhelmingly influential in human mating decisions. It is also important to note that the types of pair-bonds we see in modern environments may lead to some unusual conclusions. In the typical Western model of attraction, individuals court each other, typically with men doing more courting of women than the other way around, then date, and then enter into a long-term union that might culminate in marriage. As we will revisit in the final chapter, dating and marital patterns today are changing rapidly, owing to a variety of social factors. Yet the nature of romantic attachments is not necessarily experiencing the same shifts. Marriage equality acts that allow for same-sex marriages will also begin to change the playing field. There is no evidence that individuals in a same-sex romantic partnership experience anything other than the power of romantic love, but simply directed to-

ward a different stimulus (a same-sex partner rather than one of the opposite sex).

SEX BEYOND THE PAIR-BOND

As several cross-cultural surveys reveal (Ford and Beach 1951; Frayser
1985), societies vary in their views toward extramarital affairs. In some
places, such as among partible-paternity Amazonian societies, women are
given more allowance for affairs. Among Australian aborigines of Arnhem
Land, both women and men were granted sexual access to specific classes
of people apart from their spouse (for example, a spouse's siblings), a practice that may have provided social insurance in the event of a spouse's
death (Berndt and Berndt 1951). In other societies, a cheating woman or
man may be given a death sentence. Several patterns emerge across the
landscape of extramarital affairs. One is that restrictions accorded to women
and men within a given society tend to be similar; that is, a society providing some allowance for women tends to do the same for men, whereas a
society constraining women tends to do the same for men. A second pattern
is that, where there are differences in the latitude granted, these slide toward a sexual double standard whereby women are punished more severely
than men within a given society. Cross-cultural surveys do not reveal any
societies wherein men's extramarital activities are constrained more than
women's.

In Ford and Beach's (1951) coverage of extramarital liaisons from a
cross-cultural perspective, they divide societies into those that allow and
those that do not allow such liaisons. However, they also underscore that
even when extramarital liaisons are allowed, they are usually under limited
circumstances rather than actively and generally endorsed. More specifically, they note,

> With [a] few exceptions . . . every society that approves extra-mateship liai-
> sons specifies and delimits them in one way or another. There are some
> peoples, for example, who generally forbid extra-mateship liaisons except in
> the case of siblings-in-law. This is true among the Siriono, [where] a man
> may have liaisons with his mate's sisters and with his brother's wives and
> their sisters. Similarly, a woman has sexual access to her husband's brothers
> and the husbands of her sisters. . . . In some societies extra-mateship liaisons

take the form of "wife lending" or "wife exchange." Generally, the situation is one in which a man is granted sexual access to the mate of another man only on special occasions. . . . Another type of permission in respect to extra-mateship liaisons appears in some societies in the form of ceremonial or festive license . . . [and] may range from harvest festivals to mortuary feasts. (Ford and Beach 1951: 121–122)

There are a variety of reasons why one might engage in extra-mateship liaisons, and not all are considered infidelity—as we saw with the couple interviewed in the sex club. But infidelity remains influential on the evolutionary landscape. And sometimes rules are not enough to keep two lovers apart, as various legal codes make a point of addressing—in fact two of the biblical Ten Commandments deal with issues of adultery. Among the various contributors to infidelity, being involved in an arranged marriage can increase the odds of it (Abbott 2010); among Himba women of Namibia, none of those studied who had entered love marriages reported having children with other men, unlike the women in arranged marriages (Scelza 2011). In China, women who reported inadequate love in their relationships and who were partnered to low-income men were more likely to have engaged in extrapair sex (Zhang et al. 2012). Other factors that can lead to infidelity include being unhappy or unsatisfied in a relationship, being uninhibited by substances like drugs or alcohol, and a desire for revenge.

In 2010 Justin and his colleagues published a study on the relationship between a particular variant of the DRD4 dopamine receptor gene and uncommitted sexual behavior, including infidelity. While several news outlets sensationalized this study as revealing "the cheating gene" or "the promiscuity gene," with shocking news headlines to match, this was not what the study actually showed. The DRD4 gene is a polymorphism found on chromosome 11 in humans (orthologs appear to be present in other species on different chromosomes). Everyone has the DRD4 gene, but individuals vary in the number of allelic repeats they have—generally two to eleven. Individuals with a particular variant of DRD4, at least one longer allele with seven repeats or more, are often classified as genetically predisposed to sensation seeking. That is, those with more repeats have stunted dopamine uptake, meaning they require more sensation-seeking or risk-taking behavior to be satiated by their brain's dopaminergic reward pathway. Genetic tests called linkage disequilibrium analysis suggest that the sensation-seeking variant of DRD4 arose roughly 30,000 to 50,000 years ago, a time in human evolutionary history characterized by high levels of migration. Having some individuals in a population with greater sensation-seeking impulses may well have been beneficial to the evolution of (new) communities. In evolutionary terms, risky behavior is about gambling big risks for big rewards. Imagine being the one who migrates to a new land, where you might starve or be killed (risks) or find new untapped resources and mates (rewards). Even today we see substantial population frequencies for the DRD4 genotype. In Western industrialized societies about one-third of the population has the sensation-seeking variant, while there are much higher frequencies among the Yanomami of the Amazon rainforest, for example; rates are much lower, however, among other populations such as the !Kung of southern Africa (Harpending and Cochran 2002).

In the study of DRD4 and uncommitted sexual behavior, Justin and colleagues found that those men and women with seven repeats or more were more likely to have engaged in promiscuous sexual encounters (one-night stands), and they engaged in infidelity much more frequently than those without the longer alleles (Garcia et al. 2010). This is interesting for several reasons. First, it is unusual for one gene to have substantial influence on human behavior; indeed, we recognize the importance of multiple genes working in concert with a particular environment. Yet in this case, when the sample was divided into two groups, behaviors were different among those with the longer sensation-seeking variant of DRD4 than those without. While there was no difference between the groups in overall number of sex partners, the context of partners mattered. As an example, nearly 20 percent of individuals who were not genetic sensation seekers had a history of committing infidelity, yet in the group with seven repeats or more, nearly 50 percent of individuals had committed infidelity before. Moreover, participants with seven repeats or more reported a more than 50 percent increase in instances (number of occasions) of infidelity. For one gene to account for so much variation is substantial. Further, that this gene is associated with sensation seeking suggests that a number of individuals who engage in infidelity and uncommitted sex may do so not because of a high libido or because they are unhappy in their current relationship, but because they crave novel stimuli. So when asked if someone can love their partner if they also commit infidelity and cheat on them, the answer appears to be yes. This is partly because the mechanisms motivating romantic attachments and those motivating desire for new stimuli are not the same. This is also consistent with the psychological literature on infidelity, revealing a wide array of self-identified reasons for committing infidelity (Tsapelas, Fisher, and Aron 2010).

Ultimately, the reasons for infidelity may be quite different from the psychosocial reasons that might come to mind. While a species may be socially monogamous and form pair-bonds, an individual may also attempt to mate guard a partner to prevent infidelity for a good evolutionary reason. Extrapair copulations (EPCs) can allow an individual to achieve more variation among offspring—different genetic combinations, potentially increased offspring quality, different environments, and extra fertilizations. For humans and probably other long-lived social animals, EPCs have functions linked to relationships and parental investment; a woman may engage in an EPC to garner resources from another mate, or with an eye toward switching to a different, more desirable long-term mate (mate swapping). In one study of fifty-three different nations, mate poaching, or the active seduction of a mate away from another partner, was documented through the International Sexuality Description Project to occur nearly universally (Schmitt 2004). In a series of psychological studies conducted in the United States, Schmitt and Buss (2001) found that more than 50 percent of participants had attempted to poach another's mate, and nearly 80 percent had themselves received a poaching attempt from someone who tried to deter them from their mate. Moreover, more than 25 percent of participants reported having had a previous partner leave them for what they believed to be another partner (having been poached away).

To combat this potential invasion of a pair-bond, humans often guard their mates and become jealous of potential extra-mateships. Cross-culturally, a variety of practices have the explicit intent of preventing cuckoldry. Some research suggests wives may attempt to initiate sex to promote further attachment before their husbands leave the home during times when EPCs might be possible. Beverly Strassmann (1992) has proposed that religious customs and menstrual taboos among the Dogon of Mali, where women are sequestered in menstrual huts while menstruating, serve to prevent cuckoldry (the raising of offspring genetically sired by another male). Based on more than two years of field data, Strassmann argues that because Dogon men are concerned that their wives might become pregnant by other males, going to the menstrual hut is a signal of nonpregnancy.

We humans are not alone in our attempts to prevent a partner's infidelity: many organisms engage in mate guarding and will fight, sometimes to the death, with other individuals that pose a threat to paternity or resources. Consider the female burying beetle, who attempts to keep her mate monogamous through coercion (Eggert and Sakaluk 1995). Burying beetles will bury the carcass of a small animal by digging a hole underneath it and then covering it with soil once it drops down. Females will then lay their eggs near the carcass. If it is a large enough food source, males will attempt to attract a second female by using pheromone emissions to chemically call other females. However, in an effort to prevent this and preserve the resources for themselves, the first female burying beetle will attempt to halt shared paternal investment by physically interfering with the male's attempts to signal to additional females.

Historically, our society's laws have been somewhat understanding about the extreme reactions a person can have to a partner's infidelity. Across many legal codes, a "crime of passion" has been an acceptable excuse for a man who kills his wife—perhaps her lover too—after uncovering her adultery (Wilson and Daly 1993). But these laws are biased toward men, as men's reactions tend to be more immediate and more violent. Using cross-cultural ethnographic records from sixty-six societies, Jankowiak and Hardgrave (2007) found that both genders almost always react in some way to sexual betrayal—with 88 percent of men and 64 percent of women resorting to physical violence, and 29 percent of men and 50 percent of women withdrawing emotionally or physically from their partner. However, men and women react differently, often depending on the degree of the sexual double standard in their society and the culturally permissible public reactions.

Men and women are not alone in their penchant for extrapair copulation, nor in their attempts to prevent their partner from doing it. While a vast majority of birds form pair-bonds, according to genetic paternity tests at least 70 percent of bird species reviewed show genetic evidence of extrapair copulations. In the socially monogamous swift fox, nearly 50 percent of offspring are the result of cuckoldry. And in humans, studies have suggested that 9 to 10 percent of children in the United Kingdom are raised by fathers who are not their biological parent but presume they are. However, a mediating factor in that finding is the likelihood of paternity certainty (see Anderson 2006): when fathers are unsure if they are the father, about one-third of the time they are right to worry; if fathers are sure they are the father, they are almost always right (98 percent of the time). In the United States, studies suggest that roughly 50 to 80 percent of individuals have maintained fidelity in their current relationship, while the rest have either once or repeatedly committed infidelity. While both men and women engage in infidelity, it occurs at a substantial cost. If a partner finds out, they may react in any number of ways, including violently or with dissolution of the relationship. While women are always 100 percent sure of their maternity (after all, they gestate for nine months), men are less certain of their paternity. One way to reinforce or discourage the presumption of paternity is to consider how much the offspring looks like the father. Across cultures, mothers often reinforce the idea of paternal resemblance by telling fathers how much offspring look like them, and her relatives will do this too, more so than the father's (Daly and Wilson 1982; Regalski and Gaulin 1993). As Malinowski wrote of Trobriand Islanders, for example, "Every child resembles its father. Such similarity is always assumed and affirmed to exist" (Malinowski 1929: 175). By verbally confirming it, the presumption is that fathers will agree and act with a high presumption of paternity. Among a polygynous community in rural Senegal, facial and odor resemblance are related to increased paternal investment, and positive health and growth outcomes are seen for those children who receive more paternal investment (Alvergne, Faurie, and Raymond 2009).

PHYSIOLOGY OF PAIR-BONDING

What happens to our physiology when we pair-bond? In some species, monogamy is the result of physiological tricks. Take, for example, the ever popular fruit fly, where males produce a sex peptide in their seminal fluids. The sex peptide promotes a brief "monogamy" in the females that have been inseminated thus attempting to commandeer a female's reproductive potential (Yapici et al. 2008). But a number of animals are known to have similar reactions, mating exclusively with each other but only for a relatively short period of time. Among African elephants, for example, a pair may mate exclusively with each other, repeatedly, for several days and then separate. Elephants are also known to stick their trunk into the mouth of another, a behavior rather reminiscent of kissing. Although other animals do indeed kiss, chimpanzees are known to embrace each other and mouth kiss. Some have argued that the initiation of the physiological cascades before arousal, sex, or love (in no particular order) requires some form of kiss. Many (but not all) human societies have some form of romantic kiss that draws two potential lovers close together, their faces in near contact. In the case of human mouth kissing, some have argued that the exchange of saliva is important for mate-choice and overall quality and health detection, particularly with open-mouth and tongue kissing. One study of U.S. college students found that roughly 60 percent of young men and women had, at least once, kissed someone they were initially attracted to and then lost interest after an unpleasant kiss (Hughes, Harrison, and Gallup 2007).

But the physiology of mate choice runs much deeper than kissing. By using functional magnetic resonance imaging (fMRI) to measure oxygenated blood flow, studies have demonstrated that attraction activates specific parts of the brain. One area in particular that has been implicated in feelings of romantic love is a part of the brain called the ventral tegmental area (VTA) (Bartels and Zeki 2000; Fisher, Aron, and Brown 2006). The VTA is at the core of the mesocorticolimbic reward system, where the neurotransmitter dopamine functions to motivate sensation-seeking behavior. Dopamine has been implicated in a variety of activities that activate the reward pathway in the brain, from genital stimulation to novelty-seeking behavior to alcohol and other drug use. When researchers took men and women who were romantically in love and stuck them in an fMRI scanner to observe their brain while they viewed images of their beloved, they found several parts of the brain were especially active, most prominently the VTA (Fisher, Aron, and Brown 2006). Although there is some overlap, this is a slightly different part of the brain from that involved in the maternal love a mother feels toward her child, which included regions not involved in romantic love such as the orbito-frontal cortex and the periaqueductal gray (Bartels and Zeki 2004). Romantic love activates parts of our brain similar to those activated by drug addiction. The motivational drive for romantic love has a profound effect on humans; it is something our physiology craves.

We should expect to see mechanisms for sexual motivation and romantic attraction across a wide variety of species (and indeed we do), and mechanisms for the romantic attachment more defined in socially monogamous species (H. E. Fisher 2004; Ellison and Gray 2009; Nelson 2011). There has been a substantial amount of research aimed at understanding these mechanisms in an unlikely subject: a tiny and short-lived rodent called the vole.

In particular, this research has focused on differences between two species of North American voles: the montane vole and the prairie vole. The species are remarkably alike except for small variations in their neuroanatomy and some striking behavioral differences. The montane vole has a promiscuous mating system; they do not pair-bond following mating, there is no apparent paternal investment, and the mothers abandon their offspring soon after birth. The prairie vole, on the other hand, is socially monogamous, forms strong pair-bonds, has fathers that provide parental investment and mothers engaged in early maternal care, and each parent spends several weeks caring for their young and remaining closely huddled together. Female prairie voles remain sexually immature until they detect the presence of a chemical in the urine of an unrelated male; this stimulates pubertal development and within a day she becomes receptive and copulates repeatedly with the male, facilitating the formation of an enduring, socially monogamous pair-bond. C. Sue Carter was the first to identify the physiological mechanisms underlying the development of this socially monogamous bond. The physiological difference between these two species of voles is the role played by the peptide hormones oxytocin and arginine vasopressin, specifically in the hypothalamus. Interestingly, it is not the amount of these hormones that makes the difference but the concentration of oxytocin and vasopressin receptors in the female and male brains, respectively: the location and density of the receptors in the brain of the prairie vole, and the resulting dopamine association that causes a natural feeling of reward, are the factors believed to support social monogamy in this species.

In humans, less is known about the role of oxytocin and vasopressin in mating and pair-bonding (see Ellison and Gray 2009; Nelson 2011). However, a series of recent studies have demonstrated that oxytocin is peripherally released during affiliative behaviors such as cuddling with a partner and playing experimental games that foster trust; even physical touch such as massage can initiate some of these cascades. This is why oxytocin is sometimes called "the cuddle hormone." In one of the best-designed studies of this type, individuals administered oxytocin via an intranasal spray (which is believed to reach the brain) were found to engage in relatively more positive behaviors during a lab-based discussion with their romantic partners than individuals who sniffed a placebo spray (Ditzen et al. 2009). Other recent genetic studies have pointed to the role of these hormones in shaping how satiated individuals are in their romantic and sexual lives.

Variation in the vasopressin receptor gene AVPR1A that results in less vaso-pressin uptake was shown to influence pair-bonding behavior and marital satisfaction in men (Walum et al. 2008). Likewise, a genetic polymorphism for the oxytocin receptor gene was associated with empathy for others (Ro-drigues et al. 2009). Interestingly, some scholars have argued that coopera-tive breeding has allowed empathy to evolve in humans, setting the stage for allowing humans to form a deep sense of love. Cooperative breeding, in this sense, is the social system that characterizes how humans around the globe willingly care for offspring (see Hrdy 2009).

When it comes to hormones associated with monogamy in humans, another has taken center stage: testosterone. Most testosterone studies have been framed in terms of the challenge hypothesis (Archer 2006; Wingfield et al. 1990), where it is suggested that testosterone levels are elevated in male-male competition occurring within reproductive contexts, such as over mates. As support for this, several studies have shown that single men in the United States (who are actively competing) have higher testosterone levels than partnered men (Gray et al. 2004; Gray and Campbell 2009). Among Swahili men of Kenya, polygynously married men had higher testosterone levels than monogamously married men (Gray 2003). And a study of single, partnered, and polyamorous men and women found that both polyamorous men and polyamarous women had higher baseline testosterone levels than singles of their same gender (van Anders, Hamilton, and Watson 2007).

At the same time, there are also dramatic physiological changes associ-ated with ending a monogamous relationship. In a spectacular study con-ducted by Helen Fisher and colleagues, fifteen individuals received fMRI brain scans after being rejected in love. What the researchers found was that regions similar to those activated in romantic love were also activated in the brain after rejection, including the ventral tegmental area, ventral striatum, and cingulate gyrus. At the same time, parts of the forebrain more typically associated with drug-use motivation (specifically, cocaine use) were also activated (H. E. Fisher et al. 2010). This may help explain the intense physiological and psychological reactions many individuals have to relationship dissolution, ranging from loss of appetite to minor depression to severe mental states (Morris and Reiber 2011). Individuals are even known to engage in forms of stalking behavior following a romantic breakup. These intense responses are indicative of the enormity of long-term pair-bonds. It will be interesting to study how these post-breakup behaviors change in our technological era, when many can simply keep tabs on a previous partner through the Internet (via Facebook or other social-networking media). An interesting finding from Justin's recent research highlights the difficulty of ending a relationship: in a nationally representative sample of single Americans, roughly 50 percent of men and women said they have been in a cyclical yo-yo relationship, where they have broken up with someone but then gotten back together.

But just as some relationships must end, so must this chapter. While we could discuss dozens upon dozens of topics on the nature of love, we now turn our attention to sex differentiation. While we have reviewed the long-term pair-bond context of much of human sexual behavior, and the significance of pair-bonds to human behavior, let us consider how we come to be the men and women who develop into sexually capable (and sometimes loving) adults.

CHAPTER FOUR

Raising Human Sexuality

Processes of Sex Differentiation and Sociosexual Expression

> For there are eunuchs who have been so from birth, and there are eunuchs who have been made eunuchs by men, and there are eunuchs who have made themselves eunuchs for the sake of the kingdom of heaven.
>
> —the Bible, Matthew 19:12

PLEASE SHIFT YOUR gaze momentarily from these words (eunuchs and all) to your fingers. In particular, look at your pointer finger and your ring finger. In terms of so-called digit ratios, these are your 2D and 4D, respectively. If you measure the length of your 2D (pointer finger) from the crease at the base of your finger to the finger tip, then divide it by the length of your 4D (ring finger), you can calculate your digit ratio for both your right and your left hand. What is the meaning of this, and what does this exercise have to do with evolution and human sexuality?

Males tend to have lower digit ratios (around 0.95) compared with females (around 0.97), although these ratios demonstrate pronounced within-sex variation, meaning that they do not strongly predict a given individual's behavior (McIntyre 2006; Puts et al. 2004). These are small differences indeed, but they have been found in all of the various human populations in which they have been measured, and in various nonhuman animals where the equivalent traits (in rat paws to baboon hands) have also been quantified. Several lines of evidence suggest that the digit ratio reflects relative androgen and estrogen exposure during early development. In the case of humans, these digit ratios appear to be defined primarily before birth, owing to prenatal hormonal exposure. The early established sex difference in these ratios is consistent with the known differences in steroid hormone exposure perinatally (before and shortly after birth) in humans. Girls with a clinical condition—congenital adrenal hyperplasia, or CAH—who are known to be faced with atypically high perinatal androgen exposure, tend to have more masculinized digit ratios (Brown et al. 2002). Recent experimental research on mice indicates that their digit ratios are shaped by steroid hormone exposure early in life (Zheng and Cohn 2011).

What do these digit ratios point to regarding human sexuality? In a meta-analysis of eighteen studies, which included around 6,000 participants, Teresa Grimbos and colleagues (2010) investigated whether digit ratios differed according to sexual orientation. They found that lesbians tended to have more masculinized digit ratios compared with heterosexual women, but that digit ratios did not differ by sexual orientation in men. As one illustration of the kinds of studies incorporated in this review, Hall and Love (2003) measured digit ratios in a small sample of seven monozygotic female twins who differed in sexual orientation; there were no differences in digit ratios in two sets of twins, whereas for the other five sets of twins, the twin with a same-sex orientation had more a more masculinized ratio compared with their heterosexual sister. Such findings suggest that some cases of at least female sexual orientation appear to trace developmentally to high perinatal androgen hormone exposure. Some females may be shunted toward a different sexual orientation before they are even born.

In this chapter, we consider the mechanisms by which females and males develop. We cover the genetic bases of human sex determination; the roles of hormones perinatally and in differentiating reproductive anatomy, brains, and behavior; the ways that development tends to unfold both typically and atypically; and how an interaction between heritable physiology and social context shapes the sexual beings we become. Along the way, we'll revisit the eunuchs too, as they illustrate quite dramatically connections between hormones and human sexuality. Altogether, we will find that neither nature nor nurture wins (though scholars and lay audiences will continue bickering indefinitely over this dead issue), but that our sociosexual selves are unfolding interactions. Unpacking this story requires more physiological details than most other chapters in the book, but we hope you will agree that fully weighing this body of work makes for a more sophisticated understanding.

MAKING MALES AND FEMALES

Suppose you are watching a nature program on crocodiles and alligators. You watch the footage of one of these beasts sloshing in the water with dinner in its jaws, or maybe, more surprisingly, you see an adult delicately carrying its babies in its mouth, an unusual instance of parental behavior among reptiles. A question it may not occur to you to ask is, how did one of those alligators or crocodiles become a female or male? What factors initiated the process of sex determination and differentiation in these animals? There is no higher power to call upon in this process, no mythical story. Instead, the process of distinguishing between females and males starts with something much more mundane: temperature. Indeed, the temperature at which the eggs of a crocodile or an alligator incubate during a particularly sensitive period shunts the developing offspring toward being females or males (Nelson 2011; Rhen and Schroeder 2010). With the American alligator, for example, eggs incubated at lower temperatures hatch females, whereas eggs incubated at higher temperatures hatch males. This is a case of environmental sex determination in which the primary environmental factor kicking off sex differences is temperature.

Nature has also hit on other ways to determine and differentiate the sexes. For birds, this process begins with sex chromosomes, but in a different way than it does in mammals (Graves and Shetty 2001; Nelson 2011). Bird females are the heterogametic sex, meaning that they have a mixture of sex chromosomes: WZ. Bird males, conversely, are the homogametic sex, meaning that they have two similar sex chromosomes: ZZ. Those different sex chromosomes underlie additional steps in sex differentiation. For bird sexual behavior, it appears that unspecified genes on the W sex chromosome lead the initially undifferentiated avian gonad to secrete relatively higher concentrations of estrogen, which in turn leads the left gonad to develop as an ovary. The estrogens produced by the left avian ovary, in turn, demasculinize sexual behavior. As an example, among Japanese quail—a research star of studies of avian sex determination and differentiation studies—the males typically strut their stuff in view of mates and then mount the female.

But males who are given estrogen during an early developmentally sensitive period mate like females, while ZW females in which the effects of the estrogen have been blocked behave like males. For other sexually dimorphic bird behaviors, such as territorial aggression or song, the hormonal mechanisms differ, indicating that different systems can, to some degree, be decoupled when specifying their genetic, hormonal, and developmental basis.

While sex chromosomes are absent in some sexually reproducing species (such as some reptiles), it is nonetheless interesting that when sex chromosomes arise they tend to be hotbeds of evolutionary activity (Vallender and Lahn 2004). Genes playing roles in sex determination and differentiation are often sequestered on the sex chromosomes, perhaps because they are transmitted in sex-biased ways (for example, since only female birds have a W chromosome, stacking it with genes involved in sex determination and differentiation helps prevent their impact from being watered down by other genes). We will look at various examples of mammals, including humans, in which genes implicated by natural selection in reproductive physiology and behavior are disproportionately located on the sex chromosomes, and for related reasons (sex-specific effects are less diluted). Still, it is worth underscoring nature's rainbow of mechanisms by which sex differences arise. As one further example of this variation, in summer 2011 Peter came eye to eye with bearded dragons while visiting a friend who had two of the Australian reptiles as pets. A discussion ensued about how one can identify females and males behaviorally and morphologically, which in turn raised the question of how sex differentiation occurs in these creatures. Interestingly, it appears that bearded dragons rely on both genetic and temperature cues to developmentally differentiate females and males (Quinn et al. 2007). The sex chromosomes largely dictate whether an individual will become a female (WZ) or a male (ZZ), but eggs incubating at a high temperature can "switch" sex: apparently because of a deactivated enzyme, would-be ZZ males develop as females instead.

HUMAN SEX DETERMINATION: GETTING STARTED

We do not become male or female based on the temperature inside our mom's womb. The mechanisms initiating human sex determination are not environmental but genetic. A number of genes, with one grandstanding above all others, are responsible for human sex determination (Knobil and Neill 2006; Nelson 2011; Pfaff 2010). Most of these key genes rest on the sex chromosomes, the X and Y chromosomes in mammals like us. Human females typically have a sex chromosome complement of XX, having acquired an X from both mom and dad. Human males typically have a sex chromosome complement of XY, having acquired an X from mom and a Y from dad. Note how in mammals, unlike birds, females are the homogametic sex. One speculation for the sex chromosome difference between birds and mammals is that in birds, females often disperse further than males at puberty, whereas in mammals males typically disperse further than females (Greenwood 1980); the idea is that male mammals may face stiffer male-male competition and courtship challenges than male birds, with fewer benefits possible from kin selection (altruism between relatives), resulting in male mammals being subject to greater pressure to harbor male-beneficial genes on their sex chromosome complement (see Haig 2000).

The gene most central to human sex determination rests on the Y chromosome. It is referred to as the SRY, for sex-determining region of the Y chromosome, an allusion to its general location on the Y, and yet with new technology we can now specify exactly where this gene resides (Knobil and Neill 2006; Nelson 2011; Pfaff 2010). When the SRY gene is expressed, yielding a protein, a cascade of events ensues. Most important, the protein encourages the undifferentiated gonads to develop as testes; in the absence of this protein (when the SRY gene is not expressed), those same gonads will become ovaries by default. The timing of this gonadal differentiation occurs late in the first trimester: prior to SRY's expression, the gonads are openminded, as it were, but based on the molecular signal they receive (or not) around this time, they commit to maleness or femaleness.

When the gonads develop as testes, those testes are responsible for the next stage of sex differentiation. They secrete two molecules that masculinize and defeminize an individual's reproductive anatomy. The testes release testosterone and Müllerian inhibiting hormone (MIH). The MIH released by the testes inhibits, as the name implies, the development of the Müllerian ducts. The Müllerian ducts wither in males, whereas in females, in the absence of MIH, they develop into the Fallopian tubes, uterus, and the inner portion of the vagina. As for other features of internal reproductive anatomy, they are affected by testosterone rather than MIH. Testosterone causes the Wolffian ducts to develop as the vas deferens and seminal vesicles in males; without the testosterone released by the testes, those same Wolffian ducts wither in females. Testosterone released by the testes also facilitates development of so-called accessory structures: the prostate gland and the bulbourethral gland, both of which regress in females in the absence of testosterone.

Testosterone masculinizes various previously undifferentiated structures in the external genitalia. The timing of these differentiating effects occurs early in the second trimester. This is a case of, once again, undifferentiated structures becoming male- or female-like, depending on the signals (testosterone) they do or do not receive. Under the influence of testosterone, the genital tubercle becomes the glans penis, or head of the penis, in a male; without the testosterone released by testes, that same genital tubercle becomes the clitoris in a female. Under the influence of testosterone, the urogenital ridge becomes the shaft of the penis in a male; without that same testosterone, the ridge develops as the labia minora in a female. Under the influence of testosterone, genital folds become the scrotum in a male; without that same testosterone, the genital folds become the labia majora in a female.

The process of sex differentiation in human reproductive anatomy thus draws upon shared precursors—undifferentiated structures like the genital tubercle—and eventually yields male- or female-typical structures. The process of human sex differentiation begins with the sex chromosomes, which lead to gonadal sex differences, which in turn lead to differences in the hormones released by the gonads (testosterone, MIH) that masculinize and defeminize gonadal males. According to this view, females are the default sex, the sex that will develop in the absence of an SRY gene and the release of high concentrations of testosterone prenatally by the testes. Why should this be so? One evolutionary interpretation is that this captures the derived nature of males. Once sexual reproduction and the sexes were "invented," nature crafted mechanisms to spin off males from females. The development of females may more closely resemble nature's original design, with males effectively carved off female ribs. Note how this is the reverse of the biblical story in which woman was born from the rib of man. Males needed to develop a different reproductive anatomy to aid fertilization, and this construction product relied on the SRY and testosterone effects noted above. Another noteworthy aspect of human sex differentiation in reproductive anatomy concerns the shared structural starting points of female and male characteristics. For example, the process of differentiation reveals how the clitoris and glans penis are homologues, built from an originally undifferentiated structure; these observations inform an understanding of the function of the clitoris, female orgasm, and sensory innervations, as we will see in later chapters.

ADDING MORE COMPLEXITY TO MALE AND FEMALE UPRISINGS

While this story of building masculinity (XY, SRY, and testosterone, with females as the default sex) captures much of the process by which human sex differentiation arises, there are additional complexities. Those complexities involve other genes as well as the endocrine system. We will describe some of the most important complexities here, and note how they are important to capturing both the variation in both normal and atypical human sex differentiation.

We know that other genes besides the SRY gene play roles in sex differentiation, although most of the critical work has been conducted on mice rather than men (McCarthy and Arnold 2011; Pfaff 2010; Vilain and McCabe 1998). As an example, non-SRY genes on the Y chromosome appear to play a direct role (rather than indirect, via effects on testes and testosterone secretion) in the differentiation of a vasopressin system in male brains (DeVries et al. 2002). There is also the issue of female XX sex chromosomes. To avoid double dosing on gene products from their paired X chromosomes, typically one of the chromosomes is "silenced," with the specifics of which one is silenced varying by cell lineage. The result is that a given female is a genetic mosaic resulting from the expression of genes on her two X chromosomes. Still, studies indicate that the process of X inactivation may sometimes lead to variable dosages, yielding variation in femaleness depending on the specifics of X-chromosome gene expression. Further, a gene referred to as DAX-1, which resides on the X chromosome, encourages the development of ovaries. Its effects are outweighed by SRY in an XY individual, but in an XX individual (a genetic female) the expression of this gene fosters the development of the undifferentiated gonads into ovaries.

If genes get the gonads started, we need to consider in great detail how the gonads work—how they release hormones, and how those hormones exert their effects. This opens up the field of endocrinology (Adkins-Regan 2005; Knobil and Neill 2006; Nelson 2011). Endocrinology is concerned with the set of ductless glands (such as ovary, testis, adrenal, thyroid) that release hormones that travel through the body's circulatory system until bound by specific receptors to exert their effects. The key to all of this terminology is information: the hormones (such as testosterone, estradiol, and cortisol) serve as information about the body's age, sex, health, status, and other contextual considerations; that information can be exploited to generate coordinated, integrated outcomes that are often adaptive. For example, elevated levels of the hormone estradiol may be released by the ovary during the most fertile phase of the ovulatory cycle, with the estradiol binding to its receptors (estrogen receptors alpha and beta) in specific tissues in order to exert adaptive effects, such as thinning the cervical mucus to facilitate sperm transport should intercourse occur, or altering motivation to interact with potential mates. In another example, the testosterone released by the human fetal testis can exert all of the masculinizing effects on internal and external reproductive anatomy, as discussed above.

Much of our focus in this chapter and indeed throughout the book is with the so-called sex steroids—the androgens (which include testosterone, dihydrotestosterone (DHT), androstenedione, and others), estrogens (which include estradiol, estrone, and estriol), and progestins (progesterone). These are hormones derived from cholesterol, hence their classification as steroids. Other hormones are produced through the expression of genes; these may be molecularly small (as in peptides such as oxytocin and vasopressin) or large (as in prolactin). The smaller steroids can readily pass throughout the body, including across the blood-brain barrier, a barrier designed to protect the brain from toxins circulating in the periphery. The larger peptide and protein hormones may be stopped by the blood-brain barrier, or in some cases they may exert their effects across it through active transport mechanisms, exerting effects on receptors at the boundary (for example, the choroid plexus), or more distally by influencing nerve activity (of the vagus nerve, for example) with those effects transmitted to the central nervous system. All hormones bind to receptors. The androgens have their greatest affinity for androgen receptors, the estrogens for two types of estrogen receptors (alpha and beta), and, as one further example, vasopressin for one of three different types of receptors (referred to as 1a, 1b, and 2). Interestingly, the androgen receptor is located on the X chromosome, a chromosome that is disproportionately held in females (who have two X's compared with one in males); this is consistent with a sexual conflict model in which females may "harness" some of the effects of androgens in males through selective effects on the androgen receptor.

In the case of sex steroids, they are released from the gonads (ovaries and testes) and also from other tissues (such as in the brain and by adipose tissue, or fat, the latter of which serves as an important source of estrogen). The release of sex steroids from the gonads occurs in both males and females via the HPG axis, referring to its hypothalamus (H), pituitary (P), and gonadal (G) components. Interestingly, in females and males the HPG axis builds on similar foundations, which is also a reflection of the shared evolutionary heritage of the sexes (that they emerged from a long-ago, sexless world). The hypothalamus, a small structure at the base of the brain, releases GnRH (for gonadotropic-releasing hormone), which in turn stimulates the pituitary gland, a small structure just below it, to release LH (luteinizing hormone) and FSH (follicle-stimulating hormone), which in turn travel through the circulatory system until exerting effects on the gonads to stimulate steroid hormone release (as well as gamete release). By cuing HPG-axis activity to age and other factors (producing elevated sex steroid hormones around the time of birth and at puberty, for example) the body can coordinate morphological, physiological, and behavioral outcomes appropriate for an individual's age, sex, and other contextual factors (including social context). The head (central nervous system) and the rest of the body (periphery) can work in tandem in adaptive ways. And to build in the potential for tissue- and organ-specific compartmentalization, different hormone pathways and receptors can be activated. A hormone traveling throughout the circulation, such as testosterone, can serve as a prohormone, converted to other steroids (estradiol, DHT) through which it exerts its effects in only certain tissues. This is the case for testosterone's effects on male reproductive anatomy and physiology: these effects are typically exerted after testosterone is converted by the enzyme 5-alpha reductase to DHT, which in turn binds to the androgen receptor, in order to grow the prostate gland or lengthen a man's penis. In the case of rodent brain masculinization, testosterone is converted by the enzyme aromatase to estradiol, through which it exerts its effects.

While we feature discussions of sex steroids here, there may also be synergistic effects between sex-specific steroid hormone concentrations and peptides (Adkins-Regan 2005; Ellison and Gray 2009). More specifically, in females estrogen appears to exert potentiating effects on the oxytocin system, whereas for males androgens similarly appear to potentiate the vasopressin system. In Chapter 3, we noted findings linking oxytocin with pair-bond formation in nonhuman animals as well as people. Oxytocin appears to dampen stress responses and to foster development of social attachments with offspring, mates, and even others. Oxytocin's effects on smooth-muscle contraction, including at orgasm, birth, and during lactation, appear to have been piggybacked with complementary emotional effects, helping to generate positively reinforcing experiences (well, at least with sex and lactation). The potentiating effects of estrogens, which are revealed in the modulation of oxytocin concentrations and receptor levels by estrogen (for example, across the ovulatory cycle, and with age: postmenopausal women have both lower estrogen and lower oxytocin levels), make sense in adaptively attuning female perceptions, motivations, and behavior for their age, reproductive status, and in other contexts. If oxytocin's effects appear more highly attuned to female sociosexual behavior, this peptide also appears to exert similar effects in males, indicative of cross-sex pleiotropy (or crossover in function between the sexes).

In a related vein, modulations in male testosterone (for example, through castration) also modulate vasopressin concentrations. Vasopressin appears to influence male motivation to respond to acute behavioral challenges, such as aggressive challenges in humans (in voles, pup defense and courtship), while also altering fluid dynamics and blood pressure, with the latter effects sensible as physiological adjustments to fighting (for example, elevated blood pressure, reducing blood loss during injury). The wider point is that sex steroids may work synergistically with other hormones in sex-specific ways to foster adaptive outcomes. Some work in voles suggests that early-life administration of vasopressin influences adult male aggression (Stribley and Carter 1999), but there is inadequate work in humans to determine whether oxytocin and vasopressin have significant early-life roles in sex differentiation that resemble demonstrated effects of these same peptides among adults.

One additional complexity, typically drawn on for behavior but with relevance for other traits, is a distinction between so-called organizational and activational effects of hormones. A given hormone, like testosterone, may exert permanent, irreversible effects (organizing effects) as well as more transient and reversible ones (activational effects). Traditionally, it was thought that organizing effects of hormones were focused on the perinatal period, but it is now clear in humans and many other animals that puberty represents a second organizational period (Schulz, Molenda-Figueira, and Sisk 2009). For example, estrogens in humans are likely playing organizational roles on adolescent female brains, contributing to shifts in attention, motivation, and behavior upon the female's entrance to the reproductive realm. The distinction between organizational and activational effects is important because tracks need to be laid (the organizing effects) in order to later run the train on them (activational effects), which is quite pertinent to discussions of sexual orientation, sexual response, and aging. At any rate, we belabor these and other complexities of human sex determination and differentiation not to scare or induce sleep in our readers, but to help gain an appreciation for the realities undergirding the mechanisms and development of our sexual systems.

SEXUALLY DIFFERENTIATING THE BRAIN

It is easy to visualize human sex differences in reproductive anatomy. The differences between a penis and vagina are readily apparent, as anyone familiar with the human body appreciates. But what about the brain? Are there sex differences in the brain? And if there are, can we unpack the genetic and hormonal mechanisms by which they emerge, and are they possibly related to sex differences in sexual behavior?

Anatomists and neuroscientists have documented a variety of sex differences in the human brain. These are not the kinds of observable differences you might scan in a potential partner's head, however. By examining and quantifying the size of specific structures in the brains of deceased humans, or more recently through brain imaging techniques such as magnetic resonance imaging, scientists can tell us some of the ways that the average female and male brains differ. Of course, like any sex differences, these are average differences; there is considerable variation between the respective sexes, meaning that for any given individual one still could not specify her or his specific neuronal dimensions without looking at them specifically.

Of the sex differences in human brains, perhaps the most accessible is brain size (Hines 2004; Nelson 2011). We cannot see each others' brains directly, but the skull that surrounds a human brain reflects, in part, the underlying brain size (Lieberman 2011). The average male human brain is larger than the average female brain. That may come as no surprise, though, since the average male human body is also larger than the average female human body (all else being equal, we would expected larger bodies to have all kinds of larger structures). However, if one adjusts statistically for average body size, then females have higher encephalization quotients (or EQs), a measure of relative brain size. So whether male or female humans have larger brains depends on whether one prioritizes absolute or relative brain size.

Smaller structures within the brain also exhibit human sex differences (Hines 2004; Nelson 2011). These include the sexually dimorphic nucleus in the preoptic area, the interstitial nuclei of the anterior hypothalamus, and the suprachiasmatic nucleus, all of which are larger in males. These structures may be implicated in courtship and sexual behavior, as similar regions in other species are also larger in males, and ablation of them in other species has been found to compromise male mating behavior. Females, conversely, exhibit larger planum temporale, dorsolateral prefrontal cortex, and superior temporal gyri, all structures that may be implicated in language, and perhaps in turn this demonstrates advantages in female verbal fluency. By several measures, such as the posterior corpus callosum, females appear to have relatively larger connections between brain hemispheres. This sex difference may underlie the more lateralized brains of males. A female experiencing a stroke, for example, is less likely than a male to suffer deficits in cognition and behavior, as the greater connectivity between her hemispheres enables more redistribution of information processing. Thus, all said, there are measurable sex differences in human brains, even if they are typically hidden under the hood (or within the cranium), and some of these may be linked with sex differences in human social and sexual behavior.

The best way to spot sex differences in human brains is indirectly—in behavioral output. For every emotion, thought, and eventually behavior, there is a physiological substrate underlying it. When those cognitive and behavioral patterns—such as in verbal fluency or male courtship—are sufficiently robust and differentiated between the sexes, there may also be measurable differences in the brain between females and males. At this juncture, we ask: what are the primary sex differences in human sociosexuality that have been found? Fortunately, many scholars, from evolutionary, comparative, and neuroendocrine backgrounds, have addressed this very issue, enabling us to draw from an impressive literature in answering this question.

As Nelson notes, "The most prominent behavioral sex differences between men and women are observed in gender role, gender identity, and sexual orientation/sexual preference" (Nelson 2011: 174). This may seem self-evident, and yet these differences are indeed pronounced, and they presumably have some kind of measurable brain differences underlying them. Other major sex differences in human behavior include juvenile males engaging in more physical play; males engaging in more physical aggression; males engaging in more sexual coercion; juvenile males maintaining higher activity levels; females displaying more verbal fluency; females engaging in more group socioemotional behavior; females displaying greater investment in childcare; sex differences in toy preference (with females more oriented toward doll-like toys that can be nurtured, and males more oriented toward objects that can be moved around, like trucks); and juveniles tending to prefer same-sex play partners (Cohen-Bendahan, van de Beek, and Berenbaum 2005; Hines 2004; Lippa 2005; Nelson 2011). There are also demonstrated sex differences in cognition that are likely related to these same sex differences in behavior. As examples, females display higher perceptual speed, while males score higher on mental rotation tasks (rotating an object in space). These cognitive and behavioral differences may even contribute to higher juvenile male mortality: in the United States, boys are more likely to die than girls from "external causes" such as accidents (Kruger and Nesse 2006), though sex-biased cultural practices in other societies, such as female-biased infanticide or providing access to health care can produce higher juvenile female mortality (Hrdy 1999).

From an adaptive perspective, we can situate these sex differences in human sociosexuality and cognition within evolved sex-specific reproductive strategies. It makes sense that a genetic and gonadal female should identify as and feel female (cases of discrepant gender identity help illustrate why): doing so facilitates the development of sex-appropriate beliefs and behaviors. Among mammals, including humans, females have long provided the bulk of parental care, even if human males display unusually high investment, by mammalian and primate standards. The greater female investment would presumably favor emotional capacities fostering attachment to and investment in her offspring, thereby aiding their survival and her own reproductive success. As a juvenile, orienting toward baby stimuli (from dolls to real babies to cute puppies) could aid the development of a female's caregiving capacities. From a male's standpoint, the overarching gradient of sexual selection would presumably favor developing the competitive nature and physical prowess necessary for male-male competition, eventually over mates, and sometimes also for employing sexual coercion to increase the likelihood of fathering offspring. Further, a male's greater brain lateralization, his orientation toward objects and things (favoring activities like throwing or kicking balls; evincing less emotional empathy), including objects situated in three-dimensional space, may be part of honing the competitive fighting abilities that male ancestors once relied on. Fights play out in three dimensions, entail attacks, require judging distances rapidly—just ask an Ultimate Fighting Championship (UFC) contestant or a playground bully—and males may have the mental machinery to match.

GENES, HORMONES, AND HUMAN SEX DIFFERENTIATION: JUST SO STORIES, OR RICH SCIENTIFIC TALES?

We began this chapter with a biblical quote concerning eunuchs. A eunuch is a male who has had his testicles removed or otherwise rendered nonfunctional. During the era in which Bible stories were first passed on and recorded, eunuchs played roles in what were increasingly complex societies. While some eunuchs were born with a culturally atypical sexual anatomy and identity, most were created by their fellow men. Eunuchs frequently served as court servants, especially to women living in harems, the thinking being that eunuchs would be less likely to attempt to mate with a despot's consorts (Betzig 1986). In fifteenth-century Ming China, Zheng He, who was made a eunuch at age eleven, eventually rose to command the largest ships in the largest fleet ever assembled at that time. The last of the eunuchs in imperial China served until the early twentieth century, when they were still sharing information that scientists and doctors would use for decades after (Wilson and Roehrborn 1999). Among the lessons learned from eunuchs is that the age of castration matters for sexual performance: men castrated before puberty had more greatly compromised sexual performance than their counterparts castrated after puberty, a finding that highlights the importance of puberty as an organizational phase. The effects of castration also illustrate the power of testosterone, and of its loss: eunuchs castrated before puberty never developed the deep voice pitch of intact men, lacked measurable prostate glands, and frequently had gynecomastia (breasts), among other characteristics.

If the experiences of eunuchs tell convincing tales of the power of sex steroids on human sexuality, what other lines of evidence speak to this same subject—specifying the genetic and endocrine mechanisms of human sexual differentiation? Ethical and logistical considerations prevent transgenic or endocrine interventions on human fetuses. So how can we ensure we are not just spinning "just so" yarns here, like the Rudyard Kipling children's tales in which every trait has some greater purpose, but that we are instead elucidating a scientifically grounded understanding of human sex differentiation?

At least five lines of evidence suggest that the stories of human sex chromosomes, SRY, DAX-1, androgens, organizing and activational effects, and so forth all hold water (Gray 2010; Nelson 2011). The first is experimental research conducted on other organisms, particularly those most closely related to us. In practice, this often means rodents, such as mice and rats, and primates, including rhesus macaques. Indeed, classic mid-twentieth-century castration and regrafting experiments were conducted on rabbits by Alfred Jost and guinea pigs by William Young. In these experiments, the removal of testes (and the testosterone they secrete) resulted in female-typical anatomy and behavior in genetic males, as would be expected if females were the default sex and males were "built" by testosterone. Further, females given high doses of androgens early in development were shunted toward maleness, while females whose ovaries were removed still developed as females. Closer to home, several groups working with rhesus monkeys have conducted studies in which pregnant rhesus monkeys were given extremely high or more modest doses of androgen at different stages in their pregnancy. Females given the really high doses had daughters whose genitals and behavior were masculinized, with behavioral effects including higher rates of rough-and-tumble play and mounting behavior, although these same studies revealed that the timing in utero of genital and brain differentiation differed (Goy, Bercovitch, and McBrain 1998). More recent studies using lower doses yielded less pronounced effects. Daughters of female monkeys who received a modest androgen dose during pregnancy exhibited a few more masculinized traits, including more male-like vocalizations when separated from their mothers, but no dramatic effects on physical play or mounting were observed (Wallen and Hassett 2009).

We must extrapolate with caution from nonhuman findings to humans (Ellison and Gray 2009; Insel 2006). Tempted by the kinds of experimental research designs on nonhuman animals that simply cannot be justified for people, we might wish to extrapolate from findings in rats or rhesus monkeys to ourselves. However, phylogeny, in this case, is a mixed bag: there are indeed deep similarities across taxa (among all mammals, for example) in the mechanisms of sex differentiation, but there are also lineage-specific effects, for which the best guides to humans are the creatures most closely related to us. This is important because we (and rhesus monkeys, more so than rats) exhibit greater cortical control over behavior than most animals, meaning that the power of hormones in impacting sexual motivation is more subtle in us than it is in rats or mice. As another illustration, in rodents, apparently unlike primates, brain masculinization typically entails the conversion of testosterone to estradiol (see Wallen 2005); for rats, then, brains are masculinized by estrogens bound to the estrogen alpha and beta receptors. Female rodents apparently avoid masculinization of the brain through the binding of estrogens by alpha-fetoprotein, leaving them with a default female brain rather than one masculinized by estrogen.

A second line of evidence addressing the mechanisms of human sex differentiation draws from correlational studies. The digit-ratio findings with which we began the chapter illustrate this method. In such studies, scholars test hypotheses for relationships between some genetic or endocrine variable and some cognitive or behavioral variable. In the case of digit ratios, there was support for the anticipated association between masculinized digit ratios and female sexual orientation. As other examples, according to meta-analysis there is a positive but weak association between adult male testosterone levels and measures of aggression (Archer 2006), and a study recently found an association between androgen-receptor polymorphisms and measures of male dominance but not sexuality (Simmons and Roney 2011). The best of this sort of work is embedded in the wider phylogenetic and mechanistic literature, while admittedly it is subject to the limitations inherent in any correlational design. A strength of this approach is that the data originate with humans rather than rats or cats.

A third line of evidence involves clinical cases of atypical human development. We will delve deeply into a set of these cases in the next section of this chapter. For now, though, it is worth noting that these cases have the strength of originating with people (rather than, say, mice), but they also have many limitations, such as small sample sizes, the fact that various co-morbidities commonly occur alongside the targeted cognitive or behavioral variables, and that individuals with these patterns (and their family and friends) may be fully aware of their condition, which can yield an altered life-course experience distal to more direct effects of hormones or genes on sex differentiation.

A fourth line of evidence concerns human hormonal interventions lead-ing to unintended consequences. As an example, women taking synthetic progestins in the 1960s and 1970s to alleviate pregnancy symptoms unin-tentionally exposed their fetuses to androgenic effects; the exposed offspring reported a higher likelihood than controls did of using physical aggression. Other examples of unintended hormone exposures are presently being stud-ied, and they often draw public attention. The compound bisphenol A in plastic products, including some baby bottles, has been found to have estro-genic properties. So many women take the Pill that we have recycled public water supplies in which estrogenic hormones occur in measurable amounts. Providing children with soy milk, which contains high concentrations of plant estrogens, raises questions about the potential effects of this interven-tion. While it is clear that the impacts of such "experiments" in estrogenic supplementation are greater in frogs or fish (swimming in the stuff is a dif-ferent equation, as is having a type of skin that more readily incorporates environmental molecules), among which alterations in sexual development have been recorded, it is difficult to quantify the specific effects of these sorts of ongoing interventions in humans. Studies of Lake Mead, which provides much of the water for Las Vegas, indicate that the levels of estrogen found in it are quite low compared with concentrations in food or other sources, suggesting that levels in the water do not pose a risk of dramatic effects on human sex differentiation (see Stanford et al. 2010).

A fifth line of evidence concerns human hormonal interventions, usually undertaken for clinical, cosmetic, or quality-of-life reasons. Within this body of evidence, numerous studies have reported on the effects on women of taking various estrogen formulations, or estrogen plus progesterone, both during reproductive years and postmenopausally. Studies of men also assess the impacts of testosterone supplementation on body composition (on muscle and fat) and other outcomes, including libido and coital frequency. These studies rarely include children, a fact that limits the insights concerning mechanisms of sex differentiation perinatally, but some studies do include adolescents, and findings from adults to some degree may apply to pubertal transitions. In this vein, some of the most dramatic effects concerning effects of hormones on sexually differentiated traits have been documented among transsexuals, or individuals who undergo hormone supplementation and possibly also associated surgeries as part of a cross-gender transitioning process. Some men transitioning to women describe a "lightened libido," while some women transitioning to men note the "electric libido" they newly feel. As one other anecdote in this vein, when Peter was doing postdoctoral work in clinical endocrinology, he attended a presentation that included a photo of a highly muscular small child. What had happened to this young super-man was that some of the testosterone gel his dad had applied to his own chest and shoulders had rubbed off on his young son, in turn having some of the predicted effects on the child's muscularity.

THE RISE OF SEXUALLY DIFFERENTIATED VARIATIONS

In Chapter 2, when discussing cross-cultural variations in human sexuality, we touched on male same-sex sexuality among the Sambia, in the highlands of New Guinea. We return now to the Sambia, but with a different focus, this time on the so-called turnim-men (Herdt and Davidson 1988). This is the term the Sambia used for individuals who were more female-like prior to puberty, but who at puberty developed more pronounced masculine characteristics, including the growth of their initially small penis. Where do the turnim-men come from? Are they the developmental product of environmental endocrine disruptors, or some other natural human experiment yielding unintended consequences? No, it turns out. Instead, the small number of Sambia turnim-men were eventually diagnosed with 5-alpha reductase deficiency. This is a condition in which an inadequate amount of the enzyme 5-alpha reductase results in a failure to convert testosterone into dihydrotestosterone. Without that DHT, the typical pattern of male genital differentiation does not occur, so individuals are born with more ambiguous genitals. However, at puberty, high concentrations of other androgens bypass, to some degree, the lack of DHT, leading to partial masculinization of an XY individual's genitals. Most turnim-men lead productive adult lives as men, often marrying, though the vague ethnographic accounts of them suggest they may be stigmatized to some degree.

The 5-alpha reductase condition among the Sambia is but one case of atypical human sexual differentiation (Levay and Baldwin 2009; Nelson 2011; Roughgarden 2004). Given the complexities of the process of sex differentiation, there are numerous steps during which the process might go stray, yielding considerable variation within the typical range of femaleness and maleness, and also in the form of more extreme atypical phenotypes, such as 5-alpha reductase deficiency. This is one of the important, take-home messages of this chapter. There is no essential female or male. Rather, there are degrees of feminine and masculine traits. This variation is important both in its own right and for other reasons. There is variation on which selection can act, should evolutionary pressures push in new directions. There is also the latitude given to "normal," which requires careful consideration when a physician or a government body contemplates its own definitions of what is normal and what phenotypes that might invoke intervention. We also suspect, however, that selection has until recently acted—and perhaps still does act—strongly in favor of the development of reproductively viable females and males. An individual born with a sociosexually atypical phenotype who is infertile or who has very low fertility would find others' genes propagated at higher levels than her or his own. So we should not get so carried away with variation that we reach a point where, for example, Roughgarden (2004) proposes scrapping sexual selection theory altogether as a means of understanding the evolution of sex differences.

Sociosexual variation is all around us. While writing this, we came across a news story about an intersex dog—a dog whose genitals were neither all female nor all male, but something between. Among kangaroo rats, 16 percent "have both sperm- and egg-related plumbing, including a vagina, a penis, a uterus, and testes in the same individual" (Roughgarden 2004: 37). On the Polynesian island of Vanuatu, intersex boars have long been prized, with 10 to 20 percent having mixed female and male reproductive anatomy. These may be the extremes, however, in intersex prevalence, with lower occurrences seen more typically in other species in the wild, including animals outside the confines of human artificial selection. Classical Greek stories of hermaphrodites (part male Hermes and part female Aphrodite) are partially born out in humans, in the sense that intersex individuals may have partially female, partially male reproductive physiology, but of course there is no person who can simultaneously play reproductively viable female and male roles (like an earthworm, a true hermaphrodite).

There are several types of atypical sex chromosome complements in humans. While the typical pattern is for females to have XX sex chromosomes and males XY sex chromosomes, other varieties occur at measurable frequencies. A reason for the viability of some individuals with atypical sex chromosome complements is that the sex chromosomes contain relatively few genes; with fewer genes, the absence or duplication of some genes in the sex chromosomes has less of a lethal impact compared with the vast majority of more gene-rich autosomal chromosomes (for example, chromosomes 1 and 2).

In Turner's syndrome, individuals are born with a single X chromosome, and are designated as XO. In approximately 1 of 3,000 live births a child has this phenotype. Because females are the default sex, XO individuals are diagnosed as girls at birth, but their ovaries do not produce typical estrogen levels, partly accounting for a host of phenotypic differences. Individuals with Turner's syndrome tend to be shorter, often have webbed necks, and may have reduced cognitive function, but some of these effects can be offset through the administration of exogenous estrogen at puberty. The flip side of Turner's syndrome, however, is apparently not viable: individuals with a single Y (YO) do not survive to be born.

Klinefelter's syndrome occurs among approximately 1 of 600 live births, and is characterized by an XXY sex chromosome complement. Because they possess both X and Y chromosomes, individuals with Klinefelter's experience masculinizing effects of the SRY gene, but they may also have reduced androgen and sperm production, leading to an inability to father children. Many individuals with Klinefelter's grow relatively tall, and they may exhibit some degree of breast development, or gynecomastia. There are also individuals born with an XYY sex chromosome complement, at a frequency of approximately 1 in 850 live births. Such individuals tend to be relatively tall and have some compromises in cognitive function.

Congenital adrenal hyperplasia (CAH) is a condition in which hormone metabolism is altered, leading to exposure of unusually high androgen levels. The effects are not pronounced in males, but for females the exposure to excess androgen has masculinizing effects that vary in their severity. The condition occurs among approximately 1 in 16,000 live births, but is higher in some populations, likely owing to effects of genetic drift and inbreeding depression (for example, there is a higher prevalence among Ashkenazi Jews). In these individuals, the genitals may be partially masculinized, with the clitoris enlarged, for example. In studies of the sociosexuality of CAH girls, some have been found to exhibit higher rates of childhood physical play (much like boys), indicating that the additional androgen affects both reproductive physiology and the brain. Another study found that CAH females score higher on the kinds of mental rotation tasks in which males typically excel (Puts et al. 2008).

One last case of atypical human sex differentiation to consider here is androgen insensitivity syndrome (AIS), which refers to XY individuals who have an atypical androgen receptor. In cases of complete AIS (or CAIS), which occurs in approximately 1 in 10,000 live births, the androgen receptor is altogether nonfunctional, with the consequence that the testosterone released by the testes is not bound and does not exert its typical masculinizing effects. CAIS individuals look like females: they have female external genitalia, develop breasts at puberty, and typically identify as females. The feminine characteristics of individuals with CAIS reflect the concept that females are the default sex, and indicate that such individuals have enough estrogen for its effects to be manifest (through breast development at puberty, for example). However, CAIS individuals are infertile (they lack ovaries, Fallopian tubes, and a uterus and have a shallow vaginal barrel because the Müllerian ducts have been inhibited), and they have underdeveloped pubic and axillary hair and sebaceous gland secretions, the latter of which are secondary sexual characteristics that develop under the influence of androgens.

DEVELOPING SOCIOSEXUALITY: AN INTERACTION

Marvelous things can happen when heritable physiology and social context interact (Nelson 2011; West-Eberhard 2003). In the case of honeybees, being served royal jelly can turn a bee into a queen. Among naked mole rats, the pestering of a dominant female can suppress a subordinate female's reproductive system, leaving her smaller and sterile, but still game to take on a reproductive role if circumstances change. In the case of parrotfish, colorful inhabitants of tropical reefs, development and social context can entail a sex change, from female earlier in life to male later. Among various vertebrate species, such as plainfin midshipman fish, the males (rarely the females) take on a different form depending on the presence of rivals. By being smaller and perhaps sneakier, the alternative male morphs may find different avenues to reproduction. Among lab rats, a male's sexual performance is better predicted by the sex ratio of his littermates than the care his mother gives him or the sex ratio of his fellow rats in utero (de Medeiros et al. 2010). These few examples alone indicate that nutritional and social cues can shape an individual's developing physiology. The image we have is of a developing organism traveling slowly along a river's meandering floodplain, with many potential paths contingent on the twists and turns at each bend. This is not a straight course.

In raising human sexuality, we also find ongoing interactions between an individual's physiology and her or his environment. Subtle variations in a genotype, quantitative differences in endocrine function, epigenetic influences (alterations in gene expression: Francis 2011), influences on stress physiology—all of these are ways in which a human organism's course of development can also seem to follow a winding floodplain. A host of environmental factors, from nutritional to family variables to wider social contextual ones, can all shunt a person down one stream rather than another, with each shift affecting future directions. The sex of one's older siblings can exert subtle effects on sex-typical behaviors, as seen in a U.K. study in which children who had older brothers leaned a bit more toward more male-typical behaviors, and children with older sisters leaned a bit more toward more female-typical behaviors (Rust et al. 2000). In other studies, boys who tend to shy away from male-typical social behaviors (such as rough-and-tumble play) are more likely to express an adult same-sex sexual orientation; one possibility for this association is that the initial personality traits and emotional reactions shaping an individual's engagement with the social world become self-fulfilling aspects in terms of preferred friends and mates (see McIntyre and Hooven 2009).

One of the most notable social influences on sexuality is birth order (Poiani 2010). Having lots of older siblings, and particularly older brothers, is the best predictor identified thus far of a male having an adult same-sex sexual orientation, at least among the predominantly European and North American samples in which this subject has been researched. More recently, scholars extended and qualified the findings among Samoan *fa'afafine,* or men who have a strong feminine orientation and play alternative gender roles, including as an important member of the family. *Fa'afafine* are often identified in early childhood, and while they are sexually active, they rarely sire offspring themselves. Vasey and VanderLaan (2007) found that *fa'afafine* have more older brothers, but *fa'afafine* also have more older sisters and younger brothers too. There are various ways by which an association between birth order and sexual orientation or gender role might arise. A major proponent of the fraternal birth order effect, Ray Blanchard, speculates that maternal immune responses underlie this pattern. One could as well imagine that subtle differences in the social niche of boys who have no, some, or lots of older male siblings could be relevant, with older siblings having an easier go of pushing around a younger one.

Having examined the genetic and endocrine mechanisms that shape human sociosexuality, including both typical and atypical cases, we are ready to see how our plastic physiology is related to sociosexual behavior across the life course. In the next chapter we will tackle the juvenile years, before greeting adolescence; later chapters will address sociosexuality in the reproductive years and, eventually, with advancing age. Along the way, we will revisit some of the physiological findings of this chapter. Yet we will not lose sight of the larger picture of human sexuality: that humans exhibit sex differences; that our recent ancestors appear to have reproduced within long-term, slightly polygynous unions; that our proclivities both shape and are constrained by the social world in which they arise. The development of a sexual self is an evolved but noisy process, capable of following many channels based on environmental influences throughout the life course.

CHAPTER FIVE

Playing at Sex

Learning, Practicing, and Developing
Sociosexual Behavior in Context

> Hugging and kissing are usual in the activity of the very young
> child, and . . . self manipulation of genitalia, the exhibition of
> genitalia, the exploration of genitalia of other children, and some
> manual and occasionally oral manipulation of the genitalia of
> other children occur in the two- to five-year olds more frequently
> than older persons ordinarily remember from their own histories.
> Much of this earliest sex play appears to be purely exploratory,
> animated by curiosity.
>
> —Alfred Kinsey, Wardell Pomeory, and Clyde Martin,
> *Sexual Behavior in the Human Male*

AS CHAPTER 4 and the quotation above indicate, human sexuality begins to
unfold early in development, well before reproduction is possible. The
mechanisms of sex differentiation press females and males to enter the postnatal world prepared to engage it in different ways, and in ways ultimately
linked to sex-specific stakes measured in survival and reproductive success.
The mechanisms of sexual pleasure, though not fully elaborated until adulthood, are already on their way, even as a parent or other person might attempt to channel that individual's sociosexual behavior.

Humans are not exceptional beasts when it comes to the development of
our sociosexual behavior. Any long-lived social species faces the need to
develop species- and sex-specific patterns of sociosexual behavior, often entailing extended practice and learning. To shed more light on sociosexual
development, who better to turn to than some of our fellow long-lived cousins, indeed our closest living relatives, chimpanzees and bonobos?

In the wake of deforestation, a growing bushmeat trade, and other challenges to living in the wilds of Africa, many young chimpanzees and bonobos

have been taken in by sanctuaries designed to offer them a chance at life. As Vanessa Woods and Brian Hare have shown, these same sanctuaries can also serve as opportunities for cutting-edge research on ape sociosexual development (Woods and Hare 2011). Since the captive young chimpanzees and bonobos living at the sanctuaries share many similar features—living together with many other unrelated individuals in a cobbled-together social world, but also a world where their behavior can be closely studied—the sanctuaries offer an opportunity to compare the sociosexual development of these two close kin of ours.

Based on descriptions of adult chimpanzee and bonobo sociosexual behavior, which we touched on earlier in this book, we would expect both similarities and differences in their behavior. We would expect females and males of both species to mate with multiple partners, but we would anticipate that bonobos might exhibit a greater degree of female same-sex sexual behavior than chimpanzees. When the sexual behavior of bonobos and chimpanzees between two and four years of age was observed, none of the sixteen chimpanzee infants displayed sexual contacts with same- or other-sex partners. However, of the eight bonobos, all engaged in sexual contacts both with same-sex and with other-sex partners, and typically with multiple partners. Further, of bonobo sexual behaviors, more involved some form of ventral-ventral contact (such as females facing each other and rubbing their genitals together) than the primary bonobo sexual position employed during reproductive behavior (ventral-dorsal, with a male mounting a female from behind). The rate of bonobo sexual behavior was amplified during feeding contexts, perhaps because sexual contacts express excitement or release tension during a time of anxiety. These findings indicate species differences in infant sociosexual behavior that are consistent with previously identified adult behavioral differences, suggesting species-specific trajectories in the development of such behavior.

In this chapter, we address core features of sociosexual development. We discuss infant and juvenile development more broadly, including the general background of age-related changes in sociosexual practice. We then further consider examples of nonhuman sociosexual development, broadening that scope beyond chimpanzees and bonobos to include an emphasis on other primate examples. Last, we cover some of the quite limited data on human sociosexual development, drawing on cross-cultural surveys and more quantitative studies in largely Western societal contexts. Throughout, we attend to sex similarities and differences in sociosexual development, as well as details about the partners (alone, same-sex, other-sex, similar in age or older, and so on) and the positions employed as youth sort out the ways of their sexual selves.

TIMES TO LEARN

Infancy is the time during which a young mammal nurses from its mother, whereas the juvenile phase refers to the time from weaning until puberty. Both infancy and juvenility represent wonderful times to learn. Guided by species- and sex-specific mechanisms of sociosexual development, budding individuals represent an ongoing interaction between the operation of these mechanisms and specific socioecological contexts. This is neither nature nor nurture exclusively; rather, the developing sociosexual behavior of any individual is the product of their interaction.

Different species exhibit different rates of development. Among subadults (either infants or juveniles), some progress quite rapidly from birth to puberty, whereas others take prolonged periods to reach that same developmental milestone. The standard mammal of lab research, the mouse, may pass from birth to puberty within weeks; in that same amount of time, a vervet monkey, or a chimpanzee, has barely begun to see its outer world. In a strange example of pet keeping, renowned biologist Alfred Russel Wallace described the developmental differences between his orangutan and macaque infants: "The [orangutan], like a very young baby, lying on its back quite helpless, rolling lazily from side to side, stretching out all four hands into the air, wishing to grasp something, but hardly able to guide its fingers to any definite object. . . . The little monkey, on the other hand, in constant motion; running and jumping about wherever it pleased, examining everything around it" (Wallace 1890: 35). And in a bizarre illustration of developmental differences, Luella and Winthrop Kellogg recorded the experience of raising their son and a captive female infant chimpanzee side by side, treating the chimpanzee like a human family member (Kellogg and Kellogg 1933). That chimpanzee had more rapidly progressing motor skills and even some cognitive skills, but even during the nine months of the experiment (chimpanzees are not recommended as pets, after all), the human took off on an accelerated linguistic trajectory that left the chimpanzee speechless.

The species-specific rates of subadult development represent both phylogeny (shared ancestry) and adaptation. As for phylogeny, closely related species tend to have similar rates of development. Primates as a group mature at slower rates than most other groups of mammals. Among primates, monkeys and apes mature at slower rates than lemurs or lorises, and apes mature at slower rates than monkeys. And the adaptation component means that even closely related species may exhibit differences in development that have adaptive origins. Gorillas mature at a faster rate than orangutans, chimpanzees, or bonobos. Humans have the slowest development of all primates, suggesting that our subadult development has been derived relative to that of not only other primates but also great apes, presumably because slower development conferred some benefit to survival and reproductive success.

The primary adaptive hypotheses for slow primate growth refer to the benefits of learning (Geary 2010; Konner 2010). Some scenarios highlight the importance of learning the necessary ecological information in order to obtain the foods, and perhaps even food-processing techniques, necessary for survival. Other scenarios feature the importance of social information—that individuals may develop slowly because they need to learn the social skills necessary for maintaining relationships that themselves can be matters of life, death, and reproductive success or failure. These hypotheses do have support in the research: as an example, across primates, the length of the juvenile period is found to be positively associated with relative brain size as well as features of social organization (see Joffe 1997). These hypotheses may help account for the faster pace of gorilla development compared with other apes, as gorillas eat more leaves (requiring less learned skill in finding them) and have relatively smaller brains compared with other great apes. Yet another idea is that slow primate growth represents a byproduct of an integrated life history, wherein slow growth is just a symptom of an overall slow life course (Charnov and Berrigan 1993). In this scenario, some other life history trait, such as senescence (loss of adult function), may be slowed due to lowered adult mortality rates, in turn carrying slow subadult development along for the ride. A considerable body of theoretical and empirical work has been marshaled in the attempt to test these competing views concerning the reasons for slow subadult growth; however, no consensus has been reached (see Jones 2011).

Whatever the adaptive basis of slow subadult development may be, several notable landmarks stand out. One is the passage from infancy to juvenility with weaning. The center of an infant primate's world is its mother (Hrdy 1999). The mother provides milk, initially exclusively, later supplemented by foods that the infant largely acquires on its own. The mother provides warmth, protection, and a foundation from which to view and learn about the world, from the features of a foraging habitat to idiosyncrasies of the social world. For an infant primate, the loss of a mother can be, quite literally, devastating. As Jane Goodall observed among the wild chimpanzees she studied in Tanzania, when a juvenile (Flint) lost his mother (Flo), he displayed behaviors indicative of deep sadness and gave up eating, dying within weeks himself (Goodall 1986). In many other primates, the loss of an infant's mother heralds the death of her infants too, so central is she to their survival and well-being. As Harry Harlow's brutal 1971 studies of captive rhesus monkeys revealed, the infants deprived of their mother clung closely to a wire and cloth monkey surrogate, underscoring that mom provides not only milk but also warmth and security to a primate infant.

As infants are weaned and enter the juvenile stage, their world widens. The social world, especially in group-living monkeys and apes, entails less of mom (who by now probably has another nursing infant commanding her primary attention) and more of other juveniles and even adults. This expanded world can be adaptive, as the individual continues to learn but without the same heightened reproductive stakes that will later arise at puberty. As a cornerstone of learning, play helps individuals enhance their motor, cognitive, and social skills (Fagen 1981; Konner 2010). Juvenile play behavior conforms to features of a species' foraging and social ecology. Picture a kitten batting around a toy mouse, which helps it hone the hunting skills conferred by its ancestors. Or imagine a puppy's joy in snarling at would-be prey, and perhaps also digging a hole in the backyard to stash a tasty treasure for a later time. These are but two examples of the ways in which subadult play conforms to food-acquisition strategies; as we shall see, there are more parallels in species-specific patterns of sociosexual development.

Sex similarities and differences in sociosexual development occur. As parents of young children may be aware, boys tend to lag behind girls in reaching various cognitive developmental landmarks, such as language acquisition. The fact that boys and girls are on slightly different developmental trajectories says something about our ancestor's sociosexual behavior—that the slight polygyny they exhibited manifests in males taking longer to grow and learn, both early postpartum and also in later ages of pubertal development. It is likely that socially monogamous species, such as gibbons, owl monkeys, and titi monkeys, lack sex differences in their pace of juvenile social development, though we do not know of any formal tests of this idea.

The closest finding of which we are aware is the recent report that owl monkeys of South America do not exhibit sex differences in growth, consistent with their socially monogamous system and lack of dimorphism in body size (Huck and Fernandez-Duque, 2012).

A general principle is that females and males may exhibit differences in their subadult sociosexual development in tune with sex differences in their adult sociosexual development. In more polygynous species, where males are much larger than females, males may undergo more extended growth than females to achieve the bulk that facilitates success in male-male competition. As another example, in species where females engage in adult same-sex sexual behavior (like the bonobos), we would expect to see this manifested in juvenile sex play. This same principle should apply not just to sociosexual behaviors but to a wider array of sex-differentiated social behaviors, such as parenting and aggression. In this vein, we would expect that since female primates provide the vast amount of direct care of offspring, juvenile female primates should be more interested in babies than their male counterparts. Similarly, since adult male primates engage in higher rates of physical aggression than do females, we should expect to find more rough-and-tumble play in male juveniles than female juveniles. In the previous chapter, we discussed some of the neuroendocrine mechanisms underlying sex differentiated behaviors, and we now turn to nonhuman primate data generally consistent with these sex- and species-specific patterns of sex-differentiated subadult behavior; the nonhuman primate data are important for contextualizing the human data, which we will present last in this chapter.

PRIMATES LEARNING THE WAY

In our discussion of nonhuman primate development, we highlight findings in sociosexual behavior (such as mountings), but we also draw on data concerning sex-differentiated patterns of offspring care and physical aggression. For a visually stunning example, try to dig up, through an Internet search, an image of vervet monkeys and toy trucks and dolls. If you hit pay dirt, you will see side-by-side photos of two captive monkeys living in Los Angeles (not that their geography explains their behavior), one of which is a subadult male pushing a blue truck, and the other a female subadult holding a plush doll. These images capture the core patterns seen in a study in which these human children's toys were introduced to young monkeys (Alexander and Hines 2002). Obviously, vervet monkeys in the wild have no history of playing with objects like these toys. Yet the study found that females spent more time interacting with the plush dolls, male vervet monkeys more time with the trucks, and there was no sex difference in time spent with cooking pots. The interpretation of these findings suggested that among vervet monkey subadults, males and females are inherently oriented toward different types of stimuli in their environments, with the consequence that the introduction of novel human toys activates these preexisting biases. More specifically, many subadult female primates are likely interested in baby stimuli, which can help draw them toward learning about the babies they may have and eventually have to care for. Indeed, consistent with such views, Stephanie Meredith notes in her review of wild primate sex differences in behavior that, "Juvenile females show more interest in infants than their male peers in many species . . . [and] when sex differences in play are found, males play more frequently and more intensely than females." (Meredith 2012: 17–18).

In Alan Dixson's 1998 review of primate sexuality, he discusses the development of subadult sociosexual behavior in various primates, including these very toy-toting vervet monkeys. As he notes, "Sociosexual patterns of mounting, presentation, mutual embracing, and genital inspection or manipulation begin to develop during infancy in many monkeys and in the great apes" (Dixson 1998: 153). And he offers illustrations of these types of sex play, from male stumptail macaques playing with each others' genitals to a juvenile squirrel monkey displaying an erect penis to a male rhesus monkey engaging in dorsal-ventral mounting of another male, even employing the same foot-clasping technique employed by adults during sexual behavior with both females and males. In rhesus monkeys, male subadults engage in higher rates of mounting, physical aggression, and rough-and-tumble play than female subadults do. The rates of these behaviors also increase from infancy to early in the juvenile period, as individuals gain more independence from mom. These kinds of studies reveal that subadult primate sexual behavior is common, and that it exhibits some features similar to sexual behavior practiced among adults.

Another clear take-home message from studies of nonhuman primate subadults is that primates, such as rhesus monkeys and chimpanzees, require learning in order to display species-typical adult mating behavior. If socially deprived of opportunities to practice with mom or other subadults, an individual's sociosexual development may suffer. This is most vividly illustrated by Harry Harlow's (1971) rhesus monkey studies wherein infants were removed from their mothers and, in some cases, raised with other rhesus juveniles, while in other cases they were with no other rhesus monkeys. Later, as adults, the socially isolated rhesus monkeys typically failed to mate properly when introduced to potential mates; the mating deficits were less pronounced among those rhesus monkeys who had been raised with other rhesus subadults. When placed with a sexually experienced and indulgent female, the performance of deprived males improved. Dixson noted that, among chimpanzees, "Isolation-rearing causes deficits in sexual behavior in chimpanzees which are more pronounced in males than in females. Some chimpanzees of both sexes exhibit improvements of sexual function, however, and the capacity to recover seems to be greater than that of rhesus monkeys reared in isolation from their mothers" (Dixson 1998: 159). Such work on chimpanzees and rhesus monkeys suggests that the natural course of sexual development requires learning, especially for males.

How do nonhuman primates like rhesus monkeys and chimpanzees learn their sociosexual ways? They can hardly share their insights through a spoken language or even naughty magazines. But they can turn to the sexual world around them for observational learning opportunities as well as potential partners in play. Depending on the species, subadult primates are variably exposed to adults mating in open view of others, themselves included. For gorillas or gibbons, such matings are relatively infrequent, whereas for rhesus monkeys or chimpanzees they are quite common. The important point, however, is that observational learning opportunities are available, and they are a key component of the learning landscape for a sexually curious primate. The variable demographics (the availability of other subadults) across primate communities also present different opportunities for sociosexual experimentation: in small groups, there may only be a few other subadults with whom to play mount, so individuals make do with whichever adults are available for this purpose, including, in some instances, mom. As one illustration of this latter type of subadult sociosexual behavior, Anne Pusey (see Wolf and Durham 2004) showed that mother-son chimpanzee mating occurred occasionally, but drastically diminished when the son reached adolescence and the consequences of sex play (an actual conception) escalated.

The fact that nonhuman primate sexual behavior may occur within view of young offspring raises another dimension of subadult sexual behavior: that of potential parent-offspring conflict. Imagine an infant monkey, living in a multi-male, multi-female group, whose mother begins to cycle again after extended lactation. She draws more male attention and exhibits heightened proceptivity and receptivity to mates. Her physiology orients her toward a mixture of mating and mothering, whereas her offspring more highly prizes her ongoing mothering than her mating. The algebra of this equation conforms to the basic outline of parent-offspring conflict, as laid out by Trivers (1974) in a groundbreaking paper. While the classic example of parent-offspring conflict is typically weaning conflict, surprisingly there has been less attention devoted to mating conflict. Yet the occasional comment by a primatologist, based on field observations, suggests mating conflicts exist. Nursing offspring may attempt to disrupt their mothers' mating with adult males, at least initially, though eventually (like weaning) the offspring adjust to the realities of their sexually energized moms. According to Thore Bergman (personal communication, 2011), who has studied gelada baboons in the highlands of Ethiopia, gelada mothers begin cycling again around the time of weaning; offspring initially do protest against their mom's mating but seem to give up on this after awhile, often remaining nearby while she mates.

While sex differences in subadult social behavior are observed in various primates, few studies have been conducted on socially monogamous species. In other words, for those primate species in which adult male and female behavior is less differentiated (gibbon males and females have similar day ranges, for example, and both engage in vocalizations to deter same-sex competitors), we are short on data regarding potential sex similarities and differences in behaviors related to parenting, physical aggression, and sociosexual behavior. As an example, an older study on gibbon subadult play noted that a gibbon juvenile might only have a sibling to play with but did not go into any details about whether that entailed sex differences in play or sexual play (Ellefsen 1974). As a rare stab in this dark world, Chau et al. (2008) reviewed the wider literature on subadult play patterns and presented new findings on subadult social behavior in captive coppery titi monkeys. Earlier studies consistently indicate that sex differences in subadult play are typically found in polygynous species such as squirrel monkeys but are absent in more monogamous primates, including common marmosets and cotton-top tamarins. Among the titi monkeys, characterized by social monogamy and body-size monomorphism (male and female body sizes are comparable), there were no sex differences in subadult physical play. However, daughters preferred to play with their dads, whereas sons exhibited no preferential play orientation toward moms versus dads. What these data indicate, apart from a need for more studies on monogamous primates, is that subadult sex differences in primates are not an absolute given, but rather are contingent on the social behavior (such as monogamy) of a given species.

Among our closest of primate kin, the great apes, sex differences in subadult social behavior are manifested in various ways. Among juveniles, male chimpanzees tend to be more interested in adult males than the females are (Watts and Pusey 1993), a pattern that appears to reflect considerable juvenile male interest in the male-male political world; since that political world will later have a defining impact on a male's reproductive prospects, it makes sense that he would be interested in learning about and engaging it. There are consequences, though, to male meanderings in chimpanzees. Among wild chimpanzees, where female subadults tend to spend more time in proximity to their mothers than males do, female chimpanzees learn food acquisition techniques, such as termite fishing, more rapidly than males do (Lonsdorf et al. 2004).

Authors of an older review of wild great-ape behavior wrote, "Sex differences in chimpanzee [subadult play] have not been firmly documented. . . . Male gorillas and orangutans seem to play more than females, as expected in highly dimorphic species if play promotes fighting ability" (Watts and Pusey 1993: 160). More recent studies of subadult chimpanzee behavior in the wild suggest that males do engage in higher rates of rough-and-tumble play, as would be predicted based on their mating system and sex differences in adult behavior. Further, both chimpanzee and gorilla subadult females appear more interested than subadult males in infants, with differences for chimpanzees primarily related to interest in younger siblings rather than unrelated infants. In a recent study of wild chimpanzees, juvenile females more often carried sticks around in infantlike ways (that is, acting as if a stick were a baby, or, in human terms, a baby doll) than did juvenile males (Kahlenberg and Wrangham 2010). As for actual great-ape subadult sociosexual behavior, "Male chimpanzees show elements of sexual behavior by 1 year. They show copulation sequences complete except for ejaculation by age 4 . . . , earlier than gorillas (7–8 years)" (Watts and Pusey 1993: 162). Chimpanzee subadult males engage in higher rates of sex play, begin sex play at earlier ages, and do so with various partners that include adult females who are not capable of conceiving (non–sexually swollen females). All said, data on the social development of great ape behavior indicate sex differences in subadult behavior that are consistent with adult sex differences in behavior, suggesting that subadults are attuned to the kinds of stimuli that will enhance their prospects for future reproductive success.

HUMAN SOCIAL DEVELOPMENTS

If we have gained a primate comparative context in features of subadult sociosexual development, what of humans ourselves? Here, we point toward one of the few descriptions of sociosexual development in a hunter-gatherer society, a brief portrait provided by Frank Marlowe, based on his extensive ethnographic work among the Hadza in Tanzania: "Hadza girls and boys begin 'playing house' literally, building little huts, around the age of 7 or 8. There is some sex play when they enter the huts. Sometimes sex play among children occurs in full view of everyone; sometimes it is between two children of the same sex. Once, several Hadza and I watched two girls about 8 years old hugging and rolling around on the ground, clearly enjoying themselves in a sexual way. With increasing age, this sex play disappears; at least, it disappears from view." (Marlowe 2010: 168). What this ethnographic anecdote underscores is that sex play occurs in this foraging society, as it has been found to occur in several other hunter-gatherer societies, such as the Aka of Central African Republic (Hewlett and Hewlett 2010), and the !Kung of southern Africa (Konner 2010). Further, same-sex sex play may occur, and the age at which these activities were reported is consistent, as we shall see, with other human studies in which human subadults (hereafter we will use the term *children*) engage in more social sex play during juvenile ages (relative to the more common pattern of masturbation during infancy), and with some concern over maintaining privacy from adults at advancing ages. It is also noteworthy that, in this anecdote, Hadza boys and girls are acting out sociosexual play within imagined long-term bonds (couples with a house); as obvious as this is, it is a reminder that human childhood sociosexual development is consistent with long-term bonds rather than frequent multipartner mating, as among, say, rhesus monkeys or chimpanzees.

Although not manifest in this Hadza anecdote, sex differences in children's interest in young (for example, childcare) and physical aggression have been widely reported. We highlighted the mechanisms of these kinds of sex differences in the previous chapter. Here, we note that reviews of studies conducted in predominantly Western societies have consistently demonstrated childhood sex differences in interest in infants as well as physical aggression (Cohen-Bendahan, van de Beek, and Berenbaum 2005; Geary 2010). As an illustration, in a study of 106 Swedish children ages one, three, and five years, researchers found that girls preferred playing with feminine toys such as Barbie dolls, whereas boys preferred playing with masculine toys, such as pushing around a truck, even from the age of one onward (Servin, Bohlin, and Berlin 1999). A glance at any U.S. toy store reveals aisles of what are clearly girls' toys and boys' toys, providing insight into the market forces designed to manipulate and profit from sex differences in toy preferences. In systematic cross-cultural studies in which comparable methods have been employed in different cultural settings, sex differences in offspring care and physical aggression have been reliably discovered, even though the magnitude of these differences varies across social contexts; that is, while boys display higher rates of physical aggression than girls in each society (for example, in rural India, rural Kenya, and a small town in New England), the rates of aggression vary across societies (Whiting and Edwards 1988). These childhood sex differences also lead one to speculate about the functions of objects from the archaeological record: could those Venus figurines, about the size of a girl's hand, and exhibiting polish from use, have served as dolls?

We underscore the measurable sex differences in children's social behavior in order to plug them back into a pattern observed in the comparative context discussed above: that sex differences in children's social behavior, such as social play, are more dramatic in polygynous species, and minimal or lacking in monogamous species. Within this comparative framework, how should we interpret the mating meaning of human childhood sex similarities and differences in these social behaviors? The fact that sex differences in physical aggression, in addition to rough-and-tumble play, are reliably observed in humans suggests we are more like other great apes than titi monkeys or common marmosets, for we fit better with a pattern of polygyny than monogamy. The fact that girls are more interested in infant stimuli (babies, dolls) than boys suggests that sex differences in parenting behavior, the norm among primates, apply to us as well. At the same time, however, if we were to embed the discussions in more specifics of sex play (as the Hadza anecdote above suggested), we would likely conclude that we are oriented toward long-term bonds even in our childhood social play.

Another feature of human social development is the shift in cognition and behavior around ages five to seven (Konner 2010; Lancy 2008). These are ages at which developmental psychologists highlight an expanded independence, as children move away from a parent or parents and engage in the wider social world to a greater degree. These are ages by which children have typically acquired a "theory of mind" (an ability to infer the mental states of others), a characteristic that humans possess to a greater degree than other primates, and that facilitates social relationships with others. As Barry Bogin (1999) and others point out, these are also the earliest ages at which children tend to be able to survive on their own (as street children, for example). Social markings of this transition may include the greater responsibility of first grade compared with kindergarten in the U.S. school system, or, among Arnhem Land Australian aborigines, having a child's nose pierced around age seven (Berndt and Berndt 1951). Among humans and at least great apes, adrenarche, or the release of adrenal androgens, also kicks in around these ages, perhaps to shape a child's social motivations to engage the larger social world (Campbell 2006). Finally, in summarizing much of the cross-cultural wisdom of this age transition, the developmental and cross-cultural psychologist Barbara Rogoff observes: "It appears that in the age period centering on 5–7 years, parents delegate (and children assume) responsibility caring for younger children, for tending animals, for carrying out household chores and gathering materials for the upkeep of the family. The children also become responsible for their own social behavior and the method of punishment for transgression changes. . . . begins to work with groups of peers, and participate in rule games. Concurrently, the children are expected to show modesty and sex differentiation in chores and social relationships is stressed."

What is the significance of this shift at ages five to seven years for human social development? We suggest that these changes in cognition and behavior are consistent with what one would expect at the age of weaning for an ape of our body size. As we saw for other primates, the transition from infancy to juvenility occurs, by definition, with weaning, and marks important shifts in social behavior from mom to other group members. While no human hunter-gatherer population, much less any urban society today, enjoys an average age of weaning around five to seven years, we suggest that our human ancestors began weaning their young at earlier-than-expected ages so recently in the course of evolution that our social cognition and biology have not had time to catch up to the earlier weaning ages, effectively leaving us with a social mind ready to be weaned at a later age than weaning actually takes place.

This proposal amounts to a human juvenility mismatch—humans today (including hunter-gatherers) enter the postweaning juvenile stage earlier than our recent forebears. This idea is consistent with other lines of evidence. In a review of hominin weaning, Katherine Dettwyler (1995a) noted that various predictors suggest our recent ancestors were weaned between ages two and a half and seven. As examples, some argue that the eruption of first molars coincides with weaning, which would be around five and a half to six years of age, while others point out that, across primates, offspring are weaned at around one-third of their adult weight, which for humans would mean at five to seven years. Dettwyler also notes that the human immune system has matured by about age six. With this in mind, we highlight the possibility of a human juvenility mismatch because, if true, it could inform several other aspects of human reproduction: how human sociosexual behavior seems to also undergo some shifts around ages five to seven (as we note in the next section), and also the idea that ages of human weaning have been shortened only recently in human evolution (with modern humans?), giving us our unusually short interbirth intervals (a topic of further discussion in Chapters 9 and 10). Direct tests of such a hypothesis may also be possible as new isotopic techniques are devised to assess the age of weaning of our recent and more distant ancestors (see Eerkens, Berget, and Bartelink 2011).

HUMAN CHILDHOOD SOCIOSEXUAL VARIATION

Working with people of the Marquesas Islands of Polynesia in the 1960s, the ethnographer Robert Suggs commented on various features of their childhood sexuality:

> Marquesan children have ample opportunity to observe much adult sexual behavior. In the typical Marquesan house, children sleep beside the parents. . . .

Only when a child is seven or eight will some attempt be made to prevent it from witnessing the parents in coitus. . . . Masturbation among males begins at about the age of three, or sometimes earlier. . . . The masturbation technique of children is exactly like that of adults—the penis and hand are moistened with saliva to increase the sensation and ease. . . . Young boys from the age of six or seven upward gather surreptitiously in the bush for masturbation contests. . . . It was nearly impossible to obtain data on female masturbation. . . . At the age of approximately seven years, other forms of group sexual activity appear, which are heterosexual. Boys and girls, playing at "mother and father," will often place their genitalia in contact for brief periods. The girl either stands against a tree or lies supine on the ground, with the boy assuming the normal position for coitus. Contact is brief, accompanied occasionally by pelvic movement with much laughter. This activity is carried out in isolated areas where adults will not be apt to surprise the gathering. (Suggs 1966)

Given the methodological and ethical challenges to studying human childhood sexuality (observing wild or captive nonhuman primate juveniles engaging in sex play elicits fewer raised eyebrows), Suggs's account rates as one of the more detailed in the cross-cultural record of human sociosexual development. Even then, however, he was short on female data. But his observations prompt us to ask more questions concerning the ages, partners, and practices of children's sexual behavior, and how similar or different these may be cross-culturally. Eventually, too, we will want to situate some of the more quantitative human studies of children's sexual behavior, largely from Europe and North America, within such a cross-cultural scope. Here, we sort through the heart of the matter: the limited information available on childhood sexual attitudes and behavior.

As children get their sexual lives revving, they already have some of the necessary equipment in place. In one report, an ultrasound of a male fetus appeared to show him masturbating, raising the idea that sexual pleasure can precede birth (Meizner 1987). Male infants can display erections, and female infants are capable of vaginal lubrication (Bolin and Whelehan 2009; Levay and Baldwin 2009). Infants of both sexes are able to experience orgasm, though without ejaculation. These facts indicate that early in their development children have the capacity for a limited sexual response and already have some of the pleasure circuitry operating. Not surprisingly, children often enjoy rewarding aspects of sexual behavior, with kids aged

two to three commonly interested in genitalia (their own, and also others')
and autoerotic behaviors (masturbation). As children grow older and their
social world expands, by age five to six they more often maintain erotic interests in others that can include sex play and fantasies. Children are clearly
developing sexual organisms, even if they hold off from fully developing
their sexual and reproductive capacities until adulthood. In an ethnographic
illustration consistent with age-related changes in children's sex play, Australian aborigines in Arnhem Land

> desire to imitate and reenact among themselves [adult] sexual activities,
> publicly (when they are very young) or secretly (when they become older
> and more self-conscious). . . . [A]s the child grows older, his (or her) sexual
> behavior resembles more closely that of an adult. . . . Small boys and girls
> may play naked together in the water, but as they grow older there is the
> tendency to choose a companion of the opposite sex. They lie together from
> time to time imitating the sexual act and talking about sexual matters, refer-
> ring to adults and calling each other "husband" and "wife." (Berndt and
> Berndt 1951: 86–87)

Cross-cultural research indicates both regularities (such as age-related
and sex-related patterns) and differences in the expression and control of
childhood sexuality. In Ford and Beach's (1951) classic cross-cultural survey, based largely on attitudes toward childhood sexuality and coded from
a standardized data set, they identified three broad categories of societies:
restrictive societies, semi-restrictive societies, and permissive societies. In
other words, they recognized cross-cultural variation in how childhood sexuality was perceived, with some societies far more open to its expression
and exploration than others. At one extreme, they included among permissive societies the Alorese of Indonesia, whose mothers would sometimes
play with a nursing child's genitals, and among whom young boys might
masturbate in view of adults. Similarly, among the Tikopia of Melanesia
young boys might masturbate, and thus induce erections, and young girls,
too, might masturbate without being punished. And among the Lepcha of
India early sex play, such as mutual masturbation for boys and girls, was
recognized, and for young girls an ethos held that maturity requires sexual
intercourse. At the other end of the spectrum, as an illustration of a restrictive
society, boys among the Murngin of Australia were removed from their
parents' dwelling to a bachelor or boys' house to prevent them from seeing

their parents having sex. In Kwoma society, in New Guinea, a woman might hit a boy with a stick if she caught him with an erection. Apart from documenting and categorizing variations in childhood sexuality, Ford and Beach recognized another important pattern: the way in which childhood sexuality was viewed within a society tended to be similar to how adult sexuality was viewed. In other words, a society with a restrictive outlook toward childhood sexuality tended to have a restrictive view of adult sexuality too. This pattern suggests that adults attempt to shape the social worlds of children in ways that prepare them for normative adult sexuality within their particular society.

As another means of viewing the international scope of childhood sexuality, the *Continuum Complete International Encyclopedia of Human Sexuality* provides various accounts by country, albeit with little attempt to systematize the findings and with variable data quality and availability (Francoeur and Noonan 2006). We learn, for example, that in Ghana parents of young children regularly wash their children's genitals, sometimes using ground ginger when washing their daughters' genitals; further, children's sex games are reportedly common. Various other national accounts, from India to Indonesia, similarly suggest that childhood sex play is common. Accounts indicate, furthermore, that children frequently engage in autosexual behavior (masturbation), which is tolerated differently across countries (in Ukraine, for example, children caught playing with their genitals are punished). In Kenya, Francoeur and Noonan report, while "living in the unmarried men's hut, a boy has ample opportunity to listen to sexual conversations and observe older boys with their sweethearts" (682). In Morocco, "from early childhood, one inculcates an implicit sexual education totally antagonistic to the consideration of the sexes. The sexuality of a boy is praised and valued. . . . The sexual education of girls is done traditionally by the women of the family, by mothers, aunts, and older sisters. The older women tell the young girl what is forbidden and what is recommended in terms of repressing their sexuality" (707). These kinds of patchy, encyclopedic accounts point both toward recurring patterns, such as the prevalence of children's autoerotic and sex play, and also to the differential societal tolerance of these forms of childhood sexual expression.

One of the patterns observed cross-culturally is that of adult-child sexual contacts. At first glance, this might seem to be a topic rife with lawsuits, given how closely regulated adult-child contacts are today in countries like the United States. In a wider view, however, specifics of human adult-child sexual contacts indicate they have various functions beyond straightforward sexual pleasure. As an example, in culling cross-cultural accounts of human sexuality, Gregersen (1994) notes that adults in some societies stimulate the genitals of infants in ways that seem to calm the child or give it pleasure. The accounts he relates do not suggest that adults performing such acts are gaining some insidious pleasure from these behaviors, but rather that they seek to soothe the child. As another example, Gregersen (1994) points to the practice of finger defloration of female infants in various societies, including some Indonesian communities, Australian aboriginal societies, and Hindu groups. Here, again, there is no sense of adult pleasure in the practice; rather, it is a practice with a local meaning for female development. In one more example, girls in various societies may be betrothed at a young age to their future husband. Among the Tiwi of north-central Australia, a girl might be betrothed at birth to a future husband who is already a grown man (Hart, Pilling, and Goodale 1988). Are girls betrothed at such young ages expected to be sexually active with their much-older fiancés? The answer to that is typically no, since sex with a spouse is usually delayed until the girl reaches puberty. As one illustration, among the Kadar of northern Nigeria, marriages might be arranged when girls were three to six years old, but the betrothed girls would not move in with their husbands until around ten years later.

As a complement to cross-cultural accounts, in-depth quantitative studies of childhood sexuality have been undertaken in the United States and several European countries. The first of these was launched by the famed Alfred Kinsey and was based largely on interviews with adults, who were asked to recall features of their childhood sexuality (Kinsey, Pomeroy, and Martin 1948; Kinsey et al. 1953). Like so many other aspects of human sexual behavior, what Kinsey found regarding childhood sexuality shocked and awed many. For ages five to thirteen, the percentage of individuals reporting any sex play, coital play, or homosexual play was quite patterned. Approximately 10 percent of women reported having engaged in any form of sex play as girls across this age span, whereas about 10 percent of men reported engaging in sex play as five-year-olds, with that percentage rising steadily to around 35 percent for age eleven to thirteen. Thus, boys and girls at around age five did not differ markedly in the reported frequency of sex play, but gradually the percentage of males engaged in sex play increased and exceeded that of females.

For coital play, 2 to 5 percent of women reported this type of sex play sometime during the age span, whereas among boys the reports for boys rose steadily from around 3 percent at age five to around 12 percent by age eleven to thirteen. For girls aged five to seven, around 8 percent reported hetero-sexual play, to around 3 percent by age thirteen. For boys aged five to seven, around 6 percent reported heterosexual play, increasing steadily to around 20 percent by age eleven. For heterosexual play and any sex play, then, but unlike coital play, girls appeared to get a slight head start on boys; then their reported rates declined while boys reportedly increased these behaviors.

In the Kinsey studies, a notable percentage of men and women reported having engaged in homosexual play as children. For girls across the age spec-trum (five to thirteen), approximately 4 to 8 percent reported homosexual play. For boys across this age spread, around 5 percent reported homosexual play at age five, with this rising steadily to nearly 30 percent at age eleven. Like the patterns for sex play and heterosexual play, these patterns of homo-sexual play also suggest similar frequencies among girls and boys at earlier ages but greater differentiation across later childhood development. These findings suggest greater motivation among males to engage in sex play, a pat-tern not unexpected in light of human sex differences in sexual desire more broadly. These patterns also make the obvious case that many mid-twentieth-century Americans, like children around the world, engaged in sexual play, both with same-sex and other-sex partners, clearly demonstrating that sexual interests develop long before puberty. There are and have been lots of child-hood doctors and nurses (and other forms of sex play).

More recent U.S. and European studies have attempted to characterize patterns of childhood sexuality by asking large samples of parents about their kids' sexual behavior. Since older children may be more aware of their own behavior and desire to shield it from parents' eyes, such studies more closely measure parentally observed childhood sexual behavior than actual childhood sexual behavior. Indeed, this methodological issue may be cru-cial to reconciling discrepant results between Kinsey's work and some of the more recent studies, to which we now turn. In a study of almost 1,000 Dutch children aged two to twelve, two overarching patterns emerged (Schoentjes, Deboutte, and Friedrich 1999). The first is that the overall frequency of sexual behaviors declined across the age spread, and the second is that males and females exhibited very similar frequencies of sexual behavior across all ages. Furthermore, the age-related declines in sexual behavior were similar to those in a large U.S. study relying on similar methods. Thus these Dutch and U.S. findings disagree with the age-related and sex-differentiated patterns of Kinsey's earlier work, and they highlight the importance of methods and the samples from which subjects are drawn. The Dutch study also indicated that children of separated or divorced parents displayed higher frequencies of sexual behavior than children of married parents; this pattern, as we will see, has some parallels with adolescent sexual behavior, but the effect size of this difference was quite small, even if statistically significant. Further, according to several individual items in the Dutch study, children exhibited increased sexual curiosity with age (for example, the percent of children asking questions about sexuality rose from 33 percent at ages two to five to 65 percent for children aged six to nine and ten to twelve), even though that finding was not manifest in the measures of reported sexual behavior.

In a U.S. study drawing on similar parent-report methods and with a large sample of around 1,100 children, but with a focus on autostimulation (masturbation), several patterns appeared (cited in Mallants and Casteels 2008). The percentage of children reported as touching their sex parts in public, at home, masturbating by hand, and masturbating with a toy/object tended to decline across childhood age groups, although these declines were less clear for several measures among ten- to twelve-year-old girls. For example, 15.8 percent of girls aged two to five girls were reported to masturbate by hand, whereas 5.3 percent of six- to nine-year-old girls and 7.4 percent of ten- to twelve-year-old girls were reported to masturbate by hand. In another pattern, younger boys (two to nine years old) were more likely than girls to be described as touching their sex parts in public or private. In describing results from other studies, some relying on individual self-reports (rather than parent, teacher, or doctor reports), the authors report that "there seems to be a gradual rise in masturbation in the prepubescent years, from around 10% at the age of 7 to about 80% at the age of 13. . . . Masturbation occurs in 90–94% of males and 50–60% of females at some point in their lives" (Mallants and Casteels 2008: 1113). These latter patterns indicate that, indeed, parental reports of childhood sexuality, such as masturbation frequency, likely miss most of the mark at advancing childhood ages; on this subject, the earlier Kinsey data and the higher frequencies of masturbation cited by Mallants and Casteels are likely closer to reality.

In discussion of prevailing U.S. childhood sexuality, Eleanor Maccoby writes, "At least by the early grade school years, children have begun to interpret cross-sex contacts in romantic or sexual terms. . . . And children sometimes invent games of romance and marriage in which other children are recruited into cross-sex role-playing." (1998: 67–68). As Maccoby notes and parents of children may observe, these kinds of dynamics sometimes entail girls engaging in proceptive behaviors (indicating their liking of a male, even by "hating" him or threatening to kiss him), eliciting male responses. In the United States and possibly elsewhere (such as the Sambia of New Guinea), as children grow a bit older they may also develop sexual attractions to same- and opposite-sex partners; McClintock and Herdt (2010) suggest these sexual attractions emerge by around age ten in both boys and girls under the influence of rising adrenal androgens in adrenarche.

The age-related changes in sexual behavior suggest that children gain considerable practice both stimulating themselves and in sexual interactions prior to puberty, at which point the sexual stakes skyrocket. Boys and girls appear to express a comparable sexual curiosity earlier in development (for example, by age five), but their sexual behavior differentiates as they grow older, with males engaging in higher rates of sex play than girls. This may mark sex differences in sexual desire at young ages that resemble those of adulthood, and likely traces to mechanisms of sex differentiation (such as organizing effects of perinatal androgens on the brain) discussed in Chapter 4. The possibly enhanced sex differentiation in sex play from around ages five to seven onward could be consistent with the hypothesis of a later ancestral age of weaning, and with a greater social awareness of appropriate behavior. Girls and boys seem to engage in more adultlike sexual behavior the closer they are to puberty, which makes sense, as they are learning and preparing for the realities of sex and reproduction. The fact that older children seek to shield their sexual behavior from adults speaks to the importance of privacy as a context for human sex, and also to parent-offspring conflict (this time from the kids' view, wanting to avoid a parent's interference). Finally, sex play featuring couples (for example, husband and wife) fits with the human evolutionary story that long-term bonds are the typical context in which sexual behavior occurs.

LEST THERE BE INCEST

Anthropologists and psychologists in days of yore seemed obsessed with incest (see Fox 1980). Sigmund Freud (1920) famously advanced the Electra and Oedipus complexes, arguing that human incest taboos were important to prevent the cataclysmic effects of a daughter's attraction to her father (Electra complex) or a son's attraction to his mother (Oedipus complex). If children aspired to mate with their parents, the taboos could intervene, preventing them from acting on these impulses and thus undermining a stable family structure. Anthropologists from Bronisław Malinowski to Claude Lévi-Strauss argued against the Freudian interpretation of incest taboos. Malinowski (1929) noted that the urban Austrian family contexts in which Freud developed his interpretations of the incest taboo could not account for more variable family structures, including the matrilineal ones among Trobriand Islanders of Melanesia, where sons expressed greater hostility toward their mothers' brothers than their fathers. Lévi-Strauss (1949) suggested that the primary benefit of incest taboos is that they motivated parents to seek marital partners for their children outside of the birth group, thereby solidifying alliances that cut across multiple groups. From Lévi-Strauss's vantage point studying kinship systems in Brazilian Amazonia, the outcomes of incest taboos could help mobilize wider social networks, should concerns—of, say, violence—arise.

Incest avoidance and childhood sexuality are related, which is the reason we are discussing incest here. The link between childhood social development and incest avoidance was articulated nearly one hundred years ago, by a Finnish anthropologist, Edvard Westermarck. His views on incest are quite simple, requiring fewer mental acrobatics than Freud's, for example, and they appear to be much closer to the truth, having support from various lines of evidence. In brief, Westermarck (1921) suggested that inbreeding (mating between close relatives) has negative effects on offspring, that early childhood association leads to sexual aversion, and that this aversion functions to reduce inbreeding. Incest taboos effectively reinforce the maladaptive nature of inbreeding.

The logic underlying the genetic disadvantages of inbreeding is that inbreeding might expose deleterious recessive alleles. Put another way, if there is a genetic variant that is rare, and that requires two copies in order to be expressed, then there are greater chances that close relatives will carry such alleles due to their common descent; matings between close relatives increase the likelihood that such recessive alleles will be expressed. This pattern does occur, particularly in small populations, and it has been linked to the manifestation of isolated genetically based disorders in human populations proscribing marriages of relatives such as first cousins (for example, among Bedouin in the Negev Desert). At the same time, large pedigree studies have also shown that these costs are relatively modest, with one review indicating approximately 4 percent excess mortality among offspring of first-cousin marriages (see Saggar and Bittles 2008). Those costs may be outweighed by other considerations: the promotion of first-cousin marriages can also consolidate land and power within a family (Saggar and Bittles 2008). Indeed, rates of consanguineous marriage today, including first cousins, remain quite high in parts of North Africa, southwestern Asia, and South Asia, and we need only recall Darwin's sanguine contemplation of the potential costs of inbreeding (with his cousin Emma) to realize that even among the evolutionarily minded landed gentry there can exist an optimal amount of inbreeding.

While the costs of inbreeding, then, are measurable but small (and in some cases outweighed by other factors), what about Westermarck's other contentions? More specifically, does early childhood association foster sexual aversion? That question has been most clearly addressed through research in Taiwan and Israel. Arthur Wolf studied the dynamics of marriage in Taiwan, including *shim-pua,* a tradition that entailed a son's future bride moving into his parents' house long before marriage, often before she had gone through puberty (Wolf and Durham 2004). *Shim-pua* marriages were characterized by higher divorce rates, and the younger the girl was when she moved in with her husband's family, the lower their fertility. A woman who was a girl of age five, then, when she moved in with her future spouse and his parents had lower fertility than a woman who was twelve years old when she moved in with her future spouse and his parents. In Israel, Joseph Shepher (1983) found the converse situation when he conducted research in the 1970s among individuals raised in communes. When children were raised together from early ages, they would be less expected to marry and have children if they had developed a sexual aversion to each other. While Shepher's work has been criticized on legitimate grounds (for instance, males tend to be a few years older than females at marriage, meaning one might not strictly expect similarly aged individuals to marry), a bottom line finding was that opposite-sex children who were raised together on a kibbutz rarely had sex with or married each other.

A few other studies, though less well known, also shed light on the potential operations of a Westermarck effect. Dan Fessler (2007) found mixed support for a Westermarck effect in nineteenth-century communal life in Oneida, New York: many individuals married commune members with whom they had grown up, but five of eight marriages of close-in-age spouses were between couples who had probably been quite segregated from each other during childhood. In Morocco, when individuals were asked to rate the desirability of a potential spouse, women evidenced an interaction between marrying a cousin and having spent lots of time with that cousin in childhood (that is, they were less interested in marrying a cousin they had spent lots of time with early in life), but men did not show this same effect (Walter and Buyske 2003). Through surveys given to U.S. students, Lieberman, Tooby, and Cosmides (2003) found that the longer an individual had coresided with an opposite-sex sibling while growing up, the more likely they were to condemn incest; that is, a girl, for example, who had spent her previous eighteen years living with a brother more highly condemned the thought of incest than did a girl who had never lived with a brother, presumably because the experience of living with a brother was enough to amplify the revulsion of such a thought. And another study flipped the situation around to ask about the childhood experiences of siblings who engaged in adult sexual behavior with each other. In this latter case, sexually disposed adult siblings had often been apart from each other during their early childhood years, suggesting that they did not develop the aversion that would have otherwise kept them apart (Bevc and Silverman 2000).

Altogether, various studies better support the Westermarck hypothesis concerning incest avoidance than the alternatives. Even a growing nonhuman primate literature makes clear that individuals go out of their way—sometimes quite literally, by emigrating around the time of puberty from their natal or birth group—to avoid mating with close relatives (see Wolf and Durham 2004). In wild baboons and golden lion tamarins, among other species, inbred offspring have also been found to die at higher rates (Pusey and Wolf 1996). The support for Westermarck's views highlights the importance of childhood sociosexual development, a time when young not only learn about their bodies, engage in sex play, and get wrapped up in sex-differentiated behaviors, but also learn to avoid mating with close relatives such as biological siblings.

Additional aspects of incest merit attention and explanation, beyond the more straightforward contributions by Westermarck. For one, the fitness costs of incest differ between the sexes, being more heavily born by females; accordingly, when incest does occur we might expect that males are more often the perpetrators of it. Indeed, national statistics in the United States suggest that adult males disproportionately engage in incest (see Giles-Sim 1998). Another consideration is that, most incestuous sexual behavior is nonreproductive, meaning that potential costs related to a conception are not at play; this helps explain why childhood sex play between close relatives, including siblings, happens far more commonly than sexual behavior between adult siblings. Further, many adult perpetrators of unwanted sex with children are not related to those children, nor have they spent considerable time with those children when they were young. This helps make sense of the greater risk of sexual contacts initiated by male step-figures compared with biological fathers, seen in studies in the United States and Finland (Gray and Anderson 2010).

With those thoughts on incest, we wrap up our discussion of subadult sociosexual development. Having also focused on the benefits of juvenile learning, comparative primate sociosexuality, and the limited data on children's human sexual behavior, we have identified several key patterns in human development. The foundations of human sociosexual development are laid early, and the patterns we have seen exhibit sex differences (in attention to infant cues, for example), suggest the evolutionary importance of long-term bonding, and also exhibit variations contingent on the socioecological environment within which an individual is raised. Continuing on the developmental course, it is now time for puberty, a transformative process in sexual behavior that will take us into the realm of reproductive potential.

CHAPTER SIX

Welcome to the Party

Puberty and Adolescent Sexual Development

> Adolescence marks the transition between infantile and playful sexualities and those serious permanent relations which precede marriage. During this intermediate period love becomes passionate and yet remains free.
>
> —Bronisław Malinowski, *The Sexual Life of Savages*

HIGH IN THE Himalayas one can find the natural world in all its splendor. One may also find the fascinating practice of "walking marriages." Along the Tibetan border, in the Sichuan and Yunnan provinces of China, the Mosuo people make their home. The Mosuo have extended matrilineal families, and marital patterns quite different from those practiced in many Western cultures. Their practice of walking marriages *(zou hun)* has gained increasing attention in the past few years (Shih 2010). When girls come of age, at around thirteen years old, they are given the (only) private bedroom in the home for a short period of time. This allows the young women to accept male suitors in their bedroom, usually privately, who may come under the cover of darkness and return home early in the morning. Suitors are chosen on the basis of (mutual) attraction, and once one is chosen an ongoing sexual relationship ensues. If a child results, fathers tend to invest in their nieces and nephews more than their own children. Yet this practice should not be mistaken for promiscuity (c.f. Ryan and Jetha 2010), as detailed accounts reveal quite the opposite (Shih 2010). In most cases, the sexual partnerships are long-term partnerships (although partners do not live together). In fact, it is a source of shame if a woman cannot identify the father of her offspring. Certainly vastly different from marital conventions in the United States, the walking marriages among the Mosuo also point to another cultural difference: sexual expression during the pubertal transition.

In this chapter, we welcome individuals to the pubertal and adolescent transition, when the stakes escalate for competition, courtship, and, potentially, rearing children. Before we draw out some of the wider evolutionary significance of these changes, we discuss life history theory as a framework for understanding shifting priorities, and also touch on comparative primate work to further enrich our understanding. We then turn to variation in human pubertal transitions, across time and populations, along with the energetic and social factors that underpin such variation. We cover the social significance of puberty and adolescence, including rites of passage and ritual demarcation, and also wider patterns in same-sex competition and alliance formation (even shopping, as a means of expressing one's status and identity in the expanding contemporary "youth culture"). Finally, we touch on key aspects of adolescent sexual behavior in the United States and in a wider cross-cultural context, as this developmental period often offers the opportunity for sexual experimentation—as Malinowski, in the quote above, alluded to for adolescent Trobriand Islanders.

CHANGING PRIORITIES AND A PRIMATE CONTEXT TO PUBERTY

In his book on sexual selection, Darwin (1871) attempted to understand the development of secondary sexual characteristics, or traits that arise at puberty to enhance reproductive success. He saw the advantages of delaying the development of secondary sexual characteristics until they could be utilized when the reproductive stakes had been raised (by metabolic costs, for example, or risks of predation). While his views focused on males, emphasizing the benefits of male traits that enhance competition, the insights apply generally to females and males alike, and make sense of the convergence between sex dimorphisms, pubertal development, and natural selection. Darwin wrote:

> But many characters proper to the adult male would be actually injurious to the young,—as bright colours from making them conspicuous, or horns of large size from expending much vital force. Such variations in the young

would promptly be eliminated through natural selection. With the adult and experienced males, on the other hand, the advantage thus derived in their rivalry with other males would often more than counterbalance exposure to some degree of danger. Thus we can understand how it is that variations which must originally have appeared rather late in life have alone or in chief part been preserved for the development of secondary sexual characters; and the remarkable coincidence between the periods of variability and of sexual selection is intelligible. (1871: 299)

Darwin recognized the escalation in reproductive effort surrounding the pubertal transition, facilitated by the development of secondary sexual characteristics. We would like to couch that observation within the wider framework of what is referred to as life history theory. Life history theory articulates the adaptive allocation of limited time and limited energy to growth, maintenance, and reproduction. One result of the allocation shift that occurs with puberty is that investment in maintenance can be compromised, though sometimes in sex-specific ways. Put another way, while a juvenile invests in immune mechanisms and behaviors to stay alive, an adolescent may scrap some of those investments in favor of reproductive investments—thereby prioritizing reproductive success over survival. This life history trade-off may be more pronounced among males in polygynous species, speaking to the sex-specific nature of adolescent life history transitions. This perspective informs us why adolescent males among many primates may suffer elevated mortality risks compared with adolescent females: those risks serve the greater end of reproductive success.

From a life history standpoint, an obvious aspect of puberty and adolescence is the shift from growth to reproduction. While some species, such as a desert tortoise, may continue to grow even during reproductive years, for mammals like ourselves, energy is shifted from growing to preparing for reproduction. Peak stature is typically reached during adolescence, for example. However, the shifts from growth to reproduction can also be sex-specific. In species with pronounced sex dimorphisms in body size, members of one sex may spend more time and energy growing than members of the other sex, leading to sex differences in growth patterns that may include one sex typically maturing before the other. For most primates, though not all, this pattern entails females shifting from growth to reproduction at younger ages than larger-bodied males. Because pubertal development is a process rather than a flipped switch, pubertal maturation takes time; in humans this process, which results in development of secondary sexual characteristics (such as gluteofemoral fat in females, upper-body muscle in males), continues into the twenties, even though puberty is initiated at far earlier ages.

The secondary sexual characteristics that develop at puberty are designed to enhance an individual's reproductive success. In females they primarily take the form of parenting effort, whereas in males the secondary characteristics primarily foster mating effort: male-male competition, courtship, and mate guarding. Functional aspects of these traits will be considered further in the next chapter, when we discuss an evolutionary perspective on human reproductive anatomy and physiology, including the mechanisms that underpin the development of these traits. In Darwinian terms, the bright coloration of a male bird or the antlers of a male deer that develop at puberty help attract mates or fend off male competitors. For females, extra fat or enlarged areolae can facilitate storage of energy to undergird reproduction and lactation (enlarged areolae around the nipples, for example, make nursing easier). Traits such as exaggerated sexual swellings in chimpanzees and bonobos evolved to enhance success in mating competition rather than direct reproduction, indicating that female secondary sexual characteristics can also arise to facilitate mating effort rather than parenting effort, even if this is less common. It is important to note that, while it goes against conventional wisdom, recently evolutionary scientists have tackled the case of female-female competition as it leads to differential mating success. In particular, female intrasexual competition clearly appears to have an impact on mate quality (compared with mate quantity in males), and on aspects of fitness across species (Rosvall 2011).

While life history shifts from growth to reproduction occur generally in mammals, there is also considerable variation across species in the timing of this transition. For human hunter-gatherers, the typical age of female menarche (first menstruation) is around sixteen years, as observed among the !Kung of southern Africa (Howell 2010; Kaplan et al. 2000). By comparison, wild chimpanzee females reproductively mature around age ten to twelve. And the great apes are comparatively slow when contrasted with monkeys from Africa and Asia: female rhesus monkeys and vervet monkeys typically reach maturity around three years of age (Campbell et al. 2011). Explanations for these kinds of species differences in pubertal maturation draw upon the socioecological pressures (social learning, diet, predation pressures) discussed in Chapter 5, and will be revisited in Chapter 11 in the context of species variation in aging (Jones 2011).

Species differences in female age of reproductive maturity have been observed across wild great apes. Age at first birth is approximately ten years in gorillas, around thirteen to fourteen years in chimpanzees and bonobos, and fifteen to sixteen years in orangutans (Robson and Wood 2008). In hunter-gatherer societies such as the !Kung, female age of first birth tends to be around eighteen to twenty years, a further indication that other apes are fast-tracked in their pubertal and reproductive transitions compared with us humans. The differences among apes also show that gorillas mature more rapidly—and orangutans more slowly—than chimpanzees and bonobos. This pattern of great ape species variation is also consistent with species differences in aging.

Apart from species differences in pubertal transitions, there is variation within species with respect to sex differences (Dixson 1998). In some species, females and males mature at similar ages. Take owl monkeys or common marmosets of South America: in these primates females and males proceed through puberty at comparable ages. In other species, females mature more rapidly than males. Among squirrel monkeys, rhesus monkeys, and the vast majority of primates, females progress through puberty at earlier ages than males. Rhesus males may be a year behind the females (reaching puberty at four years of age rather than at three, like a female). In the most extreme case of primate variation in sex differences in maturity, male orangutans are left in the dust by their female counterparts by several years, as are mandrills, whose females mature around ages three to five, while males mature around ages five to seven.

How do we explain these differences in pubertal transitions? For that answer, we must hearken back to Darwin again, and to the pressures wrought by sexual selection under different mating systems. The species with similar ages of puberty between females and males tend to be socially monogamous, and they tend to have adults of similar size. By contrast, species in which males lag behind females have greater male-male reproductive competition, either within multi-male, multi-female groups (like rhesus monkeys) or within polygyny (such as the "dispersed" forms of polygyny among orangutans or mandrills). The selection pressures of male-male competition favor males growing more, to pack on extra weight and eventually muscle that can be used during male-male competition.

Naturally these patterns of pubertal transitions and mating systems invite more questions, so we will start with this one: Are there any sex role reversals, whereby males mature more rapidly than females? In primates, the answer is a general no. But if there is an exception to that pattern it may be found in the sifaka, the largest members of the lemur group inhabiting Madagascar. Among sifakas, social monogamy is typical, and available data suggest males may mature at slightly younger ages than females (Strier 2011). Now, what about humans? Where do we fall in this world of sex differences and pubertal transitions? We can draw from personal observations and from a wealth of growth data to find the answer: female humans typically mature a year or two ahead of males (Bogin 1999). Recall the awkward seventh-grade school dances, where many girls were taller than the boys? And U.S. adolescents aren't alone: among the Ache of Paraguay, for example, adolescent females are taller than males between ages eleven and a half and fifteen (Hill and Hurtado 1996). Based on the sex dimorphism during human pubertal transitions, we humans do not look like socially monogamous animals. However, based on the degree of this dimorphism we also do not look as polygynous as an orangutan or mandrill. The magnitude of the human sex dimorphism in pubertal transition is consistent with slight polygyny/mostly monogamy. That finding is, of course, consistent with many other lines of evidence concerning the evolution of human sexual behavior.

There are other notable aspects of nonhuman primate puberty that inform our understanding of human puberty. One is the wanderlust (desire to travel) that puberty fosters, at least in members of one sex. Across primates, maturing individuals face the dilemma of finding mates. If all maturing individuals stayed and mated within the small groups in which they are members, they would face genetic costs of inbreeding, given that many of their group-mates are family. As noted in Chapter 5, one process that operates in humans against close kin mating is the Westermark effect: individuals tend to shun mating with others with whom they were raised early in their development. This process appears to operate in other primates too: "Perhaps the most striking evidence for inbreeding avoidance in monkeys comes from observations of female homosexual interactions in Japanese macaques. These interactions are quite frequent during the mating season, although the participating females also copulate with males. Females almost never choose related female partners, even though they prefer their kin for nonsexual friendly acts" (Perry and Manson 2008: 163). Another process that resolves this risk of inbreeding is dispersal, when members of at least one sex typically leave their natal (or birth) group at puberty (Strier 2011). By leaving, individuals may find new mates on the horizon. Among African and Asian monkeys, females typically remain in their natal groups (what biologists call female philopatry), while males leave around the time of puberty. Among apes, the picture is muddied: some genetic and observational studies suggest more female dispersal in chimpanzees, but there is considerable flexibility in these patterns across populations (see Campbell et al. 2011).

Among hominins, creative attempts at reconstructing philopatry and dispersal have entailed isotopic assessment, ancient DNA analysis, and analogies drawn from hunter-gatherer patterns. A study of South African gracile australopithecines suggested that females dispersed further than males (Copeland et al. 2011). Further, a sample of twelve Neandertal remains found in Spain indicates greater genetic difference among females, also suggestive of greater female dispersal (Lalueza-Fox et al. 2010). Studies of recent foragers reveal flexibility and bilocality rather than a strict male philopatry and female dispersal (Hrdy 2009; Marlowe 2010). Available data thus indicate that recent hominin evolution likely entailed considerable flexibility in philopatry and dispersal, with females and males both possibly shifting to other groups around their adolescent years. In larger samples of agricultural, horticultural, and pastoralist societies, male philopatry and female dispersal is the most common pattern, but this is likely a socially derived pattern (Hrdy 2009; Low 2000). Genetic studies corroborate the recognized population variation in philopatry (for example, patrilocal and patrilineal Turkic populations in Central Asia had localized Y chromosome variants), though other factors are also at play (Heyer et al. 2012). Ethnographic analogies point to other possible nuances among foragers. Among the Inuit and !Kung, as in other forager groups, males practice bride service, which entails staying with the bride's family early in marriage to help her parents, and later showing more residential flexibility. The mild polygyny of foragers also favors a slight tilt toward female dispersal, with a polygynous man's cowives living with him. However, the mobility and various scales of hunter-gatherer social organization (camp, network of camps, larger linguistic group) underscore how difficult it is to see clear patterns of philopatry and dispersal in humans, even if social worlds are shifting at adolescence and with marriage.

Puberty in nonhuman primates is a risky time (Campbell et al. 2011; Strier 2011). Full entry into the sexual and reproductive arena at puberty entails escalated stakes for same-sex competition and mating, sometimes with lethal outcomes. The wanderlust of, say, adolescent male vervet monkeys can translate to either reproductive output or mortality statistics. Leaving the natal group means leaving behind known social partners, including family, who can serve as allies and social buffers when challenges arise. It also means crossing potentially dangerous terrain to find new social partners and mates: there is increased risk of depredation, and potentially greater food stress (not knowing where foods can be found and lacking social means of finding it). When an individual does find a new group, he may face stiff same-sex competition while trying to settle in. And yet what alternative is there? By overcoming these obstacles, a migrating individual has at least the potential to gain the social position and mating opportunities that serve evolution's bottom line: reproductive success. These shifts in behavioral proclivities—the drive to leave one group and join another—are likely prompted by developmental changes in the brain (Spear 2000, 2010). Age-related behavioral shifts in risk taking are adaptive results of neural development. This adaptation includes the reorganization, in later development, of inhibitions in the prefrontal cortex and corticodopaminergic axis—that part of the brain affecting the dopamine "reward" system, prompting individuals to desire sensation and risk (manifested in wanderlust, alcohol and drug use, genital stimulation). Changes in sex steroid hormones also alter an adolescent's motivational world, helping shift behavior toward same-sex competition, courtship, and eventually the reproductive behavior that enhances an individual's reproductive success (Nelson 2011).

Adolescent risk is not limited to monkeys: human adolescent males are also prone to risk taking. According to the World Health Organization, the leading cause of death worldwide among males age fifteen to nineteen is automobile accidents. Jamaica and India are near the top of the pileup, having some of the highest numbers of accidents—not surprising to anyone who has traveled along the windy, narrow, mountainous roads of Jamaica, where young male drivers strive for speed, channeling the adolescent risk-taking favored among our ancestors. In a wider human scope, there are well-documented sex differences in adolescent mortality, often tracing to escalated male risk taking (Kruger and Nesse 2006; Wilson and Daly 1985). Among Aka hunter-gatherer men in their young twenties, some of the elevated male mortality is due to risks in foraging activities, such as falling out of trees while collecting honey (Hewlett 1991). Whether dangers are faced on the roads, high in trees, or in male-male fights over status and women, adolescence is a life history phase of escalated risk and high stakes, consistent with outcomes of excess male mortality. Like high rollers in a casino, adolescents are compelled to play: when the risks are high, the losses can be devastating but the winnings can be exceptional—and in evolutionary terms, the game of mating is about winning.

As if changing bodies and risky behaviors were not enough, adolescent primates face mating challenges too (Campbell et al. 2011; Strier 2011). Adolescent females are frequently less desirable as mates than older females, whether among capuchin monkeys in Costa Rica or chimpanzees in Uganda. Less fecund and without proven infant-rearing ability, young females may find that males prefer more mature mothers as mates, particularly within contexts of multi-male, multi-female groups. High-ranking males have limited time to take advantage of their preferential reproductive access, so they are better off channeling it toward more likely prospects of success (those proven mothers). Further, older females may have a competitive edge because their higher social status means they get preferred foods or have social access to mates, further limiting the sexual prospects of adolescent females. For their part, adolescent males may suffer, too, when same-sex competition and female choice (of other males) limit reproductive possibilities. Options for circumventing this outward competition can include attempting sneaky copulations, like some younger, smaller-bodied male orangutans have been known to do, or perhaps biding the time in bachelor male groups, as among gelada baboons in Ethiopia and mountain gorillas. Incidentally, male mountain gorillas are known to engage in same-sex sexual behaviors in these bachelor groups, which is one means of channeling their libido until possible matings with females.

HUMAN PUBERTAL TRANSITIONS: MEASUREMENT AND VARIATION

It is a challenge to accurately measure pubertal transitions in nonhuman primates. Estimating the age of individuals may be tricky and require long-term field research. Further, it is difficult to obtain adequate observations of diagnostic traits such as menstruation in a chimpanzee, or physiological measures (such as assessments from urine or feces of sex steroid hormones indicative of reproductive function). Still, such hurdles have been cleared often enough to yield interesting and important insights into patterns of nonhuman primate puberty and adolescence.

What about data on humans? How difficult is it to measure our pubertal transitions? The challenges are both similar (estimating an individual's age in a population where people do not have birth records or keep track of birth dates) and different in assessing human puberty compared with nonhuman primates.

Clinical assessments of female pubertal development typically rely on asking about menarche (first menstruation), measuring changes in height (peak stature), chronicling stages of breast development, and assessing changes in pubic hair development (Tanner 1989). Since humans have more pronounced pendulous breasts compared with other primates, assessments of breast development are more useful for humans than for other species. Assessments of pubic hair growth are also more useful for humans than other primates, since our general lack of thick body hair makes pubic hair stand out more readily. The fact that humans frequently cover the genitals and breasts means that assessments may be based on self-reporting, or perhaps on a physician's observations. Sometimes physiological markers (such as luteinizing hormone or estradiol) are available in pubertal assessments, but not in historic data sets; in fact, data on such markers are unavailable for most populations where pubertal transitions have been studied.

As for human males, typical measures of pubertal growth include stature, nocturnal emissions (ejaculation, perhaps during sleep), penis and testicular size changes, and pubic hair (Tanner 1989). Obtaining accurate assessments of these pubertal markers can be challenging, in part because of adolescents' sensitivity to providing such information. Physiological measures may also be available, though, as for females, these are obviously unavailable for historic populations and available in relatively few cases for international studies of pubertal transition. In one comparative study of male pubertal development, Kenyan Kikuyu boys had notable increases in serum testosterone around age thirteen and a half, whereas boys in London had notable rises around twelve and a half years of age (see Bogin 1999). A historic study of male pubertal ages tracked "breaking voices" (the age at which the voice deepened) among Italian singers, finding this took place around age sixteen, older than would now be the case in Italy (Potts and Short 1999). Interestingly, there is far more information on human female pubertal maturation, around the globe and historically, than on male pubertal maturation, probably in part because of greater interest in female fertility regulation, measurement issues, and the ability of women to more vividly remember their first menstruation (menarche) then men can remember specific pubertal milestones. Our treatment of human variation in pubertal maturation follows this pattern, featuring data on females rather than males.

Among Pume females of Venezuela, who live as forager-horticulturalists in a savanna environment, the estimated age of menarche is approximately thirteen years (K. Kramer 2008). This estimate was based on both cross-sectional and retrospective methods. From a cross-sectional survey, Kramer found that no twelve-year-olds reported menarche, whereas 83 percent of thirteen-year-olds reported menarche and all fourteen-year-olds did. How does this age of menarche compare with that of other foragers, or indeed with wider samples of women from around the world?

The age of menarche among the Pume is similar to that found in various other native South American populations, including other recently studied forager-horticulturalists (K. Kramer 2008). As an example, among the Tsimane of Bolivia the average age of menarche is around fourteen years. An age of menarche younger than thirteen years is not typically reported among hunter-gatherers, yet the age of menarche does vary with other ecological and energetic constraints. Among the Ache of Paraguay, for example, the average age of menarche during a time of exclusive foraging was estimated at fifteen years, whereas the age of menarche decreased to around age fourteen after the Ache adopted horticulture and a more settled way of life (Hill and Hurtado 1996). Among !Kung foragers of southern Africa, the estimated age of menarche was around sixteen years. Combining these kinds of findings, we take an age of menarche of approximately sixteen years to represent a human forager baseline, which helps us recognize that we have an evolved pattern of relatively late puberty compared with other great apes, but also revealing how ages of puberty are lower in most contemporary well-fed urban populations today.

For a glimpse of some of the wider population variation in female menarche—as a marker for age of puberty—we note a few data points on the landscape. At the upper end of age of menarche, Gainj women of highland New Guinea reach menarche at an average age of 18.6 years (Wood 1994). An early 1980s comparison found the average age of menarche to be approximately 14.3 years among Chinese Malays, 13.4 years among native

Malays, 13.9 years among Pakistanis, and 13.1 years among Belgians (Udry and Cliquet 1982). In a comparison of pubertal maturation of females in London and Kenya, Kikuyu girls in Kenya had an age of menarche around 15 years, while girls in London, England, reached menarche at around 13 years of age (see Bogin 1999). From historic estimates of age of menarche, samples from late nineteenth-century Europe (such as Sweden and Norway) suggest an average age of menarche around 16 years of age (Trevathan 2010). In a nationally representative study of U.S. menarche, a small number of girls had begun menstruating by age 11, and more than 90 percent had reached menarche by 14 years of age, yielding an average of around 12.4 years of age (Chumlea et al. 2003).

An important development is that the age of menarche has been declining in many populations globally. This represents a secular change, a systematic shift toward transitioning earlier to reproductive investments. This is best revealed by comparing estimated ages of menarche from the same country or population across time. In the same European countries for which we have the nineteenth-century data mentioned above, the age of menarche had dropped to around 13 by the 1960s and 1970s from the previous estimate of around 16 years (Trevathan 2010). In a study of urban Chinese girls, age of menarche declined from 13.5 years to 12.3 years from 1979 to 2003, occurring at the same time that average height increased (Ma et al. 2009). Suggesting that such secular changes may be leveling off, in the United States only a small decrease in age of menarche—from 12.8 to 12.4 years—has been observed since 1973 (Chumlea et al. 2003). Among Chinese females living in Hong Kong, age of menarche also seems to have reached a floor, after dropping to 12.4 years (Huen et al. 1997). How can the variation in age of menarche, including the secular trend of its decline, be explained? When raising this question in classroom settings, often one of the first things students propose is that it is an effect of increased use of synthetic hormones—such as those estrogenic compounds found in certain plastics, or the growth hormones in some commercially farmed food products (imagine meat-rich cattle and poultry), or even the measurable estrogen that can now be found in water supplies, from women taking the Pill. While excess estrogens from hormonal contraceptives may end up going down the toilet (excreted through urine) and eventually leech their way into the environment and drinking water, it is still unknown what consequences this may pose to population health. If such effects of exogenous synthetic hormones exist, they are likely small, and given the timing of their availability, they cannot account for the longer-standing changes like observed in Scandinavian countries. There must be something more.

The most obvious candidates are life-style factors: nutrition, energy expenditure, and disease loads (Karapanou and Papadimitriou 2010; Trevathan 2010). While diets have become more nutritionally dense (owing to more processed foods), we have also experienced reductions in energy expenditure (we do less walking and more sitting in cars or buses), and reductions in disease burdens (we spend less metabolic energy to fight off helminthes, or worms, and other pathogens). Taken together, this can allow for an organism's life history energy budget to be recalculated. We now have more energy to channel into growth (as in the secular change in Chinese growth noted above) as well as reproduction. The pubertal transition can begin even earlier— at age thirteen rather than age sixteen. Findings from a pile of studies are consistent with roles of these main life-style variables. As an example, the decline in menarche among the Ache in Paraguay, associated with settlement, is consistent with dietary and energy expenditure changes leading to earlier achievement of menarche. The later ages of pubertal development among girls with anorexia nervosa (who have inadequate nutrition) and among intensively engaged athletes (for example, gymnasts and runners) also highlight the importance of energetic factors. Jumping back to comparative primate research, even chimpanzee females mature at earlier ages when their mothers are highly ranked in the group, likely owing to preferential access to food resources during their development (Pusey, Williams, and Goodall 1997).

The Frisch hypothesis, proposed in the 1970s and often taken as accepted wisdom in textbooks, attempted to explicitly link female energetics with pubertal development (see Frisch 1984). Frisch suggested there is a "critical" level of fatness required for pubertal development to occur; more specifically, the hypothesis linked a body weight of 47 kilograms to the onset of menarche. This idea has intuitive appeal in that it links energetic reserves (fatness) with a capacity to kick-start the female reproductive engine. However, it fails to account for population variation in age of menarche (which does not have a single critical point) and in comparison with measures of longer-term growth (height, for example, which indexes energetic conditions over longer durations). As Peter Ellison (2001) points out in a critique of the Frisch hypothesis, another skeletal measure—pelvic size—serves as a better predictor of female reproductive development. Because it indexes energetic conditions during childhood and is importantly linked to capacity for giving birth, the size of the pelvis makes more sense as an adaptive gauge of reproductive development. As another line of evidence consistent with a role for childhood energetic conditions, girls from India who were adopted into Swedish households (and thus released from some of their preceding physiologically entrained energetic constraints) were found to experience accelerated menarche (Proos, Hofvander, and Tunevo 1991). There are also proposals that childhood social cues (such as the presence of unrelated males in the local environment, including a stepfather) may accelerate age of menarche. These social effects, if they exist, are small—pale in comparison with energetic factors—and are more relevant to the expression of adolescent sociosexual behavior than to age of menarche (see Ellis 2004).

Even in the wake of pubertal development, females experience a phase of adolescent subfecundity (and, initially, sterility) (Wood 1994). While a girl may have had her first period, her ovulatory cycles initially tend to be inconsistent and anovulatory (meaning no egg is released during the cycle). Rather than turning on immediately, the reproductive system takes time to get up and running. This phase, which can last several years, has implications for adolescent sociosexual behavior. As Ashley Montague (1946) long ago noted, female adolescent sterility enables sex play without pregnancy typically occurring, which helped account for how unmarried Polynesian girls could engage in intercourse without pregnancy during a time of quite fluid sexual partnerships. This phenomenon holds true widely, explaining how a girl who has recently had her first period might be able to engage in unprotected sex without an immediate high risk of conception.

As pubertal development proceeds, cycles become more functional, ovulation becomes more frequent, and fully functional cycles eventually arise. Yet throughout adulthood substantial variation exists among populations in patterns of menstruation and cultural ways in which they are contextualized (Vitzthum 2009). In some cases, adolescent girls can conceive before they themselves have finished growing. This sparks parent-offspring conflict: should a mother invest in her own growth, or in that of her fetus? The outcome of early adolescent motherhood often entails both diminished maternal growth and reduced birth weight for her offspring.

Among the Pume of Venezuela, girls conceiving before age fourteen have a fourfold higher risk of their infant dying than girls conceiving at seventeen years of age or later, one measure of the cost of early reproduction (K. Kramer 2008).

ADOLESCENTS AWAY

In a convergence of female fatness, sexuality, and fertility, seminomadic Arab pastoralists of Niger seek to fatten females from childhood onward. The value of female fatness has a cultural logic to it that entails aspects of status and economics. In a world of subsistence pastoralism, the women who can afford to fatten are of higher status, and are able to convert their energetic wealth to reproductive success. As Rebecca Popenoe puts it in her book describing this practice: "Because a heavy body with wet, meaty flesh is the paragon of sexual attractiveness, women seek to achieve it, at the same time limiting their ability to work and requiring the presence of a servant caste, which in turn supports their own status and that of their families. In order to best achieve the cultural ideal women must ingest large quantities of milk that only cows—male property—produce, making women dependent on men, men who are dependent in turn on women for transforming their production into the stuff of sexiness, female virtue, and, eventually, sons" (Popenoe 2004: 189).

This practice of preferring fat females stands in stark contrast to contemporary Western values favoring lithe female physiques. The preference for fat females among Arab pastoralists in Niger is but one of numerous cross-cultural cases of idealizing females with bulk to them, although this example stands out for inculcating it from early childhood; the most common pattern is to try fattening adolescent females, particularly prior to marriage (Brown 1991). That extra fat in young women enhances their allure, and in a world of energetic constraint translates to more offspring: the women often lose fat with each child born (maternal depletion), resulting in a thin-framed postreproductive mother of many children.

If fattening females opens up a discussion of puberty rites, what other patterns do we see in those rites, for both females and males? While puberty rites are not practiced in all societies, they do appear frequently: in one cross-cultural survey, they were recognized in 68 percent of societies of hunter-gatherers, pastoralists, horticulturalists, and so forth (Schlegel and Barry 1991). Female initiation ceremonies occur more commonly among hunter-gatherer societies than in other types of societies, and initiations for females and males alike are less common in socially stratified nation-states compared with smaller-scale societies. Beyond that broad scope, there are some overarching cross-cultural patterns in puberty rites, particularly when viewed in light of an overarching evolutionary perspective.

Female puberty rites typically emphasize female economic, marital, and reproductive capacities (Bolin and Whelehan 2009; Schlegel and Barry 1991: Weisfeld 1999). They may entail some emphasis or training of women in successful performance of subsistence and domestic activities, given a local sexual division of labor, and they are often performed with an audience of local group and family members. Among hunter-gatherer women, for example, women's economic productivity is often underscored in adolescent initiations. More to the point of sexuality, however, female puberty rites often highlight the sexual maturity, marital eligibility, and fecundity of an adolescent girl. An Apache girl may undertake a multiple-day initiation upon menarche, announcing her transition into the mature world. !Kung girls, on achieving menarche, are secluded for the duration of their menses and then ceremonially introduced as a woman. The celebration of a Latina's quinceañera recognizes the transition into womanhood at a female's fifteenth birthday, a time for celebration with family and friends. Debutant balls, a waning tradition, announce the social arrival of young women in high society, as they transition from childhood to adulthood.

For males, the nature of initiation rites is somewhat different. Male initiation ceremonies tend to involve a wider public sphere, occur at later ages, more often emphasize responsibility and achievement, are more severe, and often sublimate an individual's desires into activities that may serve a greater public good, such as involvement in warfare (Bolin and Whelehan 2009; Schlegel and Barry 1991: Weisfeld 1999). A successful Masai *moran,* or warrior, may prove his mettle by killing a lion, at least he might have in the past (changing norms, not to mention availability of lions, makes such practices less tenable today). Young Ariaal men of northern Kenya are initiated in age sets, after a painful circumcision, thereby providing a demographic of young men who can be tasked with defense against enemies and livestock raiding, and also be kept somewhat apart from the wives of older, polygynous men. A Plains Indian might undertake a vision quest, mixing starvation, isolation, and possibly the use of hallucinogens to find a guardian spirit for his path forward. Peter would equate his high school ice hockey experience to an adolescent initiation: frequent early-morning or late-night practices, an exercise regime pushing him to his physical limits (altered states of consciousness when exercise-induced endorphins kick in), an ethos of teamwork, all towards shaping him and inculcating traits that might aid success in society today (such as being able to sit and write for long stretches?). Many of today's martial arts trace back to training exercises for young males that helped them become warriors, and one can readily see the parallels with training groups of adolescent males on football fields, for similar ends.

The sex-specificity of adolescent initiation rites speaks to the wider sex-differentiated patterning of adolescent behavior. In an evolutionary context, featuring a hunter-gatherer society for inspiration, females typically transitioned relatively quickly through adolescence to adulthood. With an age of menarche around sixteen, age of first marriage around eighteen, and first reproduction around eighteen to twenty, there was little time to dwell on the process of adolescence. The time period between spending time predominantly with older female family members (mothers and aunts) and female friends, and spending time predominantly with a husband and children would have been quite short. During this brief spell, a young, unattached woman might have welcomed her sexual allure. Among the Hadza, for example, young women enjoy cultivating men's attention, fully aware of their sexual status (Marlowe 2010).

In wider cross-cultural context, the duration of female adolescence varies considerably (Bolin and Whelehan 2009). In societies in which females shift rapidly to marriage and child rearing at or after puberty, adolescence is kept to a minimum. As an example, many rural Bangladeshi women transition from puberty to marriage early, with ages of first reproduction typically in the midteens (see Wellings et al. 2006). In many more societies today, there are gaps between pubertal development and the formation of long-term reproductive bonds, allowing for a more extended adolescence phase. Where this occurs, female have opportunity to explore a shifting social landscape. The importance of parental relationships can be challenged and traded off for the greater draw to peer relationships, both supportive and competitive, as adolescent girls strive for social status. Competition often highlights attractiveness and sexual capacities (being accused of excessive promiscuity; spreading negative gossip about a competitor), indications that adolescent status competition is frequently couched in sexually relevant contexts. Competition to attract the sexual attention of desirable males also features in an adolescent world, part of females' experimenting to find their value in the mating market and ideally to find a long-term partner fit for procreation.

For males, a longer adolescence than females have is a given, but there is considerable cross-cultural variation in the length of this phase. Even after undergoing puberty, males typically require time and effort to develop the status, however it is defined locally (for example, by being a good warrior, a good hunter, a man with job prospects), that may make them desirable as long-term mates. To do this, adolescent males may push away from family relationships toward male-male relationships that can aid their rise in status (e.g., through alliances with older high-status men or with peers). Those same Ariaal age-graded warriors illustrate this dynamic nicely: the young warriors serve a useful purpose in society while also garnering status and resources for themselves, if slowly. In turn, this will eventually make them the eligible marital partners they will be—by around age thirty. Is it any wonder that today's front-line military services are largely made up of late-adolescent males working to serve some larger societal interest, perhaps for pay or for the promise of something better to come?

From an evolutionary perspective, one of the most striking aspects of contemporary adolescence is its extended duration, both in females and males. Declining ages of sexual maturation in both sexes activate sexual motivation and capacity at still younger ages. The social necessity of prolonging adolescent activities and delaying long-term partnering and reproduction in order to reach a desired "place" (having the right job, education, state of mind) continues to edge toward later ages. A result is that in the United States, for example, females and males alike may have ten to fifteen sexually active years, on average, before having a first child. This is a profound gap, especially for females. It is a gap with important implications for sexual behavior, as we discuss below, and also for other features of adolescent sexuality.

One of the dynamics affected by the growing adolescent gap is parent-offspring conflict. The well-regarded and professional advice giver Dr. Benjamin Spock long ago suggested that adolescent rebelliousness is a given. There is some theoretical evidence favoring this view: the shifting life history priorities of adolescents can indeed run up against the will of parents, yielding some tension over acceptable behaviors (having to work hard to inherit the farm, as it were). However, this view underestimates the cross-cultural variation, if difficult to quantify, in parent-offspring dynamics at adolescence. Tensions may be minimal in mobile, foraging contexts where there is no heritable wealth (land, livestock, or money) to fight over, where an adolescent, out of self-interest, may wish to work hard to establish a reputation enabling status, marriage, and reproduction; one can move to another camp if social tensions arise, where older males can serve as economic and social models. As Schlegel and Barry (1991) find across cultures, the relations between adolescents and parents are generally fairly harmonious, with adolescents frequently working with and learning from parents, and willingly heeding rather than battling parents' wishes.

The expanding adolescence gap has other ramifications. It arguably has fostered growing segments of youth culture wherever it occurs. Marked by free time, peers, and often resources (from parents or from their own employment), young adults represent a massive consumer demographic, with a focus on obtaining products that advertise their status and desirability (Saad 2011). They may have resources to throw at purchases of luxury goods, as witnessed in abundant shopping malls providing social and shopping opportunities to many adolescents (and of course air conditioning, which doesn't hurt—as Peter realized on entering many such marvels in Singapore). In another example, Laura Miller (2006) describes the beautifying practices of adolescent Japanese girls—from visits to salons to public displays as provocative school girls (*koga*)—that reinforce their relationships through consumerist activity. This is exactly the kind of adolescent female activity that can develop in a youth culture, and that would not develop in a world where females proceed quickly from menarche to marriage. Further, university dorm life, in which many adolescent females can engage in creative and competitive activities together, removed from parental or spousal oversight, has no obvious evolutionary precursor.

A relatively new term that has taken hold with respect to sociosexuality refers to a popular concept among adolescents in North America today: hooking up. Over the past several years Justin and his colleagues have been investigating aspects of the sexual "hook-up" culture among college students (see Garcia et al. 2012). The developmental period spent at college is also framed as a period of "emerging adulthood." It allows for both an appreciation of the extended adolescent period and also a realization that in many cases individuals are rather behaviorally independent. Hook-ups have been particularly studied on the unique environment of the college campus (where there is minimal age stratification in groups and little or no parental supervision per se), and they generally refer to uncommitted sexual encounters, ranging from sexual kissing to intercourse—the key concept being that they occur outside a long-term committed relationship. In one study, 81 percent of respondents reported having hooked up (a percentage of campus students consistent with findings in other studies on the topic); 58 percent said they engaged in sexual touching above the waist and 53 percent below the waist, 36 percent performed oral sex, 35 percent received oral sex, and 34 percent engaged in sexual intercourse during a hook-up (Reiber and Garcia 2010).

Yet, one should not take this to mean individuals are no longer interested in long-term romantic bonds. The evidence suggests that individuals still desire such bonds, but they are consciously pushing back their efforts to obtain such relationships. In one of Justin's studies with Chris Reiber (Garcia and Reiber 2008), while roughly 89 percent of individuals indicated physical gratification as a reason for hooking up, approximately 51 percent of both females and males indicated a desire to initiate a more traditional romantic relationship. Hook-ups have entered the fabric of mainstream culture, with abundant examples in popular entertainment, along with magazines and consumer products designed to make one more competitive on the hook-up market. As an example of the placement of sex in popular culture, consider Hobbs and Gallup's (2011) findings following a content analysis of song lyrics from *Billboard* Top 10 charts for country, pop, and R & B music. They found that, of 174 different songs in the Top 10 lists from the year 2009, 92 percent contained messages regarding reproduction or mating, with the best-selling songs containing more of such messages than less-successful songs.

In a classic study of linguistic innovation, conducted on the island of Nantucket in Massachusetts, adolescent females were highlighted in their role as sources of new words (see Boyd and Richerson 1985). This appears to be a widespread pattern, with scholars pointing to young women as propagating the now pervasive speech patterns of "uptalk" (raising the voice at the end of a phrase in a questionlike way) and "vocal fry," among other examples (see Wolk, Abdelli-Beruh, and Slavin, 2011). Any verbal person can, in theory, create new words. But the new memes that take hold and spread are inordinately those developed by adolescents, especially females. The fashion industry is also driven predominantly by adolescent females seeking to accentuate their sexuality and status, by impressing other women and also enticing men during a life history phase featuring investments in mating effort.

Companies are trying, with some success, to pull adolescent males onto the same consumer treadmill as females (just watch a commercial for Axe deodorant or male hair care products), with social success redefined in consumption patterns. In a discussion of adolescent male social lives, David Lancy (2008) covers an array of ethnographic examples. In ancient Sparta, adolescent males were subject to arduous training exercises as part of their preparation for success in battle. Among the Ache of Paraguay, adolescent males engage in obnoxious or risky behavior, presenting a swagger when in the presence of male peers or women, but they are easily cowed by older men. In a tantalizing proposal, Lancy also suggests it is possibly the artistry of adolescent males that is displayed in European cave paintings like those at Lascaux in France, depicting primarily animals and dating to around 12,000 years ago. The idea is that adolescent males are the kinds of individuals most likely to undertake risky endeavors (such as crawling into the hard-to-reach recesses of caves) and most likely to have the opportunity, which few females at the time would have had (adolescence would likely have been of short duration, and a nursing mother climbing into these same caves with her infant seems unlikely). Perhaps adolescent males were portraying the game animals and predators occupying their dreams in their quest to become good, successful hunters. At least some footprints left behind in long-undisturbed caves are consistent with imprints of adolescent males, an indication this idea may not be a far-fetched possibility, and one that would focus on a more sensible life history interpretation than a theory of general-society animal mysticism.

ADOLESCENT SEXUAL BEHAVIOR

Among Arnhem Land Australian aborigines, "A girl usually has her first sexual intercourse at about the age of nine years, although cases are known of her participating in this act when seven or eight years old. The natives recognize that she is ready to enter sexual life at a much earlier age than the boy, whose sexual organ may not be sufficiently developed till the age of twelve to fourteen years" (Berndt and Berndt 1951: 91). Among the Ache foragers of Paraguay, "Teenage girls become women, and soon become the object of the attention of older men, whilst boys of the same age are still small, thin, and immature relative to the adult men. Both boys and girls begin experimenting with sex around twelve years of age, however, in a manner very similar to that described for the !Kung [hunter-gatherers of southern Africa]" (Hill and Hurtado 1996: 224). These anecdotes open our discussion of adolescent sexual behavior. When does sexual behavior begin, and with whom (that is, who are the typical sexual partners)?

As the forager accounts suggest, adolescent sexual experimentation is a frequent element of this phase in the life course. In wider cross-cultural scope, adolescence commonly represents a time of relative sexual license, preceding more constrained long-term sociosexual partnering and reproduction. Individuals may enjoy the opportunity to play the mating field, finding mates with whom they are compatible, possibly for marriage, although adult relatives are also commonly poised to weigh in on those marital transactions. Still, there is cross-cultural variation in the nature of adolescent sexual exploration: in some societies, such as among the Ache, it is expected, while in others, especially for girls, it can lead to honor killings (in which a father or other older male relative kills an adolescent female who has tarnished her reputation and her family's by having sex before marriage). We describe some of this cross-cultural variation and present the factors that help account for it.

In Ford and Beach's (1951) scheme, they categorize adolescent cross-cultural variation into bins of restrictive, semirestrictive, and permissive societies. Among restrictive societies, they report that Chagga males of Tanzania are prohibited from engaging in sex until they have gone through circumcision, conducted during initiation rites. Jivaro males are also forbidden from having sex until undertaking initiation around the time of puberty. More often among restrictive societies, however, the sexual behaviors of females are constrained, yielding to a sexual double standard whereby males are given more sexual license than females. For example, "Among the Hopi a strong attempt is made to keep boys and girls apart from the age of ten until marriage; the girls are kept at home and are accompanied by an older woman whenever they go out. Girls are expected to be chaste until marriage. Boys, however, are not similarly restricted, and, whenever possible, they defeat the chaperonage system by crawling into the girls' house stealthily at night or by holding clandestine prearranged meetings" (Ford and Beach 1951: 190). Among these restrictive societies, the attempt to guard girls' sexuality is frequently designed, as among the Hopi, to ensure their virginity at marriage, even though the same is not asked of males at marriage.

At the other end of the cultural range, among permissive societies, adolescent sexual behavior may be normalized in both females and males, meaning there is no sexual double standard (Ford and Beach 1951). The Lepcha of India expect girls by age eleven or twelve to engage in intercourse. Trobriand Islanders are expected to engage in adolescent sexual behavior, which offers an opportunity to test for possible long-term mate compatibility and sometimes leads to marriage and child-rearing after a couple's protracted sexual experience together. In some societies, including the Lepcha, it is thought that females must engage in sex for puberty to occur. In other words, sex precedes menarche, though it also tends to be common during subsequent adolescence as well. Among the Canela of Brazil, for example, girls begin having sex before menarche, which helps explain the belief that sex causes menarche. Historically, young Canela women also engaged in "group sex," which entailed childless but partnered women sexually servicing various males, sometimes in sequence. In the ethnographic report of this practice, the deeper context appears to reflect a time when intergroup violence occurred regularly and young men served as warriors; sexually servicing the unmarried men helped keep them from becoming a social hassle for older society members and also recognized the sexual needs of these unpartnered adolescent men. The adolescent women who so freely engaged in these practices changed their behavior dramatically—restricting it—after having children, and this was also somewhat true of adolescent men (Crocker and Crocker 2004).

Given the cross-cultural variation in adolescent sexual behavior, are there any factors that help account for it? In their review of such studies, Schlegel and Barry (1991) find that premarital female sexual permissiveness is associated with "simpler subsistence technologies, absence of stratification, smaller communities, matrilineal descent, matrilocal residence, absence of belief in high gods, absence of bridewealth, high female economic contribution, little or no property exchange at marriage, and ascribed rather than achieved status" (112). To appreciate the associations between these variables and adolescent female sexual permissiveness, we can examine the converse: adolescent female restrictions. Imagine a patrilocal society (bride moves in with husband's family) in which a prospective husband's family provides bridewealth (money or livestock in exchange for the woman's hand in marriage): they may seek maximal return on their investment, with one aspect being that she is a virgin, perhaps as an indicator of her future sexual fidelity too. Among Kipsigis agropastoralists of Kenya, women with illegitimate children (one sure sign they were not virgins) commanded a lower brideprice in the mating market, also providing an incentive to her 'keepers' to ensure her virginity is closely guarded (Borgerhoff Mulder 1988). In the hallmark of virginal assurances, a number of societies have historically conducted checks for virginity (by investigating whether the hymen is intact, for example, or by testing whether the blood of a broken hymen appears on a bed sheet after conjugal relations are initiated), though these tests are hardly foolproof (Gregersen 1994).

As for male adolescent sexual restraint or permissiveness, there is some patterning to this as well. One finding, reported by Schlegel and Barry (1991), is that the degree of adolescent sexual latitude granted to males is associated with that granted to females. In other words, if members of one sex are restricted, often so are the members of the other sex. That could include the Chagga, as noted above, who prohibit adolescent males from engaging in intercourse until initiated, and who historically also constrained adolescent female sexuality and practiced clitoridectomy. Still, if there is greater sexual latitude granted to one sex, it is typically to males, leading to a sexual double standard among a fair number of sexually restrictive societies.

One of the primary political shackles on adolescent male sexuality is the group of older males in the society. Especially where polygynous marriage is permitted, an older male may have multiple wives, including some quite young wives who are a similar age to the young unmarried men (with whom those younger wives would often rather have sex). An array of practices, such as secluding adolescent men or shipping them off to war or other distant economic endeavors, can help remove these adolescent males from the immediate sexually competitive arena. In an analysis of cross-cultural male genital mutilation practices (such as circumcision in southwest Asia, or subincision among some Australian aborigines, the latter of which entailed slicing the underside of the penis), it seems that several processes of male-male sexual conflict may be at play (C. Wilson 2008). We note that the practices of secluding, initiating, subjecting to painful genital procedures, and restricting male sexual behaviors, can also serve the reproductive interests of older males, who perpetuate the practices. Recall Sambia adolescent men, who initially fellate older males, then become the fellated: this is arguably a means of keeping the adolescent males sexually occupied with each other rather than with the young wives of older men. Expressing permissive attitudes toward adolescent male masturbation may be another means of attempting to distract younger male competitors: Malinowski (1929) reported on adolescent males masturbating with use of their saliva as a lubricant.

Offering a different data-rich entry into teenage sexual behavior, Wellings et al.'s (2006) massive compendium of sexual studies helps illuminate adolescent sexual behavior in countries around the world (from Australia to Bolivia, Niger to Turkey). One of the central findings is that in countries with young ages of female marriage (for example, in the teens), females typically engage in earlier sexual intercourse than males, though often within marriage. This pattern is most prevalent in West Africa and parts of South Asia. In Nigeria, for example, the average age of first sexual intercourse for females is fifteen and a half years, while for males it is twenty and a half, in large part reflecting the sexual initiation of females within marriage and the fact that males are subject to premarital constraints on available partners (since many otherwise eligible females are already married to older men). In another pattern, in developing countries the duration between sexual initiation and marriage is shorter among females than in males; in other words, on average the onset of female sexual behavior is more closely tied to marriage than it is for males. However, this is not true in developed countries like Australia or the United States, where both females and males tend to engage in premarital sex that is not closely coupled with marriage but part of an extended adolescence. And as one additional trend, premarital sex is becoming more common globally, a reflection of the later age of marriage and first reproduction worldwide, in addition to effects of earlier pubertal development.

In the United States, at least, approximately 95 percent of Americans have premarital sex by age forty-four (Finer 2007)—although it is possible, in many cases, that the person one has sex with may ultimately be the one they marry (and possibly their only lifetime sexual partner). These trends have not appeared to change much since the 1950s. In a press release from the Guttmacher Institute, a well-respected independent agency focused on sexual and reproductive health and rights, following the release of this finding, lead author Lawrence Finer stated (December 19, 2006): "The data clearly show that the majority of older teens and adults have already had sex before marriage, which calls into question the federal government's funding of abstinence-only-until-marriage programs for 12–29-year-olds. It would be more effective to provide young people with the skills and information they need to be safe once they become sexually active—which nearly everyone eventually will." We couldn't agree more.

In the United States, nearly half of all high school students report having engaged in sexual intercourse, with no noticeable gender differences, although ethnic differences suggest rates are higher among high school students that identify as black (65 percent) and Hispanic (49 percent) compared with white (42 percent) (Hamilton and Ventura 2012). Moreover, 26 percent of female and 29 percent of male U.S. high school students have had more than one sexual partner, with approximately 15 percent having had four or more sexual partners (Abma, Martinez, and Copen 2010). Sexual behavior among adolescents often raises social and political concerns focused on unintended teenage pregnancy and sexually transmitted diseases. In some cases these fears have resulted in programs to direct behavioral change. It has long been recognized that, among industrialized populations, the United States has one of the highest rates of teenage pregnancy, abortion, and birth (Hamilton and Ventura 2012); more recently, teenage pregnancy rates in the United States have been declining. Yet in today's contemporary technological era, the means of sexual communication are expanding (something we will return to in the final chapter): one 2009 report suggested that approximately 10 percent of teenagers have sent naked photos or videos of themselves via the Internet or text messaging, a practice now often referred to as "sexting." The extended phase of adolescence is helping foster greater adolescent sexual freedom and behavior.

How does same-sex sexual behavior appear during adolescence? While we have noted Sambia male-male sexual behaviors, for example, as a life-stage phenomenon, what about wider patterns? More work on this topic is available for males than for females, in part because adolescent male-male sexual behavior is more common than female-female behavior. After all, the limited duration of adolescence for females proceeding rapidly from menarche to marriage and reproduction allows less time for exploration of any potential same-sex sexual desires, whereas adolescent males, over a longer adolescence, frequently have unmet sexual needs to cope with. Adolescence, cross-culturally, appears to be a developmental stage allowing for personal maturation, including sexual exploration.

"Estimates of lifetime prevalence of sexual intercourse with men ranged from 3–5% of men in east Asia, 6–12% for eastern Europe, 6–20% for Latin America. The prevalence is 6% in the UK and 5% in France," report Wellings et al. (2006: 1714) in their sexuality review. In the United States, the number of males who have ever engaged in same-sex sexual activity (estimates of 15 percent; 10–20 percent in other studies) far outnumber the men who identify as homosexual (estimates of 4–8 percent) (Herbenick et al. 2010a). While this comparison reflects lifetime behavior versus adolescent male-male sexual behavior, the numbers offer some quantitative range for these behaviors. With a more direct cross-cultural focus on adolescent same-sex sexual behaviors, Schlegel and Barry (1991) observed several patterns. First, there is an association between tolerating adolescent same-sex sexual behavior in one sex and in the other; in other words, if same-sex sexual behaviors are permitted for males, then they often are for females too. Second, for males, tolerance of adolescent same-sex sexual behaviors tends to covary with tolerance for adult same-sex sexual behaviors. Third, for adolescent males, engaging in same-sex sexual behavior tends to be a short-lived phase that

> appears to be a substitute for heterosexual intercourse when intercourse is prohibited or access to girls is problematic [as among the Nyakusa]. . . . Homosexual practices are said to be very common in the boys' villages— they begin among boys of ten to fourteen herding cows, and continue among young men until marriage, but they are said never to continue after that, and are regarded simply as a substitute for heterosexual pleasure. . . . [T]hese relations among the Nyakusa generally take the form of interfemoral intercourse, also the common form of sexual activity with adolescent girls. (Schlegel and Barry 1991: 126–127)

While there are individuals, as among the Sambia, for instance, who have a lifelong same-sex sexual preference (as noted in Chapter 2 on cross-cultural sexual variation), much of adolescent same-sex sexual behavior represents a transient exploration and the satisfaction of sexual need, rather than the expression of same-sex partner preference.

While adolescent female same-sex sexual behaviors received little space in Schlegel and Barry's (1991) review, owing to the shortage of data, these behaviors represent a growing niche ripe for research. The expanding adolescence gap, particularly for females, entails unprecedented durations of ovulatory cycling before or perhaps without long-term heterosexual partnering and child-rearing. No forager female could have anticipated having ten to fifteen years of life in this stage, and yet this is increasingly the case in contemporary global society. Concentration of females together, as at women's colleges, may help spawn what some have called a "LUG" experience (lesbians until graduation), a time when females engage in same-sex sexual behaviors as part of exploring identity and satisfying curiosity, often followed by marriage with a man. In her book on sexual fluidity, Lisa Diamond (2008) has emphasized the plasticity of female sexual desire (toward females or males); her longitudinal sample of U.S. women includes some women who have had LUG-like experiences, but also women whose sexual behavior remained quite plastic throughout their life. An interesting question will be whether an expanding female adolescence gap gives rise to stage-specific same-sex sexual behaviors; given that adolescent females are typically in high sexual demand, they may have readily available sexual partners among males, and yet in social contexts with demographic skew (female-only schools, communities with a shortage of desirable males, who may be away at work, war, or in prison) stage-specific female same-sex sexual behaviors might be anticipated.

In summary, pubertal transitions and adolescence represent remarkable biobehavioral life history shifts. During this time, both females and males may experiment sexually, explore their newfound desires (and tools), and be subject to local cultural customs that attempt to control these shifts. Individuals begin a new chapter of their life as a reproductively capable self. Having reviewed some of the biological and cross-cultural patterns in pubertal transitions, we too are ready to begin a new chapter. We next turn to those tools that puberty has helped hone: the intricacies of reproductive anatomy. In any discussion of sexual behavior, it is essential to understand how the penises and vaginas involved have come to be, and how they operate.

CHAPTER SEVEN

Kinsey Takes Anatomy Class

Human Reproductive Anatomy and Physiology within Evolutionary Perspective

> The results of research on comparative reproductive anatomy, physiology, and copulatory patterns indicate that *H. sapiens* was originally polygynous or monogamous (or both), and did not arise from forms which had a multi-male, multi-female mating system.
>
> —Alan Dixson, *Sexual Selection and the Origins of Human Mating Systems*

HAVING GOTTEN A puppy recently (after his daughters persistently asked for one), Peter learned a lesson in comparative reproductive anatomy. This puppy (Teddy), like most, enjoyed chewing everything, so on a clerk's recommendation he bought and gave Teddy some treats called bully sticks to chew on. Teddy loves them. Yet this purchase invited the question: what is a bully stick? As Peter subsequently learned, a bully stick is a dried bull's penis. These sticks can be more than three feet long, though that's after drying, stretching, and flavoring them; a fresh bull's penis is around two feet in length. A bully stick is mostly "crude protein" in content—all tissue— though at one time the penis was connected to a baculum, or penis bone. Most male mammals have a baculum (and females often have a homologous clitoral bone), which aids in rapid achievement of erections. Walruses have the longest ones, coming in at over two feet in length; Inuit sometimes use them as clubs. Interestingly, dogs like Teddy have a baculum too, as do most primates, including chimpanzees and bonobos, but humans do not. In humans, the baculum has been lost in the past, a victim of evolutionary changes in hominin reproductive anatomy and sexuality.

As this example attests, the study of reproductive anatomy and physiology can yield some marvelous surprises. When comparing our reproductive features with those of other animals, we can also see some patterns that have inscribed our ancestors' mating ways into our own bodies. Something in our mating behavior changed to lead our ancestors to "discard" the baculum, also leaving human females without a clitoral bone. The chapter epigraph provides a general view of what this sort of effort tells us about our ancestors' mating behavior. In a rich scientific world, the inferences drawn from the study of our human reproductive anatomy and physiology should converge with all the other lines of evidence—sex surveys, studies of the human sexual response, an understanding of the proximate genetic and hormonal mechanisms underlying sex differentiation, the cross-cultural record of human sexuality, and so forth. This effort will help flesh out the evolution of human sexual behavior.

ANATOMY AND PHYSIOLOGY 101

Before we dive into the details of reproductive anatomy and physiology in females and males, respectively, we outline some of the conceptual background helpful for interpreting the evolutionary meaning of those physical details. One of the first concepts is that we should forget sex differences (briefly). Rewind the evolutionary tape to a distant past before sexual reproduction, then hit play: sexual reproduction arose in our distant ancestors, and when it did, it built on shared anatomical and physiological substrates between the sexes. This is why female and male reproductive anatomy and physiology are so similar. Really, the similarities are as striking as the differences. A useful heuristic, as noted in an earlier chapter, is that females are the "default sex," and males are a version derived from females. This heuristic breaks down dramatically at puberty, and is an admitted simplification of reality, but it does help capture the core ideas that sex evolved from an asexual world, that male and female reproductive anatomy and physiology share many features, and that male traits are derived from female ones.

Another conceptual foundation of human reproductive anatomy and physiology is that hormones such as testosterone, estradiol, and progesterone play central roles in orchestrating the development of reproductive traits (Adkins-Regan 2005; Nelson 2011). We ran through some of this material in Chapter 4. It makes a lot of sense that hormones released from the gonads (ovaries or testes) coordinate the development of reproductive anatomy and physiology. Really, the gonads have two tasks: they produce the gametes (eggs or sperm) that may become part of a new life, and they release hormones that shape all kinds of other traits that facilitate reproduction. Those other traits can seem quite distal—imagine testosterone released by the testes favoring enhanced development of male musculature. Or the traits may seem closer to home—consider subtle changes in female vaginal secretions across the ovulatory cycle that influence the likelihood of sperm movement. The key is that the traits influenced by gonadal hormones commonly play roles that ultimately facilitate reproductive success.

A distinction between primary and secondary sexual characteristics is also relevant here (Geary 2010). The primary sexual characteristics are the early-developing features of reproductive anatomy—the vagina, penis, and so forth. The secondary sexual characteristics are the traits that arise at puberty, typically under the influence of steroid hormones such as estradiol and testosterone, and play roles in intrasexual and intersexual competition. From a theoretical life history standpoint, it makes sense that these secondary sexual characteristics emerge at puberty: they might be energetically expensive to maintain (extra muscle) or they might intrude on other prereproductive priorities (using limited energy that can be channeled to either growth or reproduction, but not both), so they are best left in waiting until an organism prepares to enter the reproductive arena.

In many endeavors of life, males and females sometimes cooperate with each other and other times compete. This is also true of sex differences in reproductive anatomy and physiology. Both females and males have a vested interest in successful reproduction; under some circumstances, such as within long-term, relatively monogamous sociosexual bonds and with biparental care, there may be more mutually beneficial coordination. What's good for the goose is often good for the gander. In other contexts, such as when females mate with multiple males around the time of ovulation, the reproductive anatomy and physiology of females and males may have more competing agendas. A female may wish to screen her suitors through assessment of their success in male-male competition, personality, or other traits; that screening process may continue after copulation, with her anatomy and physiology presenting an obstacle course for sperm trying to reach the egg, thereby finding the optimal sperm among the many competitors. For his part, a male may benefit from dimorphic traits (such as extra muscula-ture) that facilitate the male-male competition, courtship, or perhaps even coercion that lead up to mating. He may try to load the dice in his gametes' favor postintercourse by producing more of them or providing other sub-stances to aid their transport, all in an attempt to bypass the barriers sup-plied by females and the challenges presented by other males.

The outcome of these conceptual possibilities is that the characteristics of female and male reproductive anatomy and physiology may be variably cooperative or competitive. When their agendas are maximally aligned, we may expect male and female traits to exhibit a "lock and key" functionality with minimal antagonism. In other cases, features may be shaped to facili-tate intrasexual selection as well as intersexual selection, with less apparent efficiency built into the system (that is, females may need to create more effective obstacle courses for sperm, as in cryptic female choice) and males may have to produce far more sperm than would otherwise seem necessary (think of sperm competition, or competition between the sperm of multiple males inside a female's reproductive tract).

If these concepts seem abstract, how about illustrating them by comparing the reproductive performance of fruit flies in experimental conditions, or the anatomy of birds' differing in mating systems? In the first case, William Rice toyed with the mating behavior of fruit flies in captivity, requiring some of the naturally promiscuous flies to be monogamous. The result was dramatic: "In the monogamy treatments male behavior and seminal fluid proteins evolved to enhance rather than reduce female survival, relative to promiscuous controls, and fitness of monogamous males declined in parallel, when they were placed back in their promiscuous ancestral population. Interestingly, females from the monogamy treatment evolved to be less resistant to male-induced harm when they were placed back in their promis-cuous ancestral populations" (Rice 2000: 12955). As for the birds, Patricia Brennan and colleagues (Brennan et al. 2007) compared the reproductive anatomy of birds differing in mating behavior, particularly with respect to whether forced copulation occurs or not. Notably, forced copulations are relatively rare in birds, but more common among some waterfowl species whose males have evolved intromittant organs (phalluses) to facilitate sex in the water. In two species of waterfowl—harlequin ducks and the African goose—females have relatively simple vaginas and males have short phalluses, and there are and no forced copulations; by contrast, in long-tailed ducks and mallards, females have remarkably elaborate vaginas and males long phalluses, and there are high rates of forced copulation. The female-male covariation could be viewed as the outcome of an "arms race" whereby males' forcing copulations has favored elaborated male phalluses as well as female responses designed to blunt the efficacy of those same instruments.

REPRODUCTIVE ANATOMY AND PHYSIOLOGY: HERS

During a lecture on primate behavior, Peter showed an image of a baboon female's exaggerated sexual swelling, that is, a female's pink perineum engorged with extra fluid. A male student shared openly his disgust at the sight of her swelling. We humans might be forgiven such a view; we're not baboons, after all, and if we were, this male student would likely have responded quite differently to the sight of such a trait.

Human females do not have exaggerated sexual swellings. Our closest living relatives, chimpanzees and bonobos do (Dixson 1998), but we don't know if our common ancestor with them had sexual swellings or not. Either our common ancestor evolved this trait and we lost it, or sexual swellings arose among more recent ancestors of chimpanzees and bonobos. Comparative primate evidence indicates that the trait for exaggerated sexual swellings has arisen at least three times: in the human/chimpanzee case just noted, and also among ancestors of colobus monkeys, and among ancestors of some other monkeys that include baboons and mandrills (Nunn 1999). The trait also appears to have been lost in some primate groups, including some other monkeys, such as the widespread and abundant vervets in Africa.

What would favor such a dramatic female trait? The best guess is that these swellings serve as sexually enticing signals to elicit male copulations. After all, the three cases of exaggerated primate sexual swellings share a common social dimension: the females live in multi-male, multi-female groups, where females seek to mate with multiple males. In a context where desirable males may try to optimize their limited mating effort, it appears to pay for females to carry a sexual invitation on their rump, inviting male copulation. Exaggerated sexual swellings also appear to develop under the influence of estrogens, with the result that they tend to increase in size across the follicular phase of a female's cycle, and may incidentally increase in size during pregnancy when estrogen levels also escalate.

The fact that human females lack exaggerated sexual swellings is a strike against any view that our recent ancestors may have been wildly promiscuous within multi-male, multi-female groups. Some have taken this absence to mean that we have "concealed ovulation," or unadvertised fertility, in part to facilitate long-term sociosexual bonds (if a male partner does not know when the female is ovulating, then he should stick around across the entire cycle rather than just during fertile times, so the argument goes). However, we will also consider evidence in the next chapter that suggests female ovulatory-cycle influences on female attractivity and sexual behavior sometimes do exist, but these are quite subtle and thus neither loud advertisements nor fully concealed.

What features of human females' primary sexual characteristics stand out (if not sexual swellings)? Females have fairly pronounced outer vaginal labia, or lips, in part a product of shifts toward upright bipedalism that have squished together the genitals. These more pronounced outer labia also conceal the clitoris inside, a contrast with other primates where the clitoris is more readily visible. An upright, bipedal stance also pushes sexual behavior, as noted earlier, toward different positions, including more face-to-face copulatory positions rather than dorsal-ventral (or male-from-behind) positions.

The clitoris is the female homologue of the penis. The clitoris has excited some of the most vigorous scientific debate over human sexual anatomy. A central question is whether the clitoris is merely a developmental by-product in females of a functional penis in males, or whether any features of the clitoris appear to have been favored by selection in their own female-specific adaptive ways. The clitoris clearly plays a central role in female sexual response (more on this in the next chapter); like the penis, it is intricately bound with sensory nerves, facilitating pleasure and sensitivity. Like the human penis, it also lacks a bone. Across species, there are some general patterns of female clitoris–male penis covariation, whereby species with baculae (those penis bones again) also tend to have clitoral bones, so this human covariation fits within the wider pattern (Dixson 1998).

Whatever you think of the human clitoris, what would you think if you saw the clitoris of a spotted hyena, a muriqui monkey of Brazil, or some other mammal in which this organ is quite pronounced? If you were Aristotle, you might confuse female and male spotted hyenas—he thought they were hermaphrodites—because the clitoris resembles the penis. Among primatologists studying some of the New World monkeys whose females have pronounced clitorises, these too can make sex assignment difficult during observations in a cloudy forest, since the often-useful contrast between penises and clitorises is diminished. In arguments pertinent to the potential functionality of the clitoris in humans, scholars have also debated the developmental, mechanistic, and potentially adaptive bases for large clitorises when they occur, but no clear consensus exists. These may be by-products of male organs in some ways (perhaps clitoral size is related to penis size across species), but not in other ways (in developing independent of androgen hormone levels, for example). They may serve adaptive female signaling functions, facilitating visual and olfactory communication, such as when an excited female's clitoris becomes erect (as in juvenile female capuchin monkeys; Perry and Manson 2008).

BREASTS AND GLUTEOFEMORAL FAT

Female breasts are secondary sexual characteristics, developing at puberty under the influence of estrogen (Levay and Baldwin 2009). Their purpose can be most parsimoniously related to direct female reproductive benefits: they serve as fat stores, helping provide energetic substrates that can aid gestation and also a lactational economy, and in turn providing survival benefits to a female's nursing offspring. But are they something more? Peter raises this question while sitting in a Las Vegas bookstore near a woman who has just begun nursing a baby. This nursing woman covers her breast and baby with a cloth. In many places in the world, this mother would not bother with covering her breasts while nursing, but the politics of breasts involves both their immediate caregiving benefits and their sexual connotations. After all, breasts in humans can serve as important indicators of reproductive age, letting those within sight know a female is of postpubertal age. Just a few miles from this bookstore in Las Vegas, topless shows are advertised in blaring colors.

If the primary function of human breasts is to promote direct female reproductive success (through lactation, and thus offspring survival), human breasts still raise other questions. While other ape females may have enlarged breasts during pregnancy, they tend to be quite small by comparison with those of humans, and these other apes also appear to generally lack permanently enlarged, pendulous breasts after puberty. Larger, more permanent human breasts may be another trait among many (like the fact that our newborns are the fattest of any known mammal, including whales, seals, and other primates; Kuzawa 1998) indicating what an energetically rich niche we humans have carved out for ourselves (for example, we have access to extractive foods and food processing technologies beyond what other primates might only dream about; Kaplan et al. 2000). Some speculate that human breasts are analogues of primate sexual swellings, but this seems a stretch, given the shortage of human female promiscuity in our species and the more parsimonious energetic considerations already noted. Others might ask whether male mate choice for prominent breasts has influenced their morphology; this would be difficult to test, but we would see such sexual interest inspired by breasts and male mate preferences in them as incidental to the breasts' primary lactational and energetic benefits.

The cross-cultural packaging of breasts is interesting in its own right, also perhaps shedding light on the various reproductive and sexual valences of breasts. In a hunter-gatherer context characterized by frequent on-demand lactation, and where all women have children who are nursed over prolonged periods, the politics of breasts may favor toplessness for ease and benefit of lactation; the real puzzle then becomes why cultural practices would arise to cover breasts. Additionally, breasts tend to grow saggier with age and with parity (or childbearing), resulting in breast size and shape serving as a cue of female reproductive value, or potential future reproduction (Marlowe 1998). Whereas in some contexts women may have little concern over these changes in breast morphology, in others women may seek to change the equation, and that's where the use of bras, breast enhancement, and other breast modifiers enter the picture.

In the United States, several hundred thousand women each year seek to enlarge their breasts through surgical enhancements (Swami and Furnham 2008). These may be single women seeking to attract mates with aggrandized stimuli; perhaps women in ongoing partnerships, employing breast augmentation as part of relationship maintenance in the wake of having had children and aging; or women who do this simply for their own satisfaction. The history of bras is fascinating in this light too, since apart from issues of comfort or support (for women who jog, for example) bras can be used to change the shape or reproductive value cued by breasts. In a fascinating anecdote about bras and breasts, Alyssa Crittenden, who has done field research with the Hadza foragers of Tanzania, shared a story of Hadza women asking her for bras as gifts. These are Hadza women who have no apparent history of wearing bras, but who, after seeing Western researchers and neighboring Tanzanians wearing them, thought that they would like to try them too. Expressing an ambivalence about providing bras, Crittenden eventually did, and could you imagine the result? As it turns out, bras (at least the kinds not designed for nursing mothers) work very poorly for frequently nursing women; they get in the way of lactation. The nursing women gave up rather quickly on their bras. Most of those bras ended up either in the hands of older, menopausal women who lacked the day-to-day concerns over lactation, and who might wear them as status symbols during important events, or with men, who stored some of their hunting instruments in them while in the bush. Victoria's Secret would flounder among foragers, this story would suggest.

Apart from fat storage in the breasts, females also deposit gluteofemoral fat (in the butt and thighs) at puberty (Levay and Baldwin 2009). The rising estrogen in females facilitates this prominent secondary sexual characteristic. The conceptual issues underlying the evolution of extensive gluteofemoral fat in humans are quite similar to those regarding breasts: the more parsimonious line of thinking is that additional gluteofemoral fat directly benefits female reproductive success more than it reflects the product of male mate choice. In other words, this extra fat has direct female fitness benefits, perhaps by enabling earlier age of first reproduction, shorter interbirth intervals (or time elapsed between births), or even higher offspring survival. It has even been proposed that gluteofemoral fat may serve as an important reservoir of polyunsaturated fatty acids to help fuel an offspring's brain development (Lassek and Gaulin 2008). Given the importance of energetic factors to female reproductive success, the extra fat provides a fantastic reproductive buffer and reservoir to draw on across the reproductive years (Morbeck, Galloway, and Zihlman 1997). From this view, then, it also serves as an important, visible cue of female reproductive value; males may thus develop mate choice criteria recognizing the adaptive value of female gluteofemoral fat.

A large body of mate-preference and body-image research indicates that men do indeed pay attention to female fatness, including gluteofemoral fat, when sizing up potential mates. The evolutionary psychologist Devra Singh, for example, suggested that men prefer women with waist-to-hip ratios around 0.7 (Singh 1993). This is a ratio of the waist circumference to the hip circumference; picture an hourglass figure (waist considerably thinner than thighs) and you get the idea. In many studies of mate preferences in the United States and elsewhere, men often do favor women with a ratio around 0.7, a ratio also associated with high fertility and fewer health problems (Swami and Furnham 2008). However, work also suggests that in some subsistence societies, such as the Machiguenga of Peru and the Hadza of Tanzania, that men prefer women with higher waist-to-hip ratios, a preference that appears to be linked to preferences for bigger, fatter women overall in societies where people are more concerned about having adequate food (rather than all-you-can-eat college cafeterias) and female reproductive potential. The key, at any rate, is that this body of research indicates the importance of female fatness to male mate preference.

FEMALE AXILLARY AND PUBIC HAIR

We sometimes bill humans as the naked ape, given our loss of thick body fur, but this isn't entirely true. Apart from head hair, humans have retained hair in both axillary (underarm) and pubic (groin) regions (Levay and Baldwin 2009). What does this have to do with female reproductive anatomy and physiology? The basics of axillary and pubic hair, in women and men alike, suggest it is a secondary sexual characteristic: it develops under the influence of androgenic hormones, it becomes prominent at puberty, and it appears to serve visual and olfactory functions in potential reproductive contexts. From the standpoint of olfaction, or smell, the same axillary and pubic regions that have hair are also dense with apocrine glands; these glands secrete substances that, as they decay, make us the smelly beasts we can be (without, say, showering). Axillary and pubic hair may function in part to capture compounds released from these same areas to assist in communicating our scent. Further, the developmental course of axillary and pubic hair means that it can serve as a visual signal of reproductive age.

Presumably pubic and axillary hair served olfactory and visual functions in an ancestral past, but in today's world we have modified these signals (Gregersen 1994; Moalem 2009). We might imagine that these signals were more important among our forager forebears before bathing, genital covering, and even smoky fires became common, all of which diminish the potential signal value of pubic and axillary hair. Subtle cycle-related fluctuations in female scent might have been slightly more noticeable, even though scent likely faded relative to sight in recent hominin evolution. Cross-cultural hygiene and other practices confuse the ancestral picture. Showering, bathing, and use of other smell-hiding compounds mask body smells. Modification, including outright removal, of pubic and axillary hair further compromises its functions. Whether upholding Islamic codes about removing axillary and pubic hair or following a trend, apparently encouraged by pornography and Brazilian body waxing, many women today fully remove their axillary and pubic hair. In a large U.S. Internet study, women who fully removed their pubic hair tended to be younger, sexually active, and to have recently received cunnilingus (Herbenick et al. 2010b). As an interesting consequence, in the United Kingdom the incidence of pubic lice has been declining, apparently due to "habitat loss," even while rates of chlamydia and gonorrhea have been rising (Armstrong and Wilson 2006).

OF VAGINAS, FALLOPIAN TUBES, AND FEMALE PROSTATES

Moving from the surface to the inner plumbing, human female reproductive anatomy and physiology reveal marvelous design reflective of our ancestral reproductive behavior. The vaginal depth of human females syncs with male penis length, a pattern that accords more broadly across primates (Dixson 2009). The vaginal barrel can also change dramatically in size to accommodate variation in a partner's penis size and the more dramatic passage of a baby being birthed through it. The oviduct length (the Fallopian tube, which conveys an egg from the ovary to the uterus) in human females is relatively short. This pattern is consistent with relatively low sperm competition pressures; across primates, those females engaging in multi-male mating around the time of ovulation tend to have relatively long oviducts (cryptic female choice), perhaps to create a more challenging obstacle course that will thus select for the most desirable sperm (Dixson 2009).

Female reproductive physiology fluctuates in intricately adaptive ways (Knobil and Neill 2006; Levay and Baldwin 2009). The cervix, lying at the mouth of the uterus, offering a portal into the vagina, allows blood to flow out during menstruation and, around the time of ovulation, sperm to travel through, but otherwise presents a barrier to things traveling in and out. Changes in cervical mucus structure and consistency help account for these variable passage rates. If sperm do pass through the cervix, subtle characteristics of female reproductive anatomy and physiology preferentially guide the sperm toward the Fallopian tube from which the egg was released. Further, cilia in the Fallopian tube draw sperm toward an ovary around the time of ovulation, but otherwise push away from the ovary. The egg released from an ovary almost always travels into a Fallopian tube rather than wandering into the wilderness of the body cavity.

Women also appear to possess a "female prostate" or Skene's gland (Zaviacic and Whipple 1993). Located adjacent to the urethra, where urine is passed, in some women it releases an ejaculate at orgasm that is similar in composition to secretions from the male prostate gland (high concentrations of prostate-specific antigen, or PSA). Like the male prostate, the female prostate is also sensitive to stimulation, suggesting that it may be the anatomical base of the putative "G" (Grafenberg) spot. The existence of the female prostate raises parallel questions to those about the clitoris or male nipples, and arguably demands its own studies centering on the adaptive/by-product debate (as does the clitoris). A parsimonious view would be that the female prostate is a functionless by-product of the functional gland in males. An alternative view could raise female-specific adaptive functions, such as its secretions aiding in fertilization or acting as antimicrobials.

AN OVATION FOR THE OVARY

The epicenter of human female reproductive anatomy and physiology, where the eggs and important hormones are released, is the ovary (Knobil and Neill 2006; Levay and Baldwin 2009). Here the follicles reside, diminishing in number from their prebirth peak until the time they literally run out of supply, at menopause. A few select follicles are recruited for the chance of a lifetime: for ovulation and potential fertilization. This process entails the maturing of a primordial follicle until the point of release from the ovary, at ovulation. This is a cyclical process, which is why it is recognized as either the ovulatory cycle (if we highlight ovulation) or the menstrual cycle (if we highlight menstruation).

The human ovulatory cycle is remarkably similar to that in other apes, such as chimpanzees (Dixson 1998; Martin 2007). The cycle is character-

ized by follicular and luteal phases, totaling about twenty-eight days together, although the duration of the follicular phase tends to be more variable than the duration of the luteal phase. In the follicular phase, the initial
three to five days typically entail menstruation; the follicle also develops
during this phase (hence the follicular phase) until it's ready for release at
ovulation on around day fourteen of the cycle. Estrogen levels increase across
the follicular phase, promoting follicle development in addition to fueling
proliferation of the endometrial lining in the uterus, the potential home of a
fertilized egg. The rising estrogen promotes female sociosexuality, from vaginal lubrication to sexual interest, with androgens also playing roles in female sociosexuality. After all, the ovary "wants" the egg it releases to have the
sperm of a desirable mate ready for fertilization, and it "wants" a fertilized
egg to have a nourishing home in which to implant and develop.

That's where the luteal phase enters the cycle. The corpus luteum (literally, "yellow body," based on its lipid-rich base) left in the ovary by the
vacated egg remains active after ovulation, releasing progesterone. The
progesterone (so named because it, literally, protects gestation) continues
to foster development of the endometrium of the uterus for a fertilized egg to
enjoy as its home. The progesterone likely has additional effects that are less
well investigated but that could collectively make sense as facilitating pregnancy outcomes (for example, diminishing immune responses to reduce the
chances the female body will reject the foreign tissue that the male sperm
represents; promoting water retention to facilitate increased pregnancy-related
nutrient and waste transport; impinging on female libido). The human luteal
phase lasts about fourteen days, from ovulation until the potential onset of
menstruation.

The "typical" human ovulatory cycle specified here is really characterized by considerable, graded variation, and for good reason (Ellison 2001;
Jasienska 2003; Trevathan 2010). Given the importance of energetic factors
to female reproductive success, it makes utter sense that ovarian function
(egg and hormone release) should be tied to cues indicative of food availability, energy expenditure, metabolic reserves, changes in metabolic circumstance, and health status. The time must be ripe for maternal investment in egg development, sexual behavior, gestation, and maternal investment
to overcome the energetic barriers to female reproductive success. The sensitivity of ovarian function follows a continuum, rather than an all-or-
nothing model, with the result that some cycles in women are ovulatory,

others not, and that some cycles have lower hormone profiles, with potential consequences for reproductive viability.

How do we know these things? Clinicians, anthropologists, biologists, and others have documented the influences of factors such as energy expenditure on female ovarian function. In the case of anthropologists, Peter Ellison (2001), Virginia Vitzthum (2009), and others have measured hormones like estradiol and progesterone in various populations in order to assess how their levels vary across and within populations. One of the central findings has been that clinically "normal" hormone reference ranges in, say, the United States often represent the extreme high compared with populations living under greater nutritional constraints, higher energy expenditures, and higher disease loads. Such studies have also found, for example, that Polish women's hormone profiles decrease during the hard-labor agricultural season compared with the nonharvest season, even though their diets are relatively constant (Jasienska and Ellison 1996).

Age-related changes in ovarian function also occur. The first few years of ovulatory cycles are less functional than those during the prime reproductive years. As we shall see, this has implications for adolescent sexuality and fertility. The peak years of female fecundability (likelihood of conceiving during a given act of intercourse) seem to fall between ages twenty-five and thirty-five. Fecundability thus declines long before menopause occurs, a pattern certainly of importance to women attempting to conceive.

Back to the "standard" ovulatory cycle, though: what if a woman does not conceive or implant a fertilized egg? In this event, the uterine lining is shed, resulting in menstruation (Strassman 1996). Uterine contractions, stimulated by prostaglandins, help shed menstrual blood, also giving rise to potential cramps. The characteristics of human menstruation exhibit both similarities and differences with our close primate cousins. Our relatively profligate menstrual blood loss is due, in part, to the tremendous amount of effort undertaken to build the uterine lining in the expectation of implantation; that effort, when dismantled, yields the considerable blood loss, but a loss similar to the visibly apparent menstruation of other apes, such as chimpanzees and orangutans. More distantly related primates, such as lemurs, lack overt menstruation, testimony to their lower uterine buildup. Lest we wonder why the uterine lining is not simply maintained rather than cyclically shed, Beverly Strassman (1996) has suggested the most compelling explanation: it is more energetically expensive to sustain the uterine lining than to tear it down and rebuild.

An evolutionary perspective suggests that menstruation was a relatively rare occurrence among our hominin forebears, including among many women living in natural fertility populations of hunter-gatherers and other societies where women spend the bulk of their reproductive years pregnant or lactating (Eaton et al. 1994; Short 1976). Putting some numbers to this pattern, Strassman (1997a) has also characterized menstruation patterns among Dogon women in rural Mali. She found that women aged fifteen to nineteen were cycling 72 percent of the time, sometimes menstruating during these cycles (many of these were not fully functional, given the adolescent issues already noted); women aged twenty to thirty-four were cycling only 15 percent of the time (because they were pregnant the remainder of the time, or lactating but not cycling); and women aged thirty-five to fifty-three were cycling 53 percent of the time, as their cycles became less fertile with advancing age.

Apart from frequency, the experience of menstruation varies in other ways cross-culturally. In some societies, including the Dogon and some northwestern Native American societies, women spend their menstruating days in a menstrual hut, where they may be restricted from various activities (Gregersen 1994). In some societies, women amble about but are subject to menstrual taboos, such as in some highland New Guinea societies wherein menstruating women are viewed as polluting. In Bali, temples display signs stating that women may not enter when menstruating. The exact duration and menstrual flow volume may also vary across cultures, with the Hadza foragers of Tanzania having slightly shorter cycles and flow volumes compared with the North American and European women in one study (Marlowe 2010). Energetically constrained women in rural Bolivia menstruated for a median of three and a half days, whereas in the United States the median is six days (see Trevathan 2010). The responses to menstrual flows also vary, with some women, like the Yanomami women of the Amazon, letting menstrual blood flow freely. Other societies use items designed to capture the flow, such as papyrus tampons in ancient Egypt, paper tampons in historic Japan, cloths in colonial America (hence a colloquial phrase "on the rag"), and the modern pads first invented in the 1920s (Moalem 2009).

Suppose that, instead of menstruation, fertilization and implantation occur. What happens in that event? Human chorionic gonadotropin (hCG) is released by the newly established conceptus. This is a peptide hormone that appears to have arisen in the primate ancestor of monkeys, apes, and humans (it is absent in lemurs) from a close relative, luteinizing hormone, in a wonderful example of nature's tinkering (Maston and Ruvolo 2002). The hCG serves as a pregnancy signal. Effectively, the conceptus is shouting as loudly as it can, "I'm here. Please don't destroy my home. Care for me." Absent this hCG signal, the uterine lining is shed, and menstruation occurs; responding to the signal, the uterine lining is maintained, and the pregnancy continues. Notably, home pregnancy test kits measure hCG in urine, so if you ever read one of these, you can contemplate the evolutionary beauty of this molecule in another light.

Assuming the conception holds and the uterine lining remains in place, then the placenta will grow as the interface between the mother and child-in-the-making. Humans, like other apes and Old World monkeys, have a hemochorial placenta, meaning one with lots of intricate vasculature to allow nutrient and waste exchange (Martin 2007). This is especially helpful for species that gestate an energy-hungry, big-brained primate, as we and our close kin do. The duration of human gestation, around 270 days, is just a bit longer than for our close ape kin, such as chimpanzees (around 230 days), gorillas (around 255 days), and orangutans (around 250 days), but considerably longer than the lesser apes (around 200 days) or Old World monkeys such as rhesus (around 160–170 days) (Martin 2007). To find a longer gestation, we'd have to look outside of primates, perhaps to African elephants, for which a typical gestation is around twenty-two months.

HUMAN REPRODUCTIVE ANATOMY AND PHYSIOLOGY: HIS

Penises (we'll go with that term, rather than penes) are remarkably diverse. They differentiate so rapidly in evolution that they are some of the most diagnostic morphological traits for assigning males to different species (Eberhard 1996). Given the centrality of the sexual act to a male's reproductive fitness, it should not be surprising that penises will rapidly adapt if that is required to aid the sperm they launch toward a chance at genetic posterity. You can google images of penises from various animals to make this point: you will find penises with spines, penises with complex heads, penises split in two (hemipenises), among many other types.

So where does the human penis fit into this wider natural world? In his marvelous book on comparative primate reproductive anatomy and physiology, Alan Dixson (1999) catalogs human and nonhuman primate variation, and its underlying patterns, in unprecedented detail. We rely heavily on his work for placing the human penis, and other features of men's reproductive anatomy and physiology, in a wider context. With respect to the human penis, many scholars have remarked about how large it is; Dixson, however, is unimpressed. Penis sizes are positively correlated with body size, meaning that larger-bodied primates like us tend to have larger penises. But that is not the full story: some large-bodied primates, like the biggest of all, gorillas, have relatively small penises. Some of the additional variation in primate penis size clusters by mating system: species in which males live in multi-male, multi-female groups have larger penises than males of species that tend to mate with a single female. Human penis size clusters in the single-male mating category.

The same patterning—consistent with single-male mating—holds for at least two other features of the human penis (Dixson 2009): baculum length (baculae tend to be shorter in species in which females mate with a single male—and of course we have lost ours entirely) and penis distal complexity (more complex ends of penises are found in multi-male mating species, and ours is relatively bland). Even without a baculum, the penis can do its job thanks to hydraulic action: blood flowing into the penis during an erection provides the stiffness required for intercourse.

OF TESTES AND LOW SPERM COMPETITION PRESSURES

The penis has a lot of anatomical and physiological support. Developmentally and mechanistically, much of this support is regulated by androgens secreted by the testes (Nelson 2011). This story in males thus parallels that in females: the male gonads (testes) produce both the gametes and the hormones that influence the development of primary and secondary sexual characteristics to assist the gametes in getting from point A (testes) to point B (fertilizing an egg). So let's talk about testes before opening the floodgates to the other traits promoted by androgens.

Within the Sertoli cells of men's testes, millions of sperm are in production (Levay and Baldwin 2009). In a process that kick starts at puberty and
continues through advanced age, albeit with reduced efficiency men produce profligate amounts of sperm. As a classic physiology text puts it, "The
human testis produces approximately 100 million highly differentiated sperm
each day from a pool of no more than a few hundred thousand spermatogo-
nial stem cells, a level of productivity matched in output and complexity
only by the hematopoietic system" (Knobil and Neill 2006: 1196). The process of sperm maturation continues outside the recesses of the testes: sperm
proceed to the epidydymis for continued development and storage. Eventually, at ejaculation, they are rushed through the vas deferens and onward to
the penis for release.

Human testes and the sperm they produce reflect low sperm competition pressures. In other words, our recent ancestors appear to have engaged
in relatively low degrees of multi-male mating. These inferences rest on a set
of comparative analyses. The relative size (adjusting for body size) of human
testes fall within the single-male mating group, indicative of monogamy or
polygyny, but not chimpanzee- or rhesus-like multi-male mating. Humans
have relatively small testicles. And these relatively small testes have lower
Sertoli cell mass, which goes in hand with production of lower sperm concentrations overall. With low sperm competition pressures, males do not need
such large testes because they do not need such voluminous sperm production because their sperm rarely faces off with the sperm of other males inside a female's reproductive tract. Human males also have relatively small
sperm reservoirs, as if expecting ejaculations but a few times a week rather
than, say, a few times a day.

Human sperm are of poor quality (Dixson 2009). Human sperm exhibit
lots of morphological variation, due to the fact many are abnormal. In species with heavy sperm competition pressures, like chimpanzees or many of
the highly polygynous hoofed animals we have domesticated (think of cattle, sheep, and pigs), males can ill afford so many poor-quality sperm. But
humans can, given the low likelihood that one man's sperm would meet
another's in a race to the finish (the egg). Even the pipes—the vas deferens—
that channel sperm to the ejaculatory duct are designed for low sperm competition pressures in humans: they have less smooth muscle, imparting
slower traveling times compared with those in, say, chimpanzees.

While human sperm counts and quality pale beside chimpanzees and
other multi-male mating species, these same qualities stand apart compared

with some other creatures (Birkhead 2000). Some insects have fewer than 100 sperm. Among vertebrates, pipefish males may have as few as 150 sperm per testicle, a contrast with the millions produced in a single human ejaculation (Stolting and Wilson 2007). With pipefish, females deposit their eggs in males' pouches, with males then incubating the fertilized eggs. This form of gamete congress means that males face neither sperm competition nor cryptic female choice. The pipefish thus raises the question: why do human males produce so many sperm, in light of their low sperm competition pressures? We suggest two factors could be most relevant. One is that cryptic female choice may continue to drive sperm counts upward in a back-and-forth female-male mating dynamic. Perhaps females continue to test male quality through chemical and mechanical obstacles that over time drive male sperm counts upward. A second possibility is conflict between gene-based and individual-based selection, yielding intragenomic conflict. In this latter case, "selfish" genes may propagate in order to outcompete other selfish genes within a single male's testes, with the consequence being a higher sperm count than would be expected if males could make organism-level decisions about sperm counts.

But the average human ejaculate contains much more than just sperm—sperm is transported in a marvelously evolved seminal cocktail. That cocktail contains sex steroids, among many other constituents, a mix that has prompted some scholars to ask whether semen might have antidepressant effects (Gallup, Burch, and Platek 2002). So far as other key components of semen, three accessory glands secrete substances that aid fertilization (Levay and Baldwin 2009). The first of these glands, the prostate, surrounds the urethra and ejaculatory duct. As already noted in our discussion of the female prostate, it is sensitive, and it is the primary source of prostate-specific antigen (PSA). While PSA is widely used as a biomarker of prostate growth and cancer, levels of PSA also skyrocket, unsurprisingly, in ejaculate. What is this PSA doing, apart from aiding clinicians in their diagnoses? It appears to play a procoagulatory role, helping the semen stick together, and in turn perhaps helping shuttle sperm further toward an egg that might be waiting in a Fallopian tube.

The other two male accessory glands are the bulbourethral glands and the seminal vesicles. The bulbourethral glands produce a lubricant that may reduce friction during intercourse. While female glands largely solve the problem of sexual lubrication, it appears that male glands kick in a bit toward this effort. The seminal vesicles, meanwhile, produce about 60 percent of the ejaculate volume. What do they throw into the mix? They provide fructose, a quick energy source that may aid the cause of sperm transit; bicarbonates, which help neutralize the acidic vaginal environment; and prostaglandins, which may help foster smooth muscle contractions and modulate immune responses in female reproductive tracts, among other things (Robertson 2005). Incidentally, and consistent with the theme, the size of men's seminal vesicles is relatively modest, indicative of low sperm competition pressures.

New genetic data provide complementary insights to our understanding of human male reproductive anatomy and physiology. The high sperm competition pressures in chimpanzees have led to a more pronounced reduction in Y chromosome variation compared with humans, indicative of greater pressures to remove noncompetitive Y chromosome variants in chimpanzees compared with ourselves (Hughes et al. 2005). While attempts are quite nascent to compare genes implicated in sperm and semen production between species, and to estimate divergent times for genetic variants based on a "molecular clock" approach (multiplying estimated mutation rates by genetic differentiation to infer the time elapsed for those variants to have arisen), such efforts appear promising. As an example, Carnahan and Jensen-Seaman (2008) comment on three seminal proteins (SEMG1, SEMG2, TGM4): "The general pattern that has emerged is a high rate of adaptive evolution in chimpanzees at these genes, presumably driven by strong sperm competition, and loss of function in gorilla and gibbons, with their presumed absence of sperm competition. . . . Interestingly, the observation that humans are in a sense intermediate in their rates of seminal protein evolution between chimpanzees and gorillas parallels that of anatomical and physiological correlates of mating systems" (p. 946). Put another way, if sperm competition pressures have indeed been reduced during hominin evolution, these should show up in comparative genetic research, and potentially can be timed to yield novel insights into the origins of these shifts in mating behavior.

A 2011 attention-grabbing paper in *Nature* pursued similar paths, investigating the loss of an androgen-receptor regulator in human penis morphology related to vibrissae, or small bumps thought to increase sexual pleasure during copulation (McLean et al. 2011). The researchers noted that this regulator was present in rhesus and chimpanzees but absent in humans, and in vitro work indicated its role in vibrissae development. They suggested that the loss of function in vibrissae indicated in this deleted regulator coincided with shifts in hominin mating behavior toward longer-term sociosexual bonds. Although they could not time the loss of this single element, an anthropology blogger, John Hawks, noted that Neandertals had the same version as modern humans, a line of evidence suggesting this putative shift in mating behavior occurred prior to the divergence of modern humans and Neandertals (apart from a smidgen of interbreeding) around 500,000 years ago. As yet one more clever attempt to reconstruct mating behavior based on genetic data, scholars can investigate the phylogeny of the sexually transmitted beasties that inhabit our reproductive equipment. Doing just this for herpes, Glenn Gentry and colleagues found that oral and genital herpes strains had separated long ago and remained that way throughout hominin evolution, suggesting that oral sex was likely not a regular sexual behavior in our hominin forebears (Gentry et al. 1988).

Against a general backdrop of low sperm competition pressures lies a more controversial question: are there any heritable, population differences in male reproductive physiology? That question is difficult to answer. Given that human populations have separated in some cases for more than 100,000 years (a blink of an evolutionary eye, but still a blink within which differences in adult lactose tolerance, resistance to malaria, and other heritable traits have evolved; Cochran and Harpending 2010), it could be that differences induced by cryptic female choice or variable sperm competition pressures have led to slight population-specific variations. There are, for instance, population differences in testis size, penis size (something relevant to condom manufacturers), and sperm counts (relevant to clinicians attempting to assess male reproductive function), but whether these population differences have a heritable basis is not clear (Dixson 2009; Ryan and Jetha 2010). There are also individual, population, and species differences in the vaginal flora of females, with the evolutionary significance of those differences poorly specified (see Ravel et al. 2011). We would not be surprised if some subtle population differences with a heritable basis were to be found, and they would be consistent with the rapidly evolving minefield of male reproductive anatomy and physiology writ large. As one example, in a popular human sexuality book, Ryan and Jetha (2010) posit in a footnote that culturally widespread partible-paternity societies in Amazonia would be expected to exhibit indices of greater sperm competition. That's a testable hypothesis, one that could refine population-specific pressures on human male reproductive anatomy and physiology, but also one that wouldn't change the wider story of human sexuality.

SECONDARY SEXUAL CHARACTERISTICS: BEARDS, MUSCLES, AND MORE

Moving into puberty, male secondary sexual characteristics flourish under the encouragement of testosterone (Knobil and Neill 2006; Levay and Baldwin 2009). The multiple effects of testosterone synergize to promote successful male mating effort by enhancing traits like upper-body musculature, oxygen-carrying capacity in the blood (a good thing for fighting), and beard growth, among other traits. Testosterone also promotes male libido, courtship, and male-male competition within reproductive contexts, indicating that the behavioral effects of this steroid synergize quite nicely with its wider morphological and physiological influences. Let us look at these secondary sexual characteristics a bit more closely in their evolutionary and adaptive context.

Some of the classic photographs of evolutionary master Charles Darwin feature him in his bearded glory. The cofounder of the theory of natural selection, Alfred Russel Wallace, was also known to sport a beard, even proclaiming, "My beard, too, was the subject of some admiration" during his stay in Lombok, Indonesia (Wallace 1890: 128). But nowadays you will find far fewer of the leading male evolutionary thinkers—as E. O. Wilson, Robert Trivers, David Buss—growing beards (wait: Trivers has a goatee). What is going on with beards?

Beards are a male secondary sexual characteristic (Dixson 2009; Geary 2010). They develop at puberty under the influence of testosterone. They can have visible effects that make sense within past contexts of male-male competition and female choice: enlarging the male's face (like a lion's mane?), hiding emotional expressions (do not show your fear grimace to a competitor), indicating male reproductive-age status, and cueing dominance more broadly. In a comparative context, male primates living in polygynous mating systems tend to have more facial visual traits than do male primates living in monogamous or multi-male, multi-female mating systems (Dixson 2009). In other words, the bearded man, the orangutan with enlarged cheek flesh, the mandrill with the brilliantly colored snout, and the large nose of the proboscis monkey are the sorts of traits associated with polygyny. Perhaps the human beard, then, is a legacy of a polygynous past; one might imagine a beard arising in polygynous australopithecines, and being maintained throughout *Homo* within the context of slight polygyny.

If beards owe their origins to sexual selection, then they owe their trimming, removal, or glorification to various cultural factors today and across history. Shaving a man's facial hair may reduce a dominant visage, perhaps helpful when one wants to communicate more visible facial expressiveness and friendliness rather than to hide it; copying whatever high-status males do can also generate beard fashion trends. In an effort to understand changes in facial-hair fashion, Nigel Barber (2001) found, in a longitudinal study in the United Kingdom ranging from 1842 to 1971, that mustaches, and facial hair more generally, were more commonly displayed when sex ratios were male-biased, and that facial hair was less fashionable when sex ratios tipped toward more women.

While we are still into hair, then what about male axillary and pubic hair? Here, many of the same functions would seem pertinent in men as in women: axillary and pubic hair develop under the influence of androgens from puberty onward. Further, these areas in men are rich in sebaceous glands (producing oils to give such hair its sheen) and apocrine glands (producing compounds that, as they decay, leave men smelling like the beasts they are); the axillary and pubic hair capture the oil and scent for public consumption. Men have a different smell than women; it is as if their stronger scent were designed to communicate, "I'm a smelly, adult male: take note, potential male competitors and female mates." This pattern, of men having a stronger scent than women, is consistent with wider primate patterns and sexual selection. Further, it is noteworthy that women tend to have a more acute sense of smell than men, which also fluctuates across the ovulatory cycle and pregnancy, under the influence of estrogen; these fluctuations make women more acutely aware of male scents at times when it may matter more: around her fertile time and during gestation (Nelson 2011). But just as is the case with women and their axillary and pubic hair, sometimes men decide they are better off without it, or at least with less of it. Australian aborigine men often removed body hair, including pubic hair, with beeswax, whereas in the United States today the porn industry is likely fueling a growing trend of male pubic hair removal, as one aspect of what some comics have called "manscaping."

If you cannot smell a male in the forest, then often you can hear him call instead. Indeed, Darwin's (1871) extensive documentation of sexually selected insect, bird, and mammal calls provides a rich tapestry of noises largely generated by males. Male cicadas call to mates, orangutan males call to announce their presence to potential male competitors and female mates, and the male grouse calls to mates by inflating his yellow esophagus. The different vocal tone of men, deepening at puberty (unless men are castrated, as the European castrati singers clearly demonstrate), has different acoustic properties that can signal adult status and dominance (Puts 2010). Put in ecological context, we might expect this deeper male pitch in a polygynous past, when some degree of signaling at a distance to potential competitors and mates would take place instead of face-to-face interactions.

The additional musculature that men pack on at puberty under the influence of testosterone, particularly in their shoulders and chest, is a readily visible secondary sexual characteristic (Bribiescas 2006; Dixson 2009). Men have a much higher percentage of their body composition devoted to muscle than women, with this primarily explained by the extra fat in women and extra muscle in men. The extra muscle also provides men with extra strength. In comparative primate perspective, the extra muscle was likely favored in a context of male-male contest competition, or, quite literally, fighting. The fact that men still carry lots of extra muscle suggests it did not hurt to maintain it through even recent hominin evolution, even as our ancestors' mating systems shifted toward a lesser degree of polygyny. The extra strength might have incidental benefits within a sexual division of labor, in which men do most of the hunting, particularly of larger game. But the wider pattern of sexual dimorphism in hominin musculature would argue against the sexual division of labor selecting for sexually dimorphic muscle mass, rather than simply building on it.

In today's world, male muscle has variable meaning. As much work on male body image has indicated, including research Peter conducted with David Frederick in St. Kitts, a small island in the Caribbean, men often want to be more muscular than they are (Gray and Frederick, 2012). However, cross-cultural data also reveal that Taiwanese men are less focused on extra male muscle than are men in Europe and North America (Yang, Gray, and Pope 2005), a difference that can be linked with differential social constructions of masculinity (wherein muscle matters more or less in different contexts) and with a recently surging use, by athletes, actors, and men in general, of performance-enhancing substances such as synthetic testosterone to gain otherwise unachievable amounts of muscle. Women often find male musculature attractive, though a number of U.S. studies, and in St. Kitts and elsewhere, found that often men want to be more muscular than women want them to be, reflecting conflicts between male-male dominance relationships and female choice. After all, the extra muscle that men have may be an important visual and aggressive path to success among men, but it can also be used against women in contexts of sexual coercion.

With these last thoughts on male muscles, we conclude our tour of human reproductive anatomy and physiology. In our journey through the intricate details of women's and men's reproductive biology, we have explored how their workings can be understood from an evolutionary perspective. Our ancestors' reproductive behavior has left a legacy in our bodies for all of us, from the late sexologist Alfred Kinsey onward to appreciate. Yet our study of reproductive physiology continues: in the next chapter we track the basics of the human sexual response.

CHAPTER EIGHT

Turning the Key

Human Sexual Response and Orgasm

> Muscular spasms or more intense convulsions are the usual product of the sudden release of tension in orgasm. . . . The convulsions following orgasm also resemble those which follow an electric shock. This makes it all the more amazing that most persons consider that sexual orgasm with its after-effects may provide one of the most supreme of physical satisfactions.
>
> —Alfred Kinsey, Wardell Pomeroy, Clyde Martin, and Paul Gebhard, *Sexual Behavior in the Human Female*

NICOLE IS WEARING a crisp white coat and leather boots that click on the floor as she reaches the heavy metal door. She enters the dark room as quietly as possible. The cool black chamber feels sterile, with the exception of the rustling noise, much like the sound of a thousand beating hearts. She crosses the darkness with the ease of someone who has done this before. As she flips the switch, the room is bathed in soft red light. She turns on the video camera. She heads straight to the young male she has chosen. Within minutes he is licking his own genitals. Like other males in the colony, he often self-grooms after finally managing an intromission that leads to ejaculation—which can often take several mounting attempts, as males of his species, much like our own, have a learning curve when it comes to sexual technique.

No, this is not the red-light district of Amsterdam, where one can find hundreds of legalized venues for prostitutes, sex products, peep shows, and marijuana use. This is the sophisticated behavioral neuroscience laboratory directed by Justin's colleague Nicole Cameron. Nicole, a developmental neuroendocrinologist, conducts research with a rodent model and the type of work done in her laboratory and others like it around the world puts us at the forefront of understanding mechanisms of sexual behavior and reproduction. The rats are nocturnal, and so most experiments are done under red light, in a secure facility that minimizes the influence of external variables. Research labs just like these provide the basis for our understanding of the proximate mechanisms involved in sex and reproduction, including sexual receptivity and sexual response. Yet rat sex and reproduction are quite different from what we see in humans. What we know about humans is, surprisingly, limited.

PRECOPULATORY PREPARATIONS

As we discussed in Chapter 3, humans, as well as many other animals, engage in courtship practices. While the term *courtship* refers to the European courts in which royal society was feted, there is a far greater array of courtship practices to be found in nature. These may range from displays of phenotypic quality, such as showing off attractiveness, to gestures of partner investment, such as gift giving. But these acts differ from patterns of precopulatory behaviors. Prior to sex, there exist a variety of precopulatory activities that help stimulate sexual union. These behaviors may include animals engaging in kissing, individuals investigating each other's genitals, humans engaging in heavy petting and mouth-to-genital stimulation, as well as various forms of flirting, play fighting, and erotic "teasers." What are the parallels between a male dog that licks the vulva of a female in heat, a female chimpanzee handling the semi-erect penis of a male, a male monkey that grooms a female before attempting copulation, and a woman surprising her partner by wearing new lingerie? While such behaviors may certainly be pleasurable for the individual performing such acts, they can also be sexually enticing to their partners.

A variety of species engage in some version of what we might call foreplay before initiating copulation (Ford and Beach 1951; Zuk 2003). In mammals, this may involve play fighting, affiliative gestures such as holding or kissing, or inspection of the genitals. A number of mammals display the flehmen response, a smile-like curling of the upper lip, believed to facilitate the reception of pheromones. A male horse may display this response, pointing his nose in the air and curling his lip with teeth exposed, while a mare rubs up against the male and urinates. Anyone who has spent substantial time around animals, particularly on farms or with livestock, will recognize some of the dramatic behavioral changes of females when they are sexually receptive, as well as the behavioral changes among males in their company. Elephants have been known to take their long trunk and stick it in the mouth of another, and males occasionally use their trunk to stimulate the vulva of the female, while females may use their trunk to stimulate the male's penis. Many male rodents stimulate a female's vulva before attempting to copulate, and when the females are in estrous it is common for them to hop about, touching and licking a male's scrotum and penis as a prelude or intermission to copulation. Even some female fish require males to chase them to stimulate the release of their eggs.

When it comes to humans, there is considerable variation in what happens before actual intercourse (Ford and Beach 1951; Gregersen 1994). In some societies there may be extensive precopulatory engagement, while in others, any foreplay may be perfunctory. In some, this is a continuation of courtship practices, and in others, foreplay is explicitly for the purposes of increasing sexual arousal. While the biological process involved in arousal and stimulation are universal, the cultural importance placed on certain acts varies. Across this cultural landscape, let's first consider the roll of kissing. Kissing occurs in some societies but was traditionally absent in East Asian culture, for example; Asians' responses to first encounters of Europeans kissing varied from bewilderment to disgust. Or, consider the overt touching of a partner's genitals to clearly communicate sexual interest: this is one reason why Crow men have been reported to touch a woman's genitals at night while in a public sleeping space. As another display, the physiological and psychological influences of dancing and singing, whether in a group context or one on one, may be channeled to facilitate the onset of sexual behavior.

Across cultures, evidence indicates variable erotic value given to women's breasts, with breasts more closely associated with lactation than erotic value in many societies (Dettwyler 1995b). Ford and Beach (1951) found that stimulation of the breasts prior to or during intercourse was reported among the Apache, Gilyak, and Trobrianders, among other societies. Gorer (1938) describes Lepchas as engaging in minimal precopulatory behavior, without kissing or intimate embracing, although a man might fondle a woman's breast before sex. In these cases, emphasis is put on purposeful breast stimulation. In many cultures touching of breasts during sex occurs, although this may simply be a product of the sexual positions taken—such as breasts being pushed against another's chest during forward-facing intercourse. That said, it is perhaps not surprising that some societies engage in breast play and manipulation prior to or during intercourse. The breasts (particularly the nipples) are among the primary erogenous zones for both women and men (along with the genitals, mouth and lips, buttocks, perineum, inner thighs, naval, armpits, neck, and ears). Breasts, like the other erogenous zones, are laden with dense concentrations of nerve endings, which enhance the pleasure of touch and stimulation. However, several lines of contemporary evidence suggest sexualization of the breast in modern Industrialized context. In a French field experiment women with artificially enlarged breasts (accomplished by wearing different kinds of bras) were approached more often by men while sitting at a table in a bar, and on the street outside the bar (Gueguen 2007). Other cross-cultural research showed that whether or not women concealed their breasts was unrelated to whether breasts were viewed as sexual or important to mate choice (Ford and Beach 1951).

Preceding sex, manual stimulation of the penis, the vulva, or the clitoris may occur. In Kinsey and colleagues' studies (1948, 1953), a majority of men rubbed and handled a woman's genitals, and a majority of women handled a man's penis. However, much less common was oral sex as a precursor to sexual activity. This may take the form of fellatio, cunnilingus, or soixante-neuf ("69"), when two partners simultaneously stimulate each other's genitals by mouth. Kinsey and colleagues found that 10 percent of men had reported performing cunnilingus before marriage, with 49 percent having done this in marriage (these numbers for opposite-sex oral sex did not include the 14 percent of men who said they had performed fellatio on a man or the 30 percent who had received it from another man). Similarly, 19 percent of women said they had performed fellatio before marriage, with this number rising to 46 percent in marriage. Oral sex was not always commonplace in Western societies (especially outside of marriage). This is particularly striking with respect to the types of sexual behaviors we see in the United States today, with the hierarchal reordering of oral sex before intercourse (that is, many young adults have oral sex before their first intercourse experience; Copen, Chandra, and Martinez 2012).

Oral sex in humans appears to play a different function than in many other species. While a nonhuman animal may lick the penis or vulva, this is likely to initiate copulation, and in some cases the behavior may be primarily chemically induced; a female in heat may produce vaginal secretions attractive to a male. In some human populations that engage in such activities, there appears to be a sense of sexual excitement and mutual pleasure; that is, individuals may engage in oral sex activities explicitly to pleasure their partner. For instance, a number of married couples reported engaging in cunnilingus or manual stimulation of the clitoris before, during, and after intercourse to assist women in achieving orgasm (which, in most women, does not regularly occur from penetrative intercourse alone). Ford and Beach (1951) describe several different societies that engage in oral stimulation of the female genitals before coitus. In one example, among the Ponape of Micronesia, they report that "the men stimulate the labia with their tongue and teeth. In this same society the clitoris is titillated by the penis which the man holds in his hand before attempting full insertion. After a period of such foreplay the woman urinates and only then does intromission occur. Another custom reported of the Ponapeans is for a man to place a fish in the women's vulva and then gradually to lick it out prior to coitus."

We note that foreplay is not always just fun and games. In a 1991 medical case report, foreplay resulted in a young woman's death (Eckert, Katchis, and Dotson 1991). The woman was pregnant and she died after her male partner inadvertently pushed air from her vagina into her uterine cavity, resulting in an embolism. Generally, though, the number of sudden deaths that occur during sexual activity is quite low. While heart attacks may happen during sexual activity, they are believed to be fairly rare, more common among men, and to occur more often when men are with prostitutes (Parzeller, Raschka, and Bratzke 1999). As an example of another compromising sexual injury, in one review of clinical case studies, half of the men who went to the hospital with a penile fracture (technically a tissue tear, as humans do not possess a penis bone) had apparently received the injury while having extramarital sex (A. Kramer 2011).

Much of human sexual foreplay, sexual arousal, and sexual responses, occur in the context of a long-term partnership. This has implications for conceptualizing sexual response. We can deduce from the historical and cross-cultural record that many of the modern contexts of sexual arousal are likely novel. Many of us currently live in large social communities, with a surplus of attractive, fertile, individuals—not including use of the Internet, which exponentially increases the ease with which we seek and attract potential mates. Much of human evolutionary history was characterized by living in relatively small groups, where the vast majority of individuals of reproductive age were in long-term partnerships. This likely meant that the relatively few other adults who were around were mostly relatives, or unrelated individuals who were sexually off-limits.

The social context of sexual behavior might also have acted against extensive foreplay, such as within a hunter-gatherer social world: the human propensity for sexual privacy, coupled with kids or other adults being within ear shot, if not within visual range, likely limit couples' ability to carry out prolonged and expansive precopulatory behaviors. Sexual unions might be arranged in the bush, away from prying eyes, as one partial solution to this challenge. Having sex at night offers another outlet, with relatively covert sex undertaken in social sleeping situations, including when children are in proximity. For those having sex side-by-side, relatively quickly, perhaps near a campfire at night, like spouses among !Kung hunter-gatherers, the context acts against prolonged foreplay or oral sex, and might reduce the likelihood of female orgasm taking place.

SEX FOR PLEASURE

While modern Western models of sex emphasize mutual partner sexual satisfaction, that was not and is not always the case. Further, emphasis on male ejaculation and orgasm, and similarly on female orgasm, has varied greatly over time and across cultures. As an example of culturally specified sexual pleasure, the practice of "dry sex" is found in various parts of Africa and the African diaspora, including Zimbabwe and Suriname (Beksinska et al. 1999). Dry sex (not to be confused with the colloquial use of the term to mean body grinding with clothes or underwear on) involves the removal of vaginal lubrication that is produced during sexual arousal. This is often done by wiping out the vagina or using various chemical substances. This is believed to increase sexual pleasure for men, but it is reported as painful for women. Moreover, this practice can result in increased risk of disease. Without vaginal lubrication, vaginal tissue is more likely to tear, increasing the likelihood of disease transmission. This practice also removes lactobacilli, symbiotic bacteria that reside within the vagina (and gastrointestinal tract) and help fight infection. Vaginal lubrication occurs to facilitate intercourse, to prevent damage of the vagina during penetration, and to promote pregnancy. Vaginal dryness can be a complication for some women, particularly postmenopausal women. For this reason, there are a variety of lubricants that can be purchased to simulate or in some cases enhance vaginal lubrication (lubricating lotions and gels with pleasure-enhancing properties).

As mentioned earlier, the *Kama Sutra,* an ancient guide to leading a virtuous life, contains a textbook of eroticism. Particularly popular are the descriptions of tantric sex and various sexual positions designed to maximize mutual partner satisfaction. Similarly, among the Mangaians of the Cook Islands mutual orgasm (*nene*) is sought by both men and women. Marshall (1971) wrote of sex on Mangaia: "The ideal during sexual intercourse is for the male to bring his partner to climax two or three times, to himself hold back until her second or third episode, and then to achieve orgasm at the same time as his partner. Mutual climax is a learned technique and is not always achieved." In Taoist writings well over a thousand years old, sexual intercourse was often described as an act of pleasure that should be prolonged. Men were encouraged to not waste their yang essence in ejaculation but to copulate without ejaculating, and to make women orgasm and release their yin (Hatfield and Rapson 2005). Interestingly, in the evolutionary sense, this would prevent reproduction. However, if one does reproduce with a partner, and provide prolonged sexual satisfaction, this may promote continued maintenance of the pair-bond relationship. This approach is very different from anthropological reports of sexual expression in China near the end of the twentieth century, where "foreplay is often rudimentary or nonexistent—perhaps because many couples keep on as much clothing as possible. Among peasants, 34 percent engaged in less than one minute of foreplay. Partly because of the hasty and purposeful manner with which sex is conducted, 37 percent of wives experienced pain during intercourse. Only a third of urban women and a quarter of rural women said they 'very often' felt pleasure during intercourse" (Hatfield and Rapson 2005). However, those patterns may be subject to change, like other features of sexuality in contemporary China (Parish, Laumann, and Mojola 2007).

Overall, the celebration of sex as a performance for pleasure may be relatively recent. Imagery of individuals in the throes of passion, hollering names in ecstasy, may well be based on performances intended to manipulate a partner more than on typical reactions to sexual engagement; there is a notable lack of loud coital vocalizations mentioned in the cross-cultural record. In an interesting study among sexually active heterosexual women in the United Kingdom, copulatory vocalizations did not necessarily occur during a woman's orgasm (as some pornographic depictions might have one think). Rather, women's copulatory vocalizations typically occurred before orgasm or to coincide with male ejaculation (Brewer and Hendrie 2011). The researchers suggest that some women may use vocalizations during sexual behavior to manipulate male behavior, and possibly speed ejaculation, rather than as an honest indication of ecstasy.

Primate species vary widely in their average intercourse duration, with some nocturnal lorises taking their time, literally, with intercourse duration approaching an hour (Dixson 2009). For the lorises, there are several keys to prolonged intercourse: few other lorises interfere (they are relatively asocial primates), and they have sex when and where few predators threaten (at night, in trees). However, males of most Old World monkeys and apes ejaculate relatively quickly; chimpanzee and bonobo males ejaculate within seconds of intromission (substantially sooner than a "minute man") (Dixson 2009). The typically short duration of chimpanzee and bonobo intercourse, as well as that of many monkeys from Asia and Africa (rhesus monkeys, for example), likely derives from having sex in view of competitors within a multi-male, multi-female mating system. In other words, selection may have favored males who complete intromission and ejaculation rather efficiently, when others are standing by that may attempt to interrupt and prevent ejaculation (and thus deposition of sperm).

Compare nonhuman primate intercourse durations with those of humans, who on average have intercourse for five to ten minutes before males ejaculate—although that duration for intercourse can be substantially shortened or prolonged with certain sexual practices. The typical duration of human intercourse may reflect several factors. As noted above, the human proclivity to have sex in privacy (rather than in broad daylight) places limitations on extended sexual behavior, meaning that taking too long might draw attention even in a camp at night. Having sex in privacy means that other adult sexual competitors are not overtly watching, perhaps seeking to disrupt the sexual union, so completion of sexual intercourse within seconds may not be required. Indeed, given the long-term bond within which sex typically takes place in humans, sex may be somewhat more prolonged to enhance pleasure (even if not orgasm). In this case, sex may be serving as a social means by which to help maintain relationship quality (although this is not a necessary condition for pair-bond maintenance). Although it seems remote, humans' relative buffering from predation provides another reason the duration of human intercourse can be less modest: few humans have to wrap it up quickly because of an approaching leopard or predatory eagle.

The vast majority of human sexual behavior is nonconceptive, occurring at times when a female is not capable of conceiving (during the nonfertile phase of the ovulatory cycle, for example); the relatively high frequency of nonconceptive human sexual behavior may be another means by which relationships are maintained. But the comparative evidence (as in gibbons) indicates this is not a necessary condition for pair-bond maintenance. Humans are not the only animals to engage in regular nonconceptive sex, sometimes in ways that seem to serve social functions. Among bonobos, individuals frequently engage in sexual activity (with same- or opposite-sex partners) in tense situations, such as in food patches when foraging. Sex may serve as a social lubricant in these cases. Perry and Manson (2008) describe related patterns in capuchin monkeys:

> In the Lomas capuchins, the vast majority of sex that we observed was also clearly nonconceptive, involving same-sex pairs, adult-juvenile pairs, or females that were already pregnant or nursing small infants. . . . Our research showed that male-male dyads engage in sex far more often than male-female dyads do, and the majority of male-female sex . . . involves pregnant or lactating females. Male-male social relationships tend to be fraught with more tension than female-female dyads. . . . We also found that almost 80 percent of male-male mounts occur in socially tense situations, compared with only 20 percent of female-female mounts." (110)

Multiple lines of evidence seem to suggest that sexual activity is pleasurable in order to promote reproduction, but this can be co-opted to promote positive social interactions (and encourage conflict resolution).

HUMAN SEXUAL RESPONSE

In the days before PowerPoint (when science presentations were given on actual "slides"), the practice of self-experimentation was a tad more common than it is today. One of the first effective treatments for erectile dysfunction was developed by a researcher (G. S. Brindley) who self-injected various chemicals into the corporal bodies of his penis. Brindley presented his findings, on a series of 35mm slides depicting penile erections, at a 1983 medical urology conference held in Las Vegas (Klotz 2005). Worried that his peers would be skeptical of his findings, that they would think subjects may have been more aroused due to participation in the experiment, he injected his own penis with papaverine in his hotel room before the lecture and donned a track suit for the presentation. During his talk, he pulled down his own pants to show the professional audience his full erection. His point in doing this was to show the new treatment clearly worked, for one could not possibly be sexually aroused while presenting at a professional conference (rather, the stress would make an erection difficult, or so he claimed). His full erection was, in his view, a demonstration of the treatment, without confounds on arousal. That is, he was showing physical arousal (an erection) but was not necessarily cognitively sexually aroused. His studies laid the groundwork for later treatments such as Viagra, the infamous little blue pill that helps men maintain a sex life until their deathbed. And the distinction between an erection and cognitive arousal, as it turns out, has been subject to a large body of fascinating research.

In a seminal study, Julia Heiman (1977) compared the physiological sexual arousal responses of men and women with their subjective experiences of arousal. Participants were asked to listen to a serious of audiotapes, some of which contained erotic stimuli. Both men and women reported the greatest arousal from erotic tapes, compared with romantic. Women appeared to have a greater sexual arousal response to the stimuli than men. Moreover, while men were more accurate in their ability to identify their arousal, women were not. That is, men's subjective experiences correlated well with the presence of a stiff erection, whereas women's psychophysiological responses were often greater than their subjective reporting of arousal. As it turns out, these patterns have been found in many other studies: a meta-analysis found less concordance (a correlation of .26) between women's reports of subjective and "objective" (physiological) sexual arousal than between men's (correlation of .66) (Chivers et al. 2010).

Sexual responses in humans vary tremendously by individual. On the behavioral level, there is variation in what particular actions one might like to engage in and what one might like to receive, and what feels most subjectively pleasurable. Tied to this, on the anatomical level, various scholars have commented on the profound degree of variation in penile and vulvar shape. The substantial individual differences in genital morphology may indicate relatively weak selection pressure acting on genital appearance.

During physical arousal, men often obtain an erection, and women experience vaginal lubrication. But there are several other psychophysiological changes that occur in the body, some of which many individuals are not even aware of. These physiological changes may include increased heart rate and blood pressure, engorgement and elevation of the testes, and swelling along with darkening or reddening of the labia. But if these are only a few of the measurable responses to sex, how can we situate these changes within a wider framework?

In one of the most well-known and influential models of human sexual response, William Masters and Virginia Johnson (1966) distinguish between four phases of sexual response: excitement, plateau, orgasm, and resolution. Their model was built on cutting-edge physiological studies of individuals and couples in a laboratory setting. They recorded responses as varied as male heart rate and female uterine contractions. Through these measures, they were able to document the most intimate of physiological details in unprecedented, real-time, ways. For its time, their work was controversial. In retrospect, had more details of their research methods been shared (the involvement of local prostitutes; some individuals having sex outside of established relationships for the "good of science"), it would have been even more controversial (Maier 2009).

So, what happens, as measured by Masters and Johnson, during the human sexual response? The excitement phase is often characterized by initial arousal, increased heart rate and blood pressure, increased sensitivity, nipple erection, and in some individuals, a "sex flush." The sex flush is a pink or red rash that appears on the breasts and chest during sexual arousal—more common in females, it is occasionally mistaken for irritation or something problematic. In women, there is also breast swelling, clitoral swelling, vaginal lubrication, the labia minora swell and become dark red in color, and the labia majora open. In men, this excitement phase typically includes a fully erect penis, the skin of the scrotum stiffens, and the testes elevate and engorge. The duration of this stage can be highly variable. Many contemporary sex therapists recommend up to thirty minutes of foreplay, particularly for women to reach peak arousal, before attempting penetrative intercourse— one indicator of the importance of timing during sexual excitement.

The next phase, the plateau, is the period between arousal and orgasm, marked by increases in heart rate, blood pressure, and breathing rate, as well as possible involuntary muscle contractions in the feet and hands. This may last a few moments or several minutes—again, sex therapists often recommend prolonging this period as much as possible. In women, the plateau phase is often accompanied by swelling of the areolae of the breasts, continued swelling of the labia, elevation of the uterus, and the withdrawal of the clitoris from under its hood. In men, this phase may include increased sensitivity, further elevation and engorgement of the testes, and pre-ejaculatory secretions.

The third phase in Masters and Johnson's model is the shortest: orgasm. Orgasm consists of heart rate, blood pressure, and breathing rate reaching maximum levels; involuntary muscle spasms; and contractions of the anus. However, particularly among women, there is considerable variation in this phase. Some women have one orgasm before leaving this phase, some may have multiple orgasms, and some remain in the plateau phase without ever reaching orgasm. We discuss the significance of these differences in orgasmic function among women in more detail at the end of this chapter.

Following orgasm is the resolution phase. This is the return to a non-sexually excited state. Heart rate, blood pressure, and breathing rates begin to drop and return to normal, the sex flush fades, and nipple erection subsides. The genitals, once engorged with blood, begin to become flaccid and subside rapidly. Interestingly, while women are physiologically able to return to a state of sexual arousal and even to achieve orgasm again, men experience what is called a postejaculatory refractory period. This may last several minutes, hours, or even days. During this time, men cannot experience another orgasm and have difficulty achieving and maintaining an erection. From an evolutionary perspective, this refractory period in males is likely a by-product of the functional "effort" of ejaculation, with time required to reload (to build up secretions in accessory glands). However, species differences in refractory times suggest that more frequent maters (than humans) may be selected to have shorter refractory periods, thereby facilitating more frequent orgasms.

While Masters and Johnson's work provided measurable physiological details of the human sexual response, it has also been subject to criticism, and alternative models have been advanced in its place. One weakness of the Masters and Johnson model is the focus on the refractory phase, with some researchers feeling it does not warrant its status as a distinct phase of the sexual response (many instead view it as a by-product of orgasm). Another criticism concerns the marked lack of attention to the wider frame of sexual desire. Subjects were assessed in a lab, and there was little attention given to the more typical day-to-day realities of human precopulatory behaviors (such as a romantic outing preceding intercourse, or having sex in a familiar social space rather than in a neutral lab).

In light of these and other criticisms, Helen Singer Kaplan (1979) advanced an alternative model of human sexual response. In her framework, there are three phases of sexual response: desire, excitement, and orgasm. Each of these phases is viewed as having distinct psychophysiological properties. This model casts aside the refractory phase. Kaplan's model also recognizes distinct sex differences (unlike Masters and Johnson's data, which emphasized sex similarities) and embedded typical sex responses within a context of long-term social bonds. As a sex therapist, Kaplan often focused on the difficulties that one might experience in these areas, such as concerns over sexual desire. While this triphasic model has been influential, including in its advocacy for a sexual-desire phase, it later began to fall out of favor (Crooks and Baur 2005). A primary concern was the definition of desire: how is the function of sexual desire distinct from genital changes? In Kaplan's view, desire is important to initiate sexual activity, but it can be viewed as independent from the genital response. However, the considerable cross-cultural and individual variations in the activation of sexual desire (not to mention the sex differences therein) make it difficult to clearly delineate a phase of sexual desire that holds widely.

In a more recent model, Erick Janssen and John Bancroft (2006) have advocated for the dual-control model of sexual response. This model is often described akin to a car, with both a gas or acceleration pedal and a brake pedal. While we have the ability to hit the gas (sexual excitation), we are not always revving the engine, so to speak. Rather, it is the brake (sexual inhibition) that must be more regularly applied to keep sexual response in check. This is what prevents humans from being sexually aroused by stimuli throughout the day. The dual-control model presumes there must be a mechanism to prevent sexual arousal in inopportune situations, such as when one goes to work, socially engages with others, performs routine duties, attends school, or goes to the market. Indeed it would be problematic if individuals were unable to hold a conversation or prepare a meal without being sexually aroused.

Put together, these models of human sexual response have their respective strengths and weaknesses. But collectively they have informed a contemporary science of the sexual response, one that incorporates physiological measurements, the wider social variation preceding sex, the importance of problems treated by a sex therapist, and the recognition that sexual arousal entails activation and must also be prevented from overwhelming other bodily processes (such as sleeping or eating). When conceptualizing human sexual response, then, one might start with recognizing the typical long-term bonds within which sex takes place, then denote the variation in cues and contexts facilitating sexual desire (especially in women), appreciate the physiological changes occurring during sex, remember that orgasm is a variable endpoint for males (although it is typical) and females (orgasm is less consistent, but women are also more capable of being multi-orgasmic), and note that there are both accelerators and brakes on sexual arousal. To this last point, we might add a developmental perspective: much of the social entraining of human sexual arousal and behavior entails placing brakes on sexual desire. For example, as we have seen in previous chapters, children may be variably discouraged from autosexual play (masturbation), older children discouraged from engaging in sexual play, adolescents from intercourse, and so forth, representing a societal channeling of pressures against certain sexual behaviors in specific contexts. The result is a developmentally socialized sexual response that helps attune individuals to context-specific cues.

SEX DIFFERENCES IN SEXUAL RESPONSE

In a tale from ancient Greek mythology, the gods Zeus and Hera (husband and wife) had a dispute over which sex gained more pleasure from sex. Zeus argued that it was women who had greater pleasure from sex, while Hera proclaimed that men must. They turned to the prophet Tiresias, who had experienced life and sex as both man and woman (in fact, another myth even suggests he dabbled in prostitution for awhile). Tiresias, born a man, had been transformed into a woman for seven years after disturbing copulating snakes. He responded to the gods by saying, "Of ten parts a man enjoys one only, but a woman enjoys the full ten parts in her heart." Tiresias's claim suggests that the intensity of sexual pleasure is greater for women and that women are capable of a fuller range of sexual pleasure. Let us also remember the context—in ancient Greece, sex for pleasure and mutual satisfaction was often practiced. Perhaps this is the origin of the question, which inevitably comes up in a human sexuality class: is it men or women that enjoys sexual activity more? Not to disappoint, but we will not answer that question here. However, several studies have shown that when men and women are both asked to describe their orgasm experiences, their descriptions are indistinguishable, with individual differences far outweighing gender differences (Mah and Binik 2002).

One sex difference that has received attention from a variety of disciplines surrounds the issue of female flexibility in sexuality (not in the literal sense of flexibility, although one may argue the point). Researchers such as Roy Baumeister (2000), Lisa Diamond (2008), and Marta Meana (2010), have paved the way for demonstrating sex differences in erotic plasticity and sexual fluidity. They, and others, have contended that, on average, female sexuality is more flexible than that of males. This means that on average women's sexual drives are more facultative than males, responding to environmental and social cues. Male sexual desire is presumed to be more unidirectional and consistent. This helps explain the greater variation seen in female sexual desire and comfort, and may also explain why women's sexuality appears to respond more readily to cultural influence than men's.

Some of the most consistent cross-cultural sex differences (with the largest effect sizes in several meta-analyses) are sex differences in sexual desire (Meston and Buss 2007; Symons 1979; Townsend 1998). On average, men appear to have greater and more consistent sexual motivation than women. Within an evolutionary lens, this makes sense. If women are charged, on average, with being choosier in mate selection because of potential costs, men should, on average, be more motivated to initiate sexual encounters. That is, while in many societies both men and women desire and enjoy sex, men on average have a greater desire. (Notice our overuse of "on average" as a reminder that substantial individual differences always exist). This was further shown in a review of the literature, across studies and methodologies: "Men have been shown to have more frequent and more intense sexual desires than women, as reflected in spontaneous thoughts about sex, frequency and variety of sexual fantasies, desired frequency of intercourse, desired number of partners, masturbation, liking for various sexual practices, willingness to forego sex, initiating versus refusing sex, making sacrifices for sex, and other measures. No contrary findings (indicating stronger sexual motivation among women) were found" (Baumeister, Catanese, and Vohs

2001: 242). This greater desire among men may also be responsible for initiation of sexual activity more frequently. Among Arnhem Land aborigines in Australia, males appear to initiate sexual behavior, even if in reality female and male roles intertwined (given female sexual invitations through flirtation, for example) (Berndt and Berndt 1951).

Despite sex differences in desire and in many of the processes preceding intercourse, the sexes rely on similar mechanisms for sexual satisfaction that are in many ways homologous. This was indeed an emphasis of Masters and Johnson's work, in which they underscored similarities between women and men in blood pressure and heart rate changes (among other variables) during intercourse. While both men and women experience sexual satisfaction, there is both tremendous overlap and variation in experiences. One individual may be rather satisfied with one partner but completely unsatisfied with another. At the same time, there are significant individual differences in arousal and sexual satisfaction. Physiological and anatomical variation can also result in within-sex variability in stimulation during certain sexual activities. What turns one man or woman on does not work for everyone.

OVULATORY CYCLE SHIFTS

Another sex difference in sexual response is fundamentally tied to women's ovulatory cycles. The hormonal and broader physiological changes occurring across the ovulatory cycle serve several functions: preparing an egg for release at ovulation and preparing a nice home for potentially fertilized eggs, but also rebooting the system for another cycle if conception does not occur (resulting in menses). Employing Beach's (1976) scheme, researchers have asked whether women undergo changes in their proceptivity (taking sexual initiative), receptivity (responding favorably to a sexual invitation), and attractivity (erotic value to viewers, including potential mates) across the ovulatory cycle. Of further interest has been whether any potential changes in these variables relate to ovulatory cycle shifts in sexual behavior. An adaptive, evolutionary perspective suggests that women might experience maximal proceptivity, receptivity, and attractivity around the fertile phase of the cycle (several days leading up to and during ovulation), perhaps under the influence of estrogen and androgen, when that enhanced sexual behavior may translate into reproductive success. Research on female rodents demonstrates that sexual receptivity is tightly hormonally controlled, with female rats only fully sexually receptive around the time of ovulation (Blaustein and Erskine 2002), but this is generally less true of Old World monkeys and apes (Dixson 1998).

In a review of research on ovulatory cycle shifts in female measures of sexual behavior, Steven Gangestad and Randy Thornhill (2008) found that measures of female attractiveness (such as facial features) are often rated highest during the fertile phase of a woman's ovulatory cycle. Moreover, maximally fecund women also tend to express slightly higher preferences for masculine traits in males (such as robust jaws or broad shoulders). One of the most provocative studies along these lines was conducted with a small sample of topless lap dancers at a New Mexico strip club. Miller, Tybur, and Jordan (2007) found that dancers earned higher tips around the fertile phase of their cycles, though it was not clear whether this represented greater female attractivity or subtle behavioral differences (receptivity) among dancers. Less dramatic, but still revealing, a study of eighty-eight female undergraduates found that around the time of ovulation women favored wearing outfits that displayed more skin (Durante, Li, and Haselton 2008). In two naturalistic studies conducted in France, Nicolas Guegen (2009a, 2009b) found that around the time of ovulation women were more receptive to solicitations from men—whether to dance at a club, or to give their phone number to a stranger at a park. Put another way, women in these two studies displayed higher receptivity around the fertile phase of the ovulatory cycle.

While the documentation of cycle-related shifts in female sociosexuality is of interest, the findings also need to be placed in wider context. The typical effect sizes are quite modest, meaning that effects in nicely designed studies can sometimes be found, but that these do not necessarily represent adaptively attuned fluctuations (unlike the exaggerated sexual swellings of a chimpanzee or bonobo). Additionally, several studies make assumptions of "typical" menstrual cycles, yet the best biological and cultural data suggest enormous endocrine and functional variation in women's ovarian cycles across environmental settings (see Vitzthum 2009). Thus, we must be careful in suggesting these behavioral shifts are necessarily adaptations, rather than by-products of other physiological changes across a woman's cycle.

Another important consideration is the interpretation of cycle-behavior effects. In their review, Gangestad and Thornhill (2008) suggest, "Women's sexual preferences for certain male traits probably connoting male genetic quality (or ones that did ancestrally) appear to be enhanced when women are fertile in their cycles" (996). An equally viable and arguably more compelling explanation (for a long-lived, low-fertility, social animal such as ourselves) could be that females experience slightly elevated proceptivity, attractivity, and receptivity around the time of ovulation, enhancing their sense of well-being and mate value and translating into greater attention given to sexually desirable alternative mates. Put another way, women during their fertile phase may be slightly more inclined toward an alternative prospect compared with their ongoing mate. To date, research on ovulatory cycle shifts in women's behavior have reported relatively minor influences on actual changes in sexual behavior. One large study reporting on partnered women's recent sexual activity in thirteen countries found no differences in intercourse frequency around the time of ovulation compared with times before and afterward (Brewis and Meyer 2005a). The most precise of well-designed studies, including hormone assessments, sometimes find increases in partnered women's sexual behavior around the time of ovulation, but these tend to be quite subtle rather than dramatic increases (Wilcox et al. 2004). To be sure, whatever one's preferred interpretation of such ovulatory cycle studies, this is an exciting area of active research, and the debates will be ongoing for some time.

Speaking of debates, what might be the impact of women taking hormonal contraception (such as the Pill) on potential cycle-related shifts in sociosexuality? While there are variable dosages and formulations of hormonal contraceptives, a common feature of many of them is that they include chronically dosed progesterone and estrogen during the active hormone days, in turn preventing ovulation and yielding a more steady (rather than fluctuating) hormone profile of sex steroids. Consequently, women on the Pill do not experience the elevated hormone levels during a fertile phase, and the progesterone content communicates a continued nonreproductive state (as in the luteal phase or even early-stage pregnancy). In a review of relevant studies, Alexandra Alvergne and Virpi Lummaa (2010) found that women on the Pill did not show many of the same cycle-related fluctuations in attentiveness to male sexual stimuli as normally cycling women, an indication that anticipated effects are realized. In Geoffrey Miller and colleagues' study of topless lap dancers, mentioned above, those on the Pill also had flatter tip profiles across their cycles than normally cycling women (Miller, Tybur, and Jordan 2007). Suggesting that such effects may hold more widely, a study of captive lemurs in which females were given hormonal contraception resulted in those females garnering less male sexual attention, in part because female lemurs on hormonal contraception had a quite different smell to them (and if anything, lemurs like to smell each others' secretions) (Crawford, Boulet, and Drea 2010).

BIOLOGICAL RESPONSES TO SEXUAL ACTIVITY

Some of the most intriguing studies of human sexual response have sought to understand the physiological mechanisms involved. We have seen how Masters and Johnson's work helped launch our understanding of the physiological changes taking place during sexual behavior. But the availability of new technologies has spurred new questions and an expanded grasp of the neuroendocrine effects associated with sexual behavior. Studies range from investigating psychophysiological changes during masturbation in a laboratory, to observing effects of an orgasm during an fMRI brain scan, to our own hormone studies conducted on subjects in a Las Vegas sex club. Here, we present some of these newer studies to help us understand the details of human sexual response.

Sexual arousal is much more than erections, vaginal lubrication, and penetration. In Erick Janssen's volume *The Psychophysiology of Sex* (2007), contributors raise a variety of methodological and theoretical points in understanding sexual arousal and responses to sexual stimuli. Some of the most consistent features of these studies are individual differences in responsiveness, and the discrepancy between physiological measures and subjective experiences. These studies also apply a rich set of laboratory techniques. In studying male arousal, a common measurement is the strain gauge:

> Originally described by Fisher and colleagues (1965), the first circumferential measure, the mercury-in-rubber strain gauge, was adapted from a similar transducer used by Shapiro and Cohen (1965). The device consists of a hollow rubber tube filled with mercury and sealed at the ends with platinum electrodes. The operation of the mercury-in-rubber strain gauge depends

upon penile circumference changes that cause the rubber tube to stretch or shorten, thus altering the cross-sectional area of the column of mercury within the tube. The resistance of the mercury inside the tube varies directly with its cross-sectional area, which in turn is reflective of changes in the circumference of the penis (Janssen, Prause, and Geer 2007).

In women, arousal may be studied using such assessment tools as a labial thermistor (to measure temperature of the labia) or, more commonly, a vaginal probe (vaginal photoplethysmography):

> Embedded in the front end of the probe is a light source that illuminates the vaginal walls. Light is reflected and diffused through the tissues of the vaginal wall and reaches a photosensitive cell surface mounted within the body of the probe. Changes in the resistance of the cell correspond to changes in the amount of back-scattered light reaching the light-sensitive surface. It is assumed that a greater back-scattered signal reflects increased blood volume in the vaginal blood vessels (Levin 1992). . . . The vaginal photometer is designed so that it can be easily placed by the participant. A shield can be placed on the probe's cable so that depth of insertion and orientation of the photoreceptive surface is known and held constant. (Janssen, Prause, and Geer 2007)

Studies using these techniques have demonstrated men's and women's psychophysiological responses to sexual stimuli, such as responses while watching erotic films. As discussed earlier, while men and women consistently respond to sexual stimuli, they are not always cognitively aroused. That is, the body may respond in a sexual way, but that does not imply that an individual is feeling sexual or currently desiring sexual activity.

To demonstrate this, Meredith Chivers has conducted a series of studies on men's and women's psychophysiological responses to a variety of stimuli. While participants do watch the traditional erotic videos, she has also had participants view recordings of nonhuman primates copulating. In one study Chivers conducted with Michael Bailey, they found sex differences in genital arousal to cross-species sexual stimuli (Chivers and Bailey 2005). Men did not show any genital response to films of bonobos having sex, but did respond with genital stimulation to films of humans having sex in accordance with their own sexual orientation (i.e., heterosexual men were more aroused by sex featuring female participants). Women, on the other hand, were slightly more complex. Women showed small genital responses to the videos of bonobos having sex, and large genital responses to the human videos, regardless of their sexual orientation preference for the particular videos. This adds further support to the idea that women's genital arousal is a more general autonomic response to sexual stimuli, with cognitive arousal operating independently.

In another example that many find upsetting, some women who have been raped report experiencing vaginal lubrication (and in some cases orgasmic responses) during the sexual assault (Levin and van Berlo 2004). This in no way means that the woman desired or wanted this sexual activity, nor does it mean that she found it cognitively arousing. It simply means her body responded to sexual activity. In fact, in cases of sexual coercion and rape, a women's physiological response of vaginal lubrication, despite lacking cognitive arousal, likely prevents severe internal pain and damage, as vaginal penetration without lubrication can inflict further harm. Similarly, despite his personal sexual preference, a male receiving fellatio may have an erection and achieve orgasm regardless of whether he is receiving oral sex from a man, woman, or nonhuman animal. It is important to keep in mind that physiological responses to sexual stimuli need not represent desire or consent.

To investigate the hormonal correlates of copulation in humans, in a study with our colleague Michelle Escasa-Dorne we analyzed the salivary testosterone and estradiol levels of women at a popular sex club in Las Vegas (Garcia, Escasa-Dorne, and Gray, forthcoming). Justin and Michelle collected data between 11:00 p.m. and 5:00 a.m., arriving at the club every night with clipboards, an ice-filled cooler for saliva samples, and a lack of humility to conduct scientific interviews while people walked by dressed in role-play outfits and engaged in sexual behavior around every corner. Women at the sex club were interviewed when they arrived and again on their departure. Women provided a saliva sample with each interview and detailed information on their history and the activities they engaged in while at the club. Beyond giving us an eye-opening cultural experience, this study added to the very small body of literature on women's reactions to partnered sexual activity. We found no testosterone increases among those women who engaged in sexual activity. Surprisingly, we also found that testosterone did not increase more among those women who experienced orgasm, with number of orgasms (among those who were multi-orgasmic) resulting in a dosagelike effect on testosterone levels. We also looked at salivary estradiol. There were likewise no substantial increases in estradiol in any of the conditions—not among women who were only viewers, those who engaged in various sexual activities, or those who had an orgasm. Interestingly, this may suggest a complex role of testosterone in responding to sexual activity, and estradiol may play more important roles in priming baseline sexual interest and impacting fertility downstream (the potential results of vaginal intercourse). These findings both conflict and correspond with those of previous researchers with respect to testosterone, but conflict with some studies with respect to the role of estrogens in human sexual behavior. In a series of older studies, researchers found little to no relationship between acute changes in estradiol and sexual behavior (either masturbation in the lab, or sex after dinner at home), but many did find a relationship with testosterone changes and sexual activity (Exton et al. 1999; Heiman 1991; Persky et al. 1978). More recently, Sari van Anders has undertaken a series of studies to better understand the endocrine aspects of human sexual behavior. In one study of women, vaginal pulse amplitude and salivary hormone levels were measured while subjects masturbated in the laboratory (van Anders et al. 2009). In this study, women experienced vaginal arousal, no changes in testosterone, but increases in estradiol and decreases in cortisol. Further research on the hormonal responses to human sexual behavior is clearly needed to help address some of the disparate findings; methodological advances (such as being able to measure steroid hormone levels from saliva) will certainly help fuel some of these next steps.

If bringing the laboratory to a sex club seems strange, researchers have also continued to bring sex into the lab. In a series of entertaining and enlightening studies, Barry Komisaruk, Beverly Whipple, and their colleagues have looked at what happens in the brain during the experience of orgasm (Komisaruk, Beyer-Flores, and Whipple 2006). To do this, they have put women into an fMRI device and asked them to achieve an orgasm, typically by masturbating to climax. A related study conducted in the Netherlands relied on positron emission tomography (PET) scans to identify changes in women's brains while their male partners stimulated their clitoris to the point of orgasm (Georgiadis et al. 2006). These studies have shown that during sexual stimulation and arousal, a woman's brain is heavily stimulated by sensory input, and literally lights up the computer screen. Right at orgasm, though, there appears to be decreased blood flow and behavioral disinhibition (reduced ability to manage a response). Orgasm appears to have complex physiological properties throughout the body and nervous system, particularly in women.

ORGASM

Kinsey defined orgasm as "the explosive discharge of neuromuscular tension and the peak of sexual response." For many, orgasm is often considered the grand finale of sexual activity. For males that makes some evolutionary sense, since (apart from the pre-ejaculate sperm that may, albeit very rarely, result in fertilization) our male ancestors reproduced by orgasm and ejaculation. For females, the evolutionary case is subject to considerable debate. But evolutionary bottom lines aside, plenty of people enjoy sexual activity without experiencing orgasm. As discussed earlier, a number of cultures prize the ability to have prolonged sex without the cessation typically associated with (particularly male) orgasm. And, too, the number of U.S. women who regularly orgasm from penetrative intercourse alone is rather low. While 90 to 95 percent of Western women have experienced orgasm at some point, a slightly smaller percentage has experienced it during intercourse. It is estimated that roughly one-third of women rarely to never experience orgasm with intercourse, while only 25 percent of women reliably experience orgasm during penetrative intercourse, including those who engage in self- and partner-assisted manual stimulation of the clitoris with intercourse (see Lloyd 2005).

Orgasm frequencies likely vary considerably cross-culturally (without even counting, for example, the extreme case of women who are subject to clitoridectomy). On the one hand, some women experience orgasm in unexpected places, as in a recent study documenting "coregasms," or orgasms that women experience during exercise, typically from stimulation of the core abdominal muscles (Herbenick and Fortenberry 2012). On the other hand, orgasmic "dysfunction" is often treated as a clinical condition (many sex therapists, and, one might assume, many evolutionists, disagree with the very notion of classifying variation as "dysfunction"). Gynecologists may advise women who have difficulty achieving orgasm that this may change over time, with a new partner, or with concentration. Interestingly, recent studies have suggested that both men and women orgasm more in committed relationships than uncommitted sexual encounters (hook-ups), perhaps again pointing to benefits of long-term pair bonds (Armstrong, England, and Fogarty 2009, 2012; Garcia et al., in review).

Anorgasmia (the inability to orgasm) can often be treated with simple education to teach a woman and/or her partner about how genital anatomy functions and, particularly in the case of women, to empower her to focus on achieving her own pleasure and orgasm during sexual activity (Heiman and LoPiccolo 1988). And speaking of genital anatomy and orgasm: women in one study reported having greater sexual pleasure and orgasm frequency with an "intact" male partner than with a circumcised male partner (Ohara and Ohara 1999). With so much variation in women's orgasmic function, and so many factors involved, the best evidence to date would suggest that women's orgasm is unlikely to be a trait under strong selective pressure.

Women can also fake orgasm, as the famous scene in the movie *When Harry Met Sally* (1989) vocalized. Both men and women respond with greater pupil dilation from sexual auditory stimuli than non-sexual stimuli, indicating that sexual sounds can cause arousal even without visual input (Dabbs 1997). Loud female vocalizations during sex are more likely to occur in multi-male, multi-female species, such as rhesus macaques (Dixson 2009). The functional rationale is that a female's loud sexual call may draw the attention of other males, with whom she may subsequently mate; this pattern is consistent with females seeking to "confuse paternity" of the various males in the group, thereby maybe ensuring they all will be a bit kinder, or at minimum a little less nasty, toward her future offspring (Hrdy 1981). In humans, however, loud female copulatory vocalizations are not reported in the ethnographic literature (to our knowledge), and they would seem highly inconsistent if they did, in light of other features of human sexual behavior (such as a predilection for having sex in privacy so as not to attract attention of others). Instead, where they occur, human female copulatory vocalizations may function to signal pleasure to a partner, making a partner feel special in his (or her) ability to provide sexual services, whether women are faking or not. In a recent study looking at women who fake orgasm, this appeared to be the case. Women who felt more insecure about their partner's fidelity were more likely to fake orgasm as a mate-retention strategy (Kaighobadi, Shackelford, and Weekes-Shackelford 2011).

What are the origins of orgasm? In men (and nonhuman primate males), orgasm is coupled with ejaculation and reproduction. The link is quite apparent. In women, however, no such relationship appears to exist. In females of other species, from Japanese macaques to bonobos, the capacity for orgasm clearly exists and appears to be quite regularly realized (Dixson 1998; Hrdy 1981). Female orgasm may serve to enhance female-female social bonds, as in bonobos, providing a social-selective pressure to its occurrence. Or it may be a by-product of the orgasm-reproduction link in bonobo males, one that can be employed in labile social contexts. Thus, even for nonhuman primates the evolutionary case for female orgasm remains uncertain.

Yet, whether the female orgasm is, itself, an adaptation remains hotly debated by evolutionary sex researchers (see Lloyd 2005; see also Puts, Dawood, and Welling 2012). In a recent review that advocates an adaptationist stance, Puts, Dawood, and Welling (2012) present the debate succinctly: "There are two broad competing explanations for the evolution of orgasm in women: (1) the mate-choice hypothesis, which states that female orgasm has evolved to function in mate selection and (2) the by-product hypothesis, which states that female orgasm has no evolutionary function, existing only because women share some early ontogeny with men, in whom orgasm is an adaption."

Women's orgasmic history does not appear to be consistently associated with fertility, number of offspring, lifetime sex partners, or any other meaningful metric of reproductive fitness (Lloyd 2005). Elisabeth Lloyd has argued that among the nearly twenty arguments proposed for the origin of female orgasm, the most likely is Don Symons's (1979) argument that it is a by-product of selection for male orgasm; the primary evidence for the by-product view is developmental and rests with the relative consistency of men's orgasm (suggesting selection acting on the trait in men) along with the enormous variation in occurrence among women. Now, to any scientist with feminist leanings (such as Lloyd, and ourselves), this is not a very satisfying conclusion. Yet it is the one with the best empirical support. While a number of theories have proposed that female orgasm is important for fertility and sperm retention (known as the insuck hypothesis), mate choice, and pair-bonding, the evidence is inadequate to label women's orgasm an adaptation. It is possible that future evidence will better suggest that female orgasm is an adaptation, but the existing evidence does not support such a conclusion. As Lloyd argues in the final words of her book: "The history of evolutionary explanations of female orgasm is a history of missteps, misuse of evidence, and missed references. The case is still open, and it is ripe for some good scientific work." (2005: 257).

In the open case of female orgasm, Kim Wallen and Lloyd (2011) have pushed forward further insight into its variable nature. They show from two archived data sets that slight variations in female reproductive anatomy are differentially related to orgasm frequency; specifically, they find that women with shorter distances between their clitoris and urethral meatus (called the CUMD) report more orgasms. They suggest this association traces to greater capacity for clitoral stimulation due to the shorter distance, and they further propose that prenatal androgen exposure is likely related to the distance (with higher prenatal androgens associated with a longer distance, and less orgasmic frequency). That is precisely the kind of work indicating that the continuing story of the female orgasm will remain interesting, whether one accepts the by-product argument or not.

Winding down our discussion of human sexual response, we are now prepared to put this knowledge to use: with a focus on reproduction. We shift from appreciating the finer details of sexual behavior—from cross-cultural variation in courtship to arguments about female orgasm—to our next chapter, focused on the evolution of human reproduction. In this transition, we find that intimate link between sexuality and fertility, and the ways in which sexual passions may (or may not, in the event of fertility controls) be translated into reproductive successes, or offspring.

CHAPTER NINE

The Evolution of Baby Making

Mechanisms of Fertility, Infertility, and Variation in Fertility Outcomes

> As in all apes, the successful rearing of [human] young was a challenge. Mortality rates from predation, accidents, disease, and starvation were staggeringly high and weighed most heavily on the young, especially children just after weaning. Of the five or so offspring a woman might bear in her lifetime, more than half—and sometimes all—were likely to die before puberty. Unlike mothers among other African apes, who nurtured infants on their own, these early hominin mothers relied on groupmates to help protect, care for, and provision their unusually slow-maturing children and keep them on the survival side of starvation.
>
> —Sarah Hrdy, *Mothers and Others*

WHILE TEACHING IN Singapore over a couple of summers, Peter was impressed by the ease of traveling with his own children there, but also marveled at the relative dearth of kids. Singapore, like much of East Asia and Europe, has watched its fertility rate drop precipitously in recent decades. There are fewer and fewer babies. Indeed, by some estimates, Singapore's female fertility rate of around 1.1 babies per woman ranks as the lowest of any country on earth, with Japan, Korea, Italy, and Spain some of the closest competitors (Singapore Statistics 2011; United Nations 2008). Why are people in these countries having so few kids? Have people given up sex? Are people having sex but managing their fertility in order to have smaller families—if they are having kids at all? Casual conversation and observation, as well as more rigorous analysis, suggest that a host of factors are playing into the fertility lows in Singapore (Singapore Statistics 2011).

Women and men today spend more years garnering an education that will enable them to find jobs in the competitive global labor market, while also facing high housing costs and overall costs of living, factors eventually leading to late ages of marriage (on average, at age twenty-seven for women and thirty for men). The expense of rearing children favors smaller families. With readily available contraception, couples are able to keep their families small. Yet it still takes substantial parental investment to raise a few kids who might enjoy social success.

In this chapter, we address a host of issues surrounding sex and human reproduction. We consider the ways babies are made, both with respect to cross-cultural views of conception and the current biomedical understanding. We discuss the ways people control fertility, from the historical use of plants to an array of contemporary products such as the Pill. We touch on ways people respond to having fewer kids than they want to have, focusing on the causes of and responses to infertility. To put the human reproductive process in comparative and evolutionary context, we consider fertility patterns among our ape relatives, as well as fertility patterns among more recently studied human hunter-gatherers. Finally, we scan some of the past and contemporary human fertility highlights, from the record-setting twentieth-century Hutterites to the current U.S. fertility rate of around 2.1 babies per woman. There will be plenty of both procreative and recreational sex along the way, not to mention the deeper appreciation we will gain for the ways evolutionary pressures make sense of our ongoing patterns or reproduction.

MAKING BABIES

While tales of storks dropping off babies make for fun children's stories, a question anthropologists have long grappled with is whether people around the globe believe that sex is necessary for reproduction. This is a question of cross-cultural conception beliefs—how people believe babies are made. The issue has garnered the greatest of attention among those studying parts of Melanesia, New Guinea, and Australia. As an illustration, the renowned anthropologist Bronisław Malinowski (1929) contemplated at length whether Trobriand Islanders grasped the physiological basis of conception. Trobrianders told him that conception occurred when a spirit child entered a woman's head, was met by her blood, and was eventually ushered to her womb for development. This story of course leaves out intercourse, which led Malinowski to ponder, "Are the natives really entirely ignorant of physiological fatherhood? Is it not rather a fact of which they are more or less aware, though it may be overlaid and distorted by mythological and animistic beliefs?" (153).

Further probing of Trobrianders suggested that they believed a woman must be open to conception; to open that possibility, sexual intercourse was required, a reason why virgins were not capable of conceiving. Yet once a sexually experienced woman had been opened to conception, some Trobrianders insisted she could conceive without intercourse, mentioning some sexually unattractive women who had borne children but who they believed did not have intercourse. So by this account, in one of the best-studied anthropological cases of conception beliefs, it was believed that baby making could occur without a father or intercourse. A wider swath of research, however, has planted doubts about generalizing this conclusion.

Among the Tiwi of North Australia, for example, a man must dream of a spirit child first and then have intercourse with a woman, for conception to take place (Goodale 1971). The regionally prevalent "spirit child" conception belief in Melanesia, New Guinea, and Australia thus does not provide a firm, overarching answer to whether local peoples acknowledge a physiological basis for conception. In some societies those links are made, but not in others. Yet there are several other take-away points to this discussion. The regional scope of this belief indicates cultural transmission during some period in the distant past, though it is not clear how distant. Another point is that variation within the region merits closer attention. A spirit-child belief would be adaptive within a matrilineal and matrifocal (mother-centered) system, which characterized the Trobrianders and some other groups in the region. In a matrilineal system, with lower paternity certainty and greater weight resting on a mother's brother for investment in her kids, a spirit-child belief gives the mother greater power to socially designate paternity, either shunning it altogether (she knows with whom she has had sex but may not want them as ongoing suitors, and at any rate she has her brother's resources available) or allocating it as she sees fit. Conversely, this belief system would not be sensible within a social world characterized by patrilocality (mother and offspring living with the father's parents), high paternal investment, and high paternity certainty. We hypothesize that within the region where spirit-child beliefs are held, these are more likely to be entertained among matrilineal societies than patrilineal societies, and among matrifocal societies than patrilocal ones, though for now this hypothesis remains untested.

A wider cross-cultural investigation of conception beliefs found that in 62 percent of societies, a physiological basis to conception was recognized (Frayser 1985: 286). These beliefs typically recognize the importance of bodily fluids, sometimes giving primacy to those from a female, sometimes to those from a male, and in other cases appreciating the combination of male and female components. Among the Kgatla of southern Africa, for example, Schapera (1941) notes, "Young children, on inquiring where a baby comes from, are usually told that its mother fetches it from a pool in the river. But . . . all adults, and most of the other children, are well aware that conception is due to sexual intercourse. . . . They themselves hold that a child is formed in the womb by a mixture of the man's semen and the menstrual blood of the woman" (215).

As Frayser (1985) writes, "In all of the societies that believe that conception results from a mixture of body fluids, the male element is semen, while the female contribution may be blood, menstrual blood, or vaginal fluids. Most of these ideas about conception also explain the absence of the menstrual flow after conception" (287). So the Kgatla conception belief described above is quite representative of such views. Since a woman's menstrual flow is a more obvious indicator of her ovulatory cycling and pregnancy status than changes in an egg or ovary in her abdomen, it also makes sense that many conception beliefs draw on her blood. There is also an implicit recognition that conception occurs among cycling, menstruating women rather than, say, nonmenstruating women (who might be pregnant, subject to lactational amenorrhea, or infertile). In most of the remaining societies in Frayser's sample, sexual intercourse is viewed as a necessary but insufficient precursor to conception. In such accounts, intercourse is typically complemented by supernatural forces, such as spirits, meaning that the wider context must be appropriate for conception to occur.

Among cross-cultural conception beliefs, another pattern is that frequent intercourse is commonly viewed as a requirement of conception, meaning that sex between long-term partners rather than a one-time fling is a prerequisite to making a baby (Frayser 1985; Gregersen 1994). Among Solomon Islanders, for example, repeated ejaculations are believed to fill up the womb, enabling male fluids to coalesce with menstrual blood and thus beget a child. A belief that conception requires frequent intercourse reinforces the importance of long-term sociosexual relationships as the typical context within which a child is conceived, and within which it may be cared for by its parents over the long haul.

A conspicuous feature of cross-cultural conception beliefs is that virtually none highlights the importance of midcycle female fecundability. In other words, almost no conception beliefs pinpoint the days when a woman actually can conceive as the ones during which conception is especially likely. Among the Hadza foragers of Tanzania, about half of adults believe conception occurs shortly after menstruation, while the others believe that the timing of conception is spread out across the remainder of the ovulatory cycle (Marlowe 2004). In a rare exception, however, among the Huli of the New Guinean highlands, couples are supposed to restrict their sexual activity to days eleven to fourteen of a women's ovulatory cycle (Wardlow 2008). Traditionally, Huli spouses were spatially segregated and women were viewed as polluting to men; thus, in a society where contact between ambivalent spouses is designed to be minimized, they were encouraged to have sexual intercourse only during the time of optimal fecundability.

Why do most conception beliefs fail to recognize female midcycle peaks in fertility At face value, the lack of such a connection suggests that human ovulation may be "concealed." After all, any alterations in female mood and sexual behavior taking place during the fertile phase are sufficiently small that they have not shouted out for attention, demanding to be linked with conception beliefs. Yet sometimes, in some of the best-designed studies, shifts in female libido and sexuality across the cycle, including peaks during a fertile phase, can be detected, as we discussed in the previous chapter. Those shifts tend to be subtle rather than pronounced. They are more like a rising tide than a tidal wave.

There may be another reason for the lack of a link between women's fertile phase and social conception beliefs: the right experiment has rarely been conducted. Most women and men have sex within the context of long-term relationships, and within those relationships their sexual behavior is spread fairly evenly across the ovulatory cycle except during menstruation (Brewis and Meyer 2005a; Ford and Beach 1951). The interaction of a woman's subtle cycle-related changes in libido with a male partner's more static libido typically yields relatively constant coital behavior across her cycle. These circumstances can prevent observers from isolating the fertile phase from, say, the early follicular or the late-luteal phase in terms of the differential conception risk. People would likely need to have sex less frequently to generate the possibility of associating sexual behavior at specific times in the cycle with conception (as the Huli appear to have prescribed). Accordingly, the typical conception beliefs highlight the fact that a woman needs to be cycling (as indexed by menstruating) for conception to occur, but leave out of that picture when she might conceive while cycling.

BIOMEDICAL BASE FOR CONCEPTION

A biomedical understanding of conception not only clarifies the accuracy (or inaccuracy) of traditional conception beliefs but also presents the glorious physiological details of conception. These details are remarkable for the intricate design that they reveal, the product of natural selection's fine-tuning over aeons of evolutionary development. While Charles Darwin and more recent evolutionary thinkers, such as Richard Dawkins, have marveled at the ability of natural selection to fine-tune eyes, including a structure as acute as an eagle's eye, the reproductive process deserves its own accolades. It has taken dissections, microscopes, and experiments to pull back the covers on the mechanisms of conception. Further, if you never feel lucky, then having a scientific grasp of conception will convey a deeper sense of fortune than any of us normally enjoys. Among the millions of oocytes that your mom once developed during her fetal life, she ovulated a relatively small number, one of which was helped to become you; among your father's millions of sperm, shipped off during a primordial exchange, one reached that egg and helped become you too. An egg and sperm met, fertilization occurred (yet another feat), and you are proof of the consequence.

To reinforce your good fortune, just consider all the ways in which conception could have been avoided. Suppose intercourse occurred during a time when conception was not feasible, say while your mom was already pregnant, or during a nonfertile phase of her cycle. In those cases, there would have been no egg making its way from your mom's ovary, and your father's sperm would have made their last squirming movements caught in a hostile vaginal environment, with the cervix possibly blocked by impassable mucus (which is a good thing during pregnancy, helping keep out potential infectious agents). Or suppose that your mom was not at a fertile age, perhaps during the period of adolescent subfecundity or in the years of reduced fecundity leading up to menopause. Consider how many of your mom's oocytes or your dad's sperm can now live only vicariously through you (too many to count), with just you and perhaps your siblings as the small reproductive legacy of their gamete profligacy. Or contemplate the fact that an estimated 40 to 50 percent of conceptions are spontaneously aborted, meaning that conception is hardly a guarantee of birth, much less of life beyond that (with spikes in mortality seen during infancy, around the time of weaning, and at puberty). Also consider how, on average, it takes multiple copulations to reach conception (with many couples trying for months or even years to get pregnant). From this view, we are all genetic lottery winners.

So how does conception actually occur? We first consider the female perspective, then the male perspective, followed by their union; some of this material harkens back to our discussion of female and male reproductive anatomy and physiology (Knobil and Neill 2006; Segal 2003; Suarez and Pacey 2006; Wood 1994). From the standpoint of female reproductive physiology, with each cycle a single egg is recruited and develops to a state in which it is released by an ovary. That egg is viable for approximately one day. It is typically streamed into the nearby Fallopian tube, in part thanks to small movements by the fimbria, or ends of the Fallopian tube. Traveling slowly along the tube, it awaits a possible suitor. To facilitate the arrival of a desirable gamete, the egg's owner must mate and have sperm introduced in the vagina during the time when the egg is viable. During a woman's fertile phase her cervix is dilated and secretes a thin mucus in order to allow passage of sperm. (Indeed, a woman can test the consistency of her cervical mucus as a means of tracking her ovulatory cycle.) Sperm traveling from her vagina and past her cervix may continue cruising through her uterus, following chemical signals that help it continue traveling toward the egg. Cilia in her Fallopian tubes gently beckon the sperm toward the egg, where fertilization may occur.

From the standpoint of male reproductive physiology, sperm are introduced into the woman's vagina, along with other constituents, in the complex seminal cocktail that is released when the man ejaculates. The semen temporarily reduces vaginal acidity, helping to increase the survival odds of the sperm contained within it (Masters and Johnson 1966). Prostaglandins in the semen may facilitate smooth muscle contractions of the uterus and immune modulation in order to help the sperm survive and move toward an egg. Sperm have a typical life span of approximately five days after ejaculation, so depending on the timing of sex and ovulation, some sperm might be waiting to fertilize an egg newly ovulated and entering a Fallopian tube, or the sperm might have to rush toward an already-waiting egg. Once a sperm is hardy enough—and fortunate enough—to make it to the Fallopian tubes, it encounters a kinder environment compared with the earlier obstacles.

At a meeting between egg and sperm, the egg appears ponderous and imposing in comparison, far larger than the tiny sperm. All of the sperm attempt to penetrate the egg's wall. The acrosome at the tip of sperm cells contains chemicals designed to facilitate an entry through the wall, and an egg is prepared to let in a single sperm, after which it undergoes chemical changes that will block further entries. Suppose that fertilization successfully occurs. The fertilized egg, or zygote, continues its travels along the Fallopian tube. Eventually, it reaches the uterus, where it floats freely for a few days while a supportive environment continues to be built for it (the blood-rich endometrium). If the uterine construction process and viability of the zygote remain on course, then the conceptus (fertilized egg) implants in the uterine wall, from which it will develop circulatory connections, providing it with the nourishment needed to sustain development. A placenta also develops, serving as a rich interface between the new life form and the mother's uterus. The conceptus releases human chorionic gonadotropin, closely related to luteinizing hormone, which encourages the corpus luteum in the ovary to continue releasing progesterone, which in turn helps foster continued development of the uterine lining. (Remember: hCG is what is measured in home pregnancy tests.) Without that hCG signal, the corpus luteum will not continue releasing progesterone, progesterone levels will drop, the uterine lining will slough off, menstruation will occur, and the whole process will continue anew.

CONTROLLING FERTILITY

The ultimate evolutionary currency is reproductive success. The anatomical, physiological, and behavioral legacies we and all sexually reproducing organisms bear have been stamped with various adaptations that enhanced our ancestors' reproductive success. Among other traits, females evolved sensitivity to energetic variables that would attune their physiology to times when reproduction would be more likely to pay. Males evolved traits to help ensure fatherhood and fight off other males in the attempt to mate. These kinds of selective scenarios played out in a context of scarcity. Finding resources or mates was an uphill battle. It might be difficult to imagine a creature giving up a reproductive opportunity when available, yet that is a situation in which humans frequently find themselves. In various ways, people have controlled their fertility, both in contemporary circumstances and in a wider array of historical and cross-cultural contexts (Engelman 2008; Riddle 1992, 1997; Voss 2008). While selection may have favored libido for purposes of procreation, cognitive decision making among both men and women can override this link with pregnancy-prevention behaviors, allowing them to meet their sexual interests but forestall parenthood.

As a means of contextualizing human fertility controls, we might ask whether other primates engage in seemingly intentional fertility constraint. Might a chimpanzee female in the wild seek to prevent a conception with a sexually harassing male by eating a plant with "birth-control" properties, or could a male cotton-top tamarin monkey practice coitus interruptus (withdrawal) in order to avoid an unwanted conception that he knows would require investment on his part? Perhaps these scenarios seem silly enough that they answer themselves. In a few field studies, primate females have been found to ingest foods that inhibit their reproductive capacities, but these seem more like incidental consequences than intentional outcomes: for example, Nigerian baboons feeding seasonally and heavily on black plums had elevated progesterone levels as a result, inhibiting their ovulatory cycles (Higham et al. 2007). Chances are, if you find nonhuman animals on some form of birth control, from dogs and cats (via gonadectomy) to zoo animals such as giraffes (hormonal contraception), it is because humans made them that way.

Use of fertility controls is obviously pronounced and intentional in humans, including among hunter-gatherer societies, which suggests that we have long sought to adjust fertility to local socioecological conditions (Engelman 2008). Among hunter-gatherers, Australian aborigines of Arnhem Land reported a variety of techniques to induce abortion, including "pounding the belly with stones, and the tying of rope or cord round the belly and gradually tightening it" (Berndt and Berndt 1951: 45), while infanticide has been recorded among various foraging societies such as the Inuit of the Arctic Circle and the Ache of Paraguay. Further, as we will see, the bulk of fertility control techniques are employed by females, rather than males. That makes sense given that women have a greater inherent cost-benefit ratio to investing in a given offspring than men do; those nine months of gestation, years of lactation (at least among foragers), and the extended period of maternal care mean that a woman faces high stakes and must maximize her lifetime reproductive success by attuning her physiological and behavioral responses to suitable times, including by avoiding (for example, with contraceptives) the inappropriate times. As intensive as human maternal and paternal effort can be, one consequence is that shunning a current conception for a better possible future may be a life-history trade-off amplified among us compared with other animals. This also increases the incentive to employ techniques to avoid or terminate a current conception. The more-intensive energetic and social support humans have for reproduction (even among hunter-gatherers) allows us to experience shorter interbirth intervals than other apes, but it may also be another reason why women desire to control fertility. We consider here some of the main ways in which humans have controlled fertility, although we also caution a reader interested in experimenting with any of these techniques to follow up in more detail with the logistics, health risks, failure rates, and more before they try a chosen method (see, for example, Levay and Baldwin 2009).

Abstinence, whether elective or by coercion, has been employed as one fertility control. It is, on the one hand, the simplest and most effective form of birth control. Unfortunately, the method faces a severe uphill battle against human sexual will. Many examples testify to how challenging long-term abstinence can be, with many "prevention programs" (rules set by others to regulate sexual behavior) couched in terms of supernatural justification or familial coercion as a means to blunt the force of human sexual desire. As illustrations, the requirement of abstinence for members of orders and clergy in the Catholic Church helps account for the frequent recent revelations of sexual contact between priests and boy parishioners, whereas in patrilocal, patriarchal societies in which female sexual fidelity, including virginity, has been valued, females have been guarded by various techniques—effectively coercing their abstinence—to enhance their value on the mating market.

Coitus interruptus, withdrawing the penis before ejaculation, is another way to avoid a pregnancy. According to the Old Testament, Onan "spilled his seed" as a way to avoid conception, and it is likely that plenty of men before him had done the same. By avoiding ejaculation into the vagina, the vast majority of semen is lost at sea, so to speak. However, a small number of sperm may still have a shot because they escaped in pre-ejaculate fluid (Killick et al. 2011). The use of the withdrawal technique implies knowledge of the importance of semen to conception, which also leaves us wondering if any of the "spirit child" societies mentioned earlier also employed this technique. Withdrawal is the most commonly used technique that relies on male behavior to control fertility, but it also risks conception should a male fail to "interruptus" in the heat of coitus.

The use of condoms, whether made from lamb's gut or Brazilian rubber, has been the other primary means by which males have historically had a hand in fertility control. Condoms helped make possible Casanova's legendary lovemaking in the eighteenth century (without fathering many children). They enable users to avoid both fertilization and the exchange of sexually transmitted diseases. As for other male-focused contraceptive methods, traditional Chinese medicine has long recognized the potential of gossypol, obtained from cotton plants, to control male fertility, and more recent efforts to develop a hormone-based male contraceptive have had some success, but neither of these efforts has taken off (Segal 2003).

Various pessaries, or vaginal suppositories, have been used to inhibit passage of sperm through the cervix. When Peter was an undergraduate, he spent a summer working as a ranger at Bodie State Historic Park, a former California mining town now maintained as a "ghost town," and he recalls the red-light district that once flourished there, a reflection of a population that was 90 percent male. When a few visitors inquired how Bodie prostitutes avoided becoming pregnant in the 1890s, one suggestion was that they would insert a coin in their vagina, which might sound crazy but could have partially served as a pessary, and also perhaps as an irritant that helped avoid conception (as copper intrauterine devices can do). In a related vein, Robert Engleman (2008) notes, "Some classical Mediterranean pessaries contained 'misy,' believed to be a copper compound comparable to the copper used in modern intrauterine devices to kill or immobilize sperm (111)." From Egyptian papyri we learn of pessaries made from crocodile dung and acacia oil. Sailors visiting Easter Island remarked on the use of seaweed pessaries by local women who wished to avoid pregnancy. Among contemporary fertility-controlling products, the sponge has been used for a similar function, to block off the cervix, also frequently abetted by the use of spermicides.

Another approach to fertility control is to prevent ovulation. A variety of plants can influence the female HPG axis for hormone release, potentially also affecting whether an egg is released. Some plants' leaves, bark, seeds, or roots are laced with chemicals (so-called secondary plant compounds) designed to prevent insects and sometimes larger animals from consuming them—a good form of self-defense when you cannot run away. In humans, the same estrogenic plant compounds that might have deterred, say, insects from eating soy beans, may affect the human consumer instead. Queen Anne's lace (wild carrot), a flowering plant rich in estrogenic compounds, has been used to reduce conception and perhaps has some efficacy for this very reason. Among all historic references to plants employed for fertility regulation, silphium, which is in the fennel family, is as interesting as they get. This plant was used in the Mediterranean region to treat various health concerns, including for fertility control. Ingestion may have reduced conception or favored early spontaneous abortion, or both. The plant was harvested to extinction, however, and we have only written records and depictions on ancient coins of silphium in its former glory. As for other fertility-controlling plants, some of the best documented effects were achieved by facilitating menstruation, which also terminates any possible pregnancy. This brief look at the use of fertility-controlling plants indicates that today's Pill and morning-after pill have long had their parallels.

When unwanted pregnancies have nonetheless occurred, women historically and cross-culturally have sought to induce abortions (for example, through trauma), or more frequently they have resorted to abandonment of infants to which they have given birth. The fact that so many undesired babies have been born suggests the kinds of fertility-controlling techniques described above have high failure rates, and that many women have had inadequate access either to those or more effective means of avoiding pregnancy. As one indication of the demand and also the ethical concerns over ending pregnancies, the same Hippocratic Oath that every new MD recites contains a line that translates, "I will not give a woman a pessary to cause an abortion." Classical references to "exposed" infants suggest abortion might be a preferable, earlier-acting means by which to curtail reproduction.

The magnitude of services in Europe rendered to "foundlings"—babies dropped off by mothers unable to care for them—and the estimate that 25 percent of births in late 1700s Japan ended in infanticide, provide dramatic examples of the numbers of unwanted births. Women may turn to severe measures if the time is not right and effective, alternative fertility controls are not readily available.

The cross-cultural literature on infanticide suggests that it occurs primarily within a limited set of contexts (Daly and Wilson 1988; Hrdy 1999). Importantly, most human infanticide is committed by mothers, a pattern with few primate parallels, although tamarin mothers may abandon infants if the fathers are unavailable to assist them (Hrdy 2009). The typical pattern of infanticide, in species from gorillas to langurs to lions, entails unrelated males killing infants. The adaptive hypotheses for this behavior suggest that the removal of nursing offspring facilitates a more rapid return to fecundity by the mother (by ending lactational amenorrhea), and this in turn facilitates conceptive matings with the killer male. Among humans, by contrast, the three most powerful predictors of infanticide are paternity uncertainty (wherein a woman perceives that her uncertain partner's investment in a given offspring is unclear), low infant quality (which can mean twins in a place where a female cannot afford to carry or care for both, or an infant with a congenital defect such that she cannot readily raise the child), and a lack of parental resources (which can make investment in a given offspring challenging, perhaps leading to a woman's sacrificing her child in the hope of reproducing again at a more auspicious time) (Daly and Wilson 1988). As an example, on Truk, an island in the Pacific, "abnormal or deformed newborns were not defined as human but rather as ghosts; they were burned or thrown into the sea" (Bolin and Whelehan 2009). Interestingly, these are factors consistent with the demographic profile of abortions in the United States, where the most straightforward interpretation is that women seek to terminate reproductive investment (abortions in the United States or infanticides in a wider cultural scope) when they feel the time is not right for reproduction, although it could be later.

The array of fertility-controlling techniques available to couples today includes some of the above examples, such as withdrawal and condoms, but it is largely a more biomedically regulated and produced set of options (Levay and Baldwin 2009; Segal 2003). For example, many recently developed hormone-based contraceptives are based on synthetic steroid hormones, including progestins, contained in the original Pill, that originated with compounds obtained from Mexican yams. We consider here some of the most commonly used contemporary options: hormonal contraceptives, intra-uterine devices (IUDs), and sterilization, noting that while the specific techniques employed to control fertility vary to some degree around the world, they also adhere to some overarching patterns. Most fertility controls are overseen by women, and the use of irreversible methods (for example, sterilization—tubal ligation in females, in which the Fallopian tubes are blocked, and vasectomy in males, where the vas deferens is blocked) is strongly age- and parity-related, as would make sense among couples who feel they have had enough (kids). Further, all else being equal, we might expect females to display less resistance to sterilization than males, given that older males maintain reproductive capacity, and are more likely to remarry and have additional children, and there is some evidence that these expectations are upheld (more women than men in the United States have been sterilized).

The rise of hormonal contraception with the Pill (oral contraceptives) has been viewed as fostering not just a reproductive revolution but also a sexual revolution, with the idea that freedom from pregnancy enables greater sexual freedom. Since the early 1960s, when it became available in the United States and elsewhere, the ability of women to obtain oral contraceptives that prevent ovulation (and maybe also reduce fertility through effects on cervical mucus) has been monumental. Yet early on, physicians in the 1960s often refused to prescribe oral contraceptives to single women, maintaining that premarital sex should not be encouraged (Coontz 2005). As we have seen, however, there have been earlier historical examples of fertility-controlling techniques, including ones apparently exerting effects through sex steroid physiology, but those versions differed in efficacy, reliability, and availability. Formulations have changed over time for the Pill, which initially included much higher estrogen doses than current formulations, also spawning a wide array of combination estrogen-and-progestin formulations and even, more recently, versions that work over longer time spans than a month (such as an oral contraceptive that entails menstruating only every three months, and injections, such as Depo-Provera, which can prevent pregnancies for months).

Other hormone-based contraceptives function to prevent implantation or a possible conceptus. The so-called morning-after pill (emergency contraceptive) and its relatives can be taken within 120 hours after intercourse (efficacy decreases over time) to prevent ovulation or, if conception occurs, to prevent implantation; the Plan B formulation, as the brand is called, works by blocking the effects of progesterone, thereby fostering menstruation. In some ways, then, these formulations resemble earlier uses of plants also intended to foster the shedding of the uterine lining. In the case of IUDs, these are typically made of plastic or copper, they are inserted into the uterus by a healthcare provider, and they may or may not contain hormones (progestin). They can remain inserted and prevent conception for years (nearly a decade in some cases), and are removable by a healthcare provider. Most IUDs also have strings that descend into the cervix and vaginal canal, allowing a woman to feel for the string and thereby be sure of the IUD's position.

Abortions can be performed in several different ways, depending on the gestational stage when they are conducted. In the first trimester, they typically are done by way of vacuum aspiration, a technique in which the conceptus and amniotic sac are removed via the cervix. While earlier forms of induced abortion have appeared in the ethnographic and historical record, current methods are safer, especially when legally available and carried out with healthcare support. The legal framework has also shifted over time to keep up with changing views of abortion. In the United States before the nineteenth century, women who wished to end an unwanted pregnancy commonly ingested compounds intended to induce a spontaneous abortion, before a fetus's first noticeable movements ("quickening") in the fourth month of gestation; later legal developments defined "personhood" at younger ages as part of a move toward making abortions illegal, until the 1973 *Roe v. Wade* case formally legalized abortion. Scientists and advocates alike have changed the playing field, with Margaret Sanger and Marie Stopes being instrumental in the expanded availability of fertility-controlling techniques and family-planning education.

The availability of contraception varies around the world, with the lowest rates of use in Africa (27 percent), and the highest rates of use in Latin America and the Caribbean (82 percent) (Bolin and Whelehan 2009). Forms of contraception also vary globally. In Europe, the Pill is the most commonly used type of contraception (used by 18 percent of women who use contraception), whereas in Asia, female sterilization (25 percent) and IUDs (17 percent) are most often employed. As for abortions, nearly 50 million are conducted yearly worldwide, almost 20 million of them under illegal and unsafe conditions. As of 2010, abortions were legal in fifty-four countries, and an estimated one of every four pregnancies around the world ends in abortion, with abortion rates highest in former Soviet bloc countries such as Russia and Belarus, and lowest in sub-Saharan Africa. The high rates of abortion in the former Soviet bloc countries are a legacy of Soviet fertility-control policy, which favored abortions over importing the relatively pricy Pill or internal development of other technologies (Segal 2003). Abortions are most common early and late in women's reproductive years (the latter being the case when women may have already achieved their desired family size). Contraceptive use can also change across time owing to shifts in political, legal, and technical developments. This is illustrated well by longitudinal changes in Japan. When sterilization and abortion were legalized in Japan in 1948, fertility dropped sharply from around four children per woman to around two children per woman. A medical industry subsequently formed around abortion, which generated bureaucratic inertia that hindered adoption of newer hormone-based contraceptives; only very recently (1999) did Japan see legalization of the Pill (Segal 2003).

THE PROBLEMS OF TOO FEW

While a market motivated largely by women's reproductive interests has long formed around controlling fertility, the flip side of the coin occurs when individuals want children but cannot have them. Here, we consider the causes and consequences of infertility. In biomedical view, approximately 10 to 15 percent of couples in the United States, and indeed in most parts of the world, are considered infertile, having attempted to conceive and remained unsuccessful after a year of trying (Bentley and Mascie-Taylor 2000; Bolin and Whelehan 2009; Levay and Baldwin 2009). The infertility rate is higher in parts of western, eastern, and southern Africa, owing to the spread of sexually transmitted infections (STIs), such as gonorrhea and syphilis. When scientists attempt to splice the causes of infertility, as in the United States, they attribute approximately 35 percent of cases of infertility to female factors, 35 percent to male factors, and the remainder to some combination of couple factors or unspecified sources. Among female infertility factors, these typically cluster under ovulatory problems (failure to ovulate due to inadequate nutrition or hormonal dysregulation), patent Fallopian tubes (scarring that prevents passage of sperm and egg, as can be the case in the wake of untreated chlamydia), and cervical mucus (which prevents passage of sperm to an awaiting egg). Among causes of male infertility, they typically feature problems with sperm, such as low sperm counts, poor motility or form, or genetic mutations in sperm, all of which can be in part affected by STIs, drug use, or congenital effects, among other causes. An increasing number of genetic mutations have been identified that contribute to female and male infertility (Matzuk and Lamb 2008).

From a cross-cultural perspective, as Frayser's (1985) findings indicate, the causes of infertility are more often social than physiological in nature, are more often assigned to actions by others, and are more commonly attributed to the woman than the man within an infertile union. It may be thought that the higher powers have frowned on a couple's fertile desires, or that another bad-wisher may be the source, such as the Gond widows of central India, who are thought to engage in black magic to cause infertility. So far as sex-specific attributions of cause among infertile couples, Frayser (1985) finds: "Primary responsibility for a woman's barrenness is never attributed to her husband's reproductive failure or impotence (0 of 30 societies). However, a woman is not free from being blamed for barrenness. She may contribute to her own infertility if she has a reproductive dysfunction (20% of 30 societies) or is being punished for breaking a rule of some kind (10% of 30 societies)" (289).

Sometimes a couple is just molecularly incompatible. Some women develop an allergy to a partner's sperm, and their immune system puts a brake on potential reproduction. A speculative proposal suggests that when a woman has repeated intercourse with a partner it helps her immune system adapt to his semen, in turn reducing negative birth outcomes (see Robillard et al. 2008). Additionally, the pH of a woman's vagina and the capacities of a man's semen to thrive in her "secret garden" may be at odds, leading to infertility. Not so long ago, incompatibilities of the Rhesus blood group system (this is the + or - in one's blood type) could yield incompatibilities, though medical interventions, if they are available, can now bypass this constraint. This Rhesus incompatibility comes into play if a type-negative mother mates with a type-positive partner, gives birth to a positive child, and that child's blood mixes with hers during the birth process, thus eliciting an immune response against subsequent type-positive pregnancies. In the global mating market, in which most Han Chinese are type-positive and many Europeans type-negative, one can do the math and see that such incompatibilities would be more prevalent were it not for medical intervention.

Wherever the fault lines of infertility lie, the consequences can be emotionally and socially profound. For women, reproduction typically features in her full realization of womanhood and her ultimate understanding of life's meanings. For women viewed as infertile, Frayser's (1985) cross-cultural survey finds that in 29 percent of societies such women may be divorced, may find themselves with a co-wife (as the husband seeks to reproduce with another fertile wife), or may adopt children, particularly relatives. According to Schapera (1941), long ago among the Kgatla of southern Africa, "One important motive for concubinage [was] the barrenness of a wife. In the old days this was usually circumvented by the sororate, whereby she would get her parents to provide her husband with her younger sister, or some other close relative, to bear children in her place. . . . What usually happens instead [in the 1930s] is that the husband either takes an additional wife, or, more frequently, resorts to a concubine" (Schapera 1941: 206). The Bible recounts similar tales of the sororate, evidence of what was once a more common practice.

A recent volume illustrates the continuing consequences of infertility to women (Inhorn and Van Balen 2002). In northern Vietnam, "Preoccupation with infertility, especially by childless married women who are having difficulty conceiving, often involves fears regarding the consequences for married life that are tied to issues of ancestor worship, descent, and familial expectations" (134); in southern India, "Adult identity for women is normatively organized around the milestone of motherhood" (165). The negative evaluations of women with infertility are likely more pronounced the more pronatalist the society in which they live, although it will be interesting to see if these evaluations diminish as more women have no children.

Since infertility has typically been less often assigned to men, there have been fewer repercussions for them, although having alternative male partners service a wife, or adopting, could bypass male infertility. Currently, assisted reproductive technologies (ART) offer new resources to address both men's and women's infertility (Mundy 2007). An infertile man can recruit a sperm donor, for example, to aid in conception, as is done in the

United States, although in much of the Sunni Muslim world sperm donation tends to be prohibited; also, changing regulations elsewhere make this
more or less difficult, as in Australia, where sperm donors are no longer
paid and cannot remain anonymous, which has led to a precipitous drop in
donors (Inhorn and Birenbaum-Carmeli 2008). Among women, ART has
opened up new pathways to fertility by enabling induced ovulation (for
example, cycles of clomiphene are given to spur the development and release of an egg), bypassing tubal blockage (with in vitro fertilization, or
IVF), or by saving a woman's eggs or using another's (through egg donation) in order to foster reproductive success via IVF. As for the logistics of
gamete donation, eggs are more difficult to obtain than sperm and require
more preparatory work (donor women must take injections and undergo an
egg harvesting procedure, while men can simply masturbate into a sterile
cup at a fertility clinic). Thus it is no surprise that eggs cost far more than
sperm in the fertility market. While in graduate school, Justin recalls, one
of his good friends considering selling her eggs, as the going rate for women
in her demographic—an attractive, fit, blond-haired, blue-eyed, highly educated woman—was equal to 50 percent of her graduate assistantship salary. In addition to helping couples conceive, ART has been used to extend
the age spread for female fertility, more typically into the forties, but to the
extreme of age sixty-six in the case of Carmen Bousada, who gave birth to
twins (and later died at age sixty-nine). It is also worth mentioning that many
of these assisted-reproduction methods, as well as surrogates and adoption,
are currently also employed by same-sex couples seeking to have children of
their own.

APES AND HOMININS BRING UP BABIES

Referencing a sixty-six-year-old human mother takes us into the evolutionarily new, but what about the evolutionarily old? What are the patterns of
human reproduction in evolutionary context? More specifically, across what
ages did our hominin forebears have babies, how many did they have, how
many years apart did they have them, how many survived, and what factors
account for the overall patterns? To address these questions is a tall order,
but we can home in on a reconstruction of some of these reproductive parameters from at least two primary angles: with reference to contemporary

ape reproductive patterns, and with respect to recently studied hunter-gatherers. A few other lines of evidence allow additional insight: archaeological and genetic data indicate that modern humans were relatively scanty in number until recently, suggesting that our recent ancestors did not breed like rabbits or fruit flies.

Studies of fertility tend to focus on female reproductive output rather than male reproductive output, and for several good reasons (Ellison 2001; Stearns 1992; Wood 1994). Maternity is easier to infer than paternity, especially in nonhuman animals in which females engage in multi-male mating. Females are also the reproductively limiting sex, meaning that population increases or decreases are more closely tied to female dynamics, a point that any wildlife manager or overseer of domesticated herding animals readily appreciates. In one example, to help alleviate deer overpopulation (and resulting ecological concerns) in many parts of the United States, hunters are now being encouraged, during certain parts of the season, to shoot does in addition to bucks; there are so many deer that shooting does will better help constrain population growth. Males deserve attention too, of course, since the pace of male reproduction can differ from that of females. However, since human males tend to reproduce within long-term sociosexual relationships, male and female fertility are commonly intertwined, meaning the confounds are less pronounced compared with other species, like chimpanzees, that have greater reproductive skew (meaning that the reproductive variation in males exceeds that of females). At any rate, we also follow a scholarly tradition here that features female reproductive parameters, but we will add a bit more on the male side as the story unfolds.

Among wild great apes, long-term studies have helped shed light on the reproductive parameters of chimpanzees, gorillas, and orangutans, though relatively few data are available on bonobos (Jones 2011; Robson and Wood 2008). All of these species have low fertility, though there are also species differences in reproductive pacing. Orangutan females typically have around three offspring across their lives, spaced approximately seven or eight years apart, marking them as having the longest interbirth interval of any mammal on earth. Female orangutan reproduction begins around age fifteen years and can continue into the forties. Juvenile mortality traced to nonhuman causes is virtually unheard of, yielding an estimated survival rate of nearly 100 percent. Chimpanzee females also begin reproducing in their early teen years and can continue into their forties, but they have slightly shorter interbirth intervals of around five and a half years. At Gombe, in Tanzania, a typical chimpanzee female may have around four offspring across her lifespan, with two, on average, surviving. Gorillas, by contrast, are fast-tracked relative to chimpanzees and especially orangutans. Female gorillas may begin reproducing by around age ten, they can continue reproducing until around age forty, and they have interbirth intervals of approximately four and a half years. Infant deaths do occur, due to male infanticide, among other causes.

One view of the similarities and differences of great ape reproduction in the wild is that the similarities reflect a specialty of foraging on unpredictable fruits, with differences consistent with the more sporadic fruit availability among orangutans, and this least so among gorillas (Jones 2011). Female birth spacing is highly contingent on food resource availability, which also helps ensure females are not trying to handle two highly dependent offspring stacked closely to each other. Intensive and energetically costly lactation helps inhibit resumption of cycling, thereby favoring lactational amenorrhea. Females in these species spend the bulk of their reproductive years in a state of lactational amenorrhea, interspersed by pregnancies and occasionally ovulatory cycling. Individual, site, and captive differences in great ape reproduction also reinforce the importance of socioecological factors (Bentley 1999); while talking with a representative at the Singapore Zoo in 2011, which has one of the world's largest captive populations of orangutans (truly spectacular reddish-brown beauties), Peter learned that the female orangutans there (which are provisioned by people, expend less energy daily, and have access to veterinary care) have interbirth intervals of only around three to four years. In the wild, males provide no meaningful degree of paternal care, and food provisioning by others rarely occurs, meaning that the females are effectively single mothers doing it all. As for the pacing of male ape reproduction, earlier studies estimated paternity by observing mating behavior, while more recent studies have employed genetically based paternity testing. Across all of the great apes, males have greater reproductive skew than females, and male reproduction tends to be concentrated in fewer years, when a male is able to maintain higher rank and the reproductive benefits attendant to it.

Studies of recent human hunter-gatherers highlight an uptick in human profligacy compared with wild apes. The typical age of first birth for hunter-gatherer women is around eighteen to twenty years, it is around the late thirties for last births, and they have interbirth intervals of approximately three to four years (Hewlett 1991; Kelly 1995; Marlowe 2010). Juvenile survival rates of around 60 percent are likely slightly higher than or comparable to chimpanzees and gorillas, but lower than the rate among orangutans. The outcome is that a typical hunter-gatherer woman has around 5.5 children, of whom around 3 survive. Like the great apes, forager women spent the bulk of their reproductive years in a state of lactational amenorrhea, preceded by briefer phases cycling and pregnancy. The shortened interbirth intervals are the main reason for the inflation in human fertility. The energetic and social support provided by fathers (Gray and Anderson 2010; Lancaster and Lancaster 1983), grandmothers (Hawkes et al. 1998), or a more general cooperative breeding model (Hrdy 2009; Reiches et al. 2009) are key differences between humans and the great apes that makes possible the shortened interbirth intervals. The involvement of men in long-term reproductive relationships and parental care also ties male fertility more closely to that of women: the results include less human male reproductive skew, and later ages of fathering offspring in humans compared with the other great apes (see Bribiescas et al., forthcoming).

A FERTILE WORLD

Although 100,000 years ago the sizes of the human and other great ape populations may have been on the same order of magnitude, the number of humans has exploded. While writing these pages, the world population has risen to 7 billion people. Yet, this comes at expense to the other great apes, whose numbers have dropped, especially in recent decades. The stark differences in these demographic trajectories invite questions about patterns of fertility that contribute to human increases. Here, we scan some of the wider variation in recent human fertility, as well as the primary causes for fertility reductions.

In a compilation of fertility among small-scale societies of hunter-gatherers, horticulturalists, and pastoralists, Hewlett (1991) found that fertility among horticulturalists and pastoralists was slightly higher (6.1 kids per woman) than among foragers. Considerable attention has been devoted to the question of whether fertility rates increased after animal and plant domestication, with the thought that this would have been a major demographic spur to continued human expansion (see Low 2000). In one analysis of fertility across various subsistence systems, Sellen and Mace (1997) found that the greater reliance on agriculture within a population, the greater that society's average fertility. The highest fertility rates found in any population have been in horticultural, agricultural, or otherwise industrialized societies. Among them, the grand prize of fertility goes to the Hutterites in the 1940s: they had an average of 10.9 children per woman, though that rate has dropped considerably among Hutterite groups more recently. Shipibo horticulturalists of Peru had, on average, 9.9 children per woman in the early 1960s, and in 1970s women in Oman bore an average of 9.3 children, other high points that have since dropped (Gray and Anderson 2010).

Around the world today, the general picture is of declining fertility, especially compared with the unusually high levels of the previous thirty to fifty years (United Nations 2008). The highest fertility rates today are seen in sub-Saharan Africa (for example, Mali), some countries in southwest Asia and the Arabian Peninsula (such as Yemen), and Papua New Guinea. Fertility is lowest overall in Europe and East Asia, where below-replacement fertility (meaning that women are having fewer children than are required to replace population) has become the norm. In China, for example, women now have 1.4 children on average, and we saw at the beginning of this chapter that Singapore's fertility may be the lowest of any country in the world. It will be interesting to see whether fertility rates stop dropping at a level above one child per woman, a level at which many women in, say, Korea, might still be having one child rather than foregoing fertility altogether. Peter also speculates that the drop in fertility is or will be accompanied in most places by enhanced investment in pets such as dogs, with the thought that rearing a furry dependent is better than nothing (see Gray and Young 2011).

In the United States women have an average of 2.1 children, whereas just north of the border, in Canada, women have an average of 1.5 children. The United States' fertility is the highest of any developed country, even though it has dipped since the baby boomer era in which women had around four children. Within the United States, states in the northeast have the lowest fertility (for example, Vermont), whereas those with higher fertility are in the West and Midwest; Utah has the highest fertility rate of any state, though fertility has also fallen there recently (Lancy 2008). Virtually all countries exhibit lower fertility today than during the previous generation, with pronounced drops seen in places ranging from Iran to Bangladesh to Korea. Robert Engelman, taking an even longer time span into consideration, emphasizes the dramatic shifts in global fertility: "For most of the last several centuries—and possibly for most of the last 200,000 years—women have had an average of five or more live births each. Today the average is 2.6. It is among the most remarkable behavioral shifts in history" (Engleman 2008: 12).

How can we account for population variation in fertility, including its recent decline? A host of variables, such as the increasing costs of raising children, have been advanced by scholars to explain this variation in fertility. Here, we have focused on the factors that have garnered the most attention, as well as those we feel are most central to explaining fertility variation and decline. Given space constraints, we have left aside other considerations, such as the availability of adult relatives, the assistance that older siblings might provide, pronatalist religious values, or sex-specific preferences, even though these are among the relevant factors. To complete this story of fertility variation and decline, we need to revisit sex differences in reproductive strategies. We go back to basics.

Women bear the bulk of reproductive costs. The relative reproductive costs and benefits favor women cueing their fertility decisions to salient features of their socioecological context, which in practice means energetic factors (such as changes in food availability), health factors (illnesses), and social factors (availability of family support, including a prospective child's father) (Ellison 2001; Hrdy 1999; Trevathan 2010). Women also weigh life history considerations when deciding whether to invest now or later in reproduction, as sometimes it may pay to wait, a point made earlier when we discussed fertility control. The female calculus eventually yields to emotions, sex, discussions, and fertility outcomes. When contrasted with a male perspective, in which the reproductive costs and benefits differ, we see that often women seek to have fewer kids than do men. So while there may be considerable population differences in ideal and actual fertility (in some societies, men and women alike want many kids, whereas in other societies men and women both want few kids), within any given population women may be more likely to want fewer kids.

A sex difference in fertility preference (number of children desired) has been documented in many places. As Segal (2003) notes, "Surveys have been done in Africa and Asia concerning husband's attitudes toward fertility, contraceptive use, and reproductive preferences. Compared to their wives, men want more children, not fewer. In the west African countries in the survey, the ideal family size reported by men was nine, well above the upper limit preferred by women" (114). Pointing toward similar patterns, but on a smaller societal scale, among Pimbwe agriculturalists of Tanzania the modal female fertility preference was for four children, whereas the modal male fertility preference was for six children (Borgerhoff Mulder 2009). Furthermore, while males sometimes employ techniques to control fertility (such as withdrawal or sterilization), much of the male "sexual medicine" arsenal used in small-scale and larger contemporary societies is designed to enhance fertility; males are more likely to make use of plants or pills to treat their own impotence or to stimulate female sexuality than to suppress fertility, for example.

Several implications stem from these sex differences in reproductive strategies and fertility preferences. One is that, when women have greater autonomy over reproduction, fertility rates will likely be lower than when male reproductive decisions hold sway. This may be especially true concerning access to fertility-controlling options. Another implication is that women's fertility outcomes are highly contingent on socioecological context in seemingly adaptive ways. If a woman's energetic circumstance makes reproduction challenging, her ovaries may respond by not ovulating; if she has conceived under inauspicious circumstances, she may elect to shut the reproductive investment down by ingesting a plant or pill. Furthermore, a woman and her partners encounter a wider span of socioecological variation than their hunter-gatherers once did, which also requires some additional degree of fine-tuning. The economic costs of raising kids vary by context, with forager kids able to help feed themselves as juveniles, and not requiring advanced degrees in order to earn the social status that could make them eligible as mates. All-you-can-eat buffets and a global obesity epidemic (Power and Schulkin 2009) mean that the contraceptive effects of lactation differ in a well-fed, if not overweight, sample, compared with energetically strapped foragers. Indeed, unlike among foragers, women today might choose not to breastfeed and yet still have surviving offspring.

A couple of models highlighted by demographers to account for fertility differences are infant mortality and the value of children as providers of "social security" (Gray and Anderson 2010; Low 2000). The first of these factors points toward the fact that low fertility and low mortality are often found together, and in contrast to previous states of both higher fertility and mortality. Even if reductions in infant mortality have minor impacts on fertility outcomes (deciding not to have that fourth child, for example, because the previous three survived), there are plenty of examples in which fertility has changed in ways unrelated to infant mortality changes. The drop in U.S. fertility from the time of the baby boom to now is but one example in which it would make little sense to attribute that change to a drop in infant mortality. As for the social security model, the idea is that initially resources flow from parents to descendants, but that eventually the resource flow will come back to parents, who have effectively stocked their retirement portfolio with their kids. This model may hold in limited circumstances: anecdotally, Peter's dentist shared a story in which his (the dentist's) parents, new immigrants to California from Southeast Asia, funneled all of their resources into their kids' educations but now find themselves, at older ages, financially dependent on support from their kids (including Peter's dentist). More widely, however, there are plenty of small-scale societies where older adults never "retire," engaging in economically and politically important activities until their last breath. Further, as Kaplan and colleagues (2000) show, resources in a number of small-scale societies do not flow from young to old, contrary to what this model would have predicted. So if neither of these models adequately accounts for fertility variation, including fertility declines, what other factors might be up to the task?

Some scholars favor female access to fertility controls as the best predictor of fertility reductions. Other scholars point toward shifts in female "social capital," in the form of enhanced maternal education, as the best predictor of fertility reductions. In practice, these two factors may overlap, with the same countries promoting female education being the ones that also make contraceptives and other fertility controls accessible to women. To illustrate these two perspectives, compare Engelman's view (2008: 206), that "the correlation among the three—government commitment to family planning programs, contraceptive prevalence, and completed fertility size—is more consistent than the commonly cited one between girls' average years of completed education and how many children they end up having," with that of another population scientist, Segal (2003: xx): "The premise is that enabling women to achieve personal autonomy over their own reproductive aims is an intrinsic right and will reduce fertility rates as well as improve the quality of life for women, their families, and communities. Moreover, there is substantial evidence to support this conclusion. In countries where studies have been done, a higher level of mother's education correlates well with lower fertility."

Both views feature female reproductive decision making, implicitly recognizing that women will best attune their fertility to context, if they are allowed to do so.

Continuing with the social capital viewpoint (which can be collapsed into a formal education model), the world increasingly rewards formal education in a competitive labor market. In the United States and more widely, those who are better educated tend to earn higher salaries and have better economic prospects. However, the more time and energy spent pursuing education, the less time there is available for other life history pursuits, such as finding a mate and having children. The trade-offs between education and child rearing are borne out more heavily among women than men, which helps account for the inverse relationship between a woman's educational attainment and her fertility in many societies. Additionally, raising kids to compete in the same labor market entails a cost-benefit analysis that can often favor low fertility. Where once kids could contribute to the family economy by herding livestock, assisting in the family store, or maybe earning some change as a chimney sweep, they now contribute less because they spend their time in school (providing fewer economic benefits), while also requiring greater investment in their education, to ensure their social success (expenses of school fees, books, clothes, and so on). The relative cost-benefit ratio in which formal education rules thus also make sense of low fertility; it is hard enough to raise two children (or even one) when parents aspire to provide their kids with a university education that will garner a good job that pays well. It also turns out that just having those kids—not to mention raising them—can have a significant impact on a couple's sex life, which is the subject of the next chapter.

Tying up our discussion of human fertility, in this chapter we have covered a wide-ranging landscape. We considered cross-cultural and biomedical views of conception; methods by which women and men control fertility; the causes and consequences of infertility; patterns of reproduction among wild apes and recently studied human hunter-gatherers; and the wider patterns that account for human fertility variation. From an evolutionary perspective, we highlighted hominin shifts toward later ages of first reproduction, shorter interbirth intervals, higher fertility, and expanded energetic and social support for reproduction, compared with our ape cousins. We also featured adaptive female reproductive investment, and how women's fertility is highly attuned to energetic and social factors, while male contributions to fertility are highly intertwined (in long-term bonds) with those of women's, and may have a slant toward higher fertility preferences (given sex differences in cost-benefit ratios of reproductive investment). However, despite having scanned this wide and fertile terrain, we are not yet ready to leave the reproductive process behind. In the next chapter, we ask how and why human sexual behavior varies across the peripartum period, the time of pregnancy and postpartum care. While sex can beget reproduction, sometimes, as we'll see, reproductive investment impinges on sex.

CHAPTER TEN

Born to Be Less Wild

Peripartum Shifts in Human Sexuality

> The arrival of a baby has a significant impact on the frequency of sex.
>
> —David Buss, *The Evolution of Desire*

FOR THE PARENTS reading this passage, the observation that a baby profoundly alters a couple's sex life probably sounds eminently sensible, even obvious (ah, there is a science underlying those first-hand experiences). For the nulliparous, or childless, this observation is still worthy of reflection, particularly if the future holds a baby or more. Somehow even the remarkable story of Nim Chimpsky—a chimpanzee raised by humans in an attempt to test his language acquisition abilities—helps us understand the impact of childcare on sexuality. In the case of Nim, when he was an infant he was raised for about eighteen months by Stephanie LaFarge and her human family. LaFarge reported she even breast-fed Nim, which did not provide him with milk but offered some comfort. She also kept Nim close, realizing he craved maternal love: "Most nights, Nim slept in between Stephanie and [her husband] WER, which, after the first few months, began to bother the new father. They could have sex only if the chimpanzee was in a deep slumber" (Hess 2008: 78).

The observation that having a baby (even a baby chimpanzee) can impinge on a couple's sex life has a deeper theoretical and empirical significance. From an evolutionary perspective, in early humans most females' reproductive years were spent either pregnant or in a state of lactational amenorrhea. Our primate female forebears vacillated between intensive off-spring care and mating concerns throughout their reproductive careers; they were not wildly open, childless sexual swingers. As for their male counter-parts, the fact that most male sexuality occurs within long-term partner-ships means that the bulk of male reproductive years were also spent track-ing a partner's sexual and reproductive ups and downs. Males were embedded in social groups containing few unpartnered, childless females and sur-rounded by women who spent a small fraction of their peak reproductive years experiencing ovulatory cycles.

In this chapter, we emphasize the physiological and life history adjust-ments faced by females during the peripartum (pregnancy and postpar-tum) period. There are good reasons for doing this. One reason is that the life history trade-offs faced by females during transitions from ovulatory cycles to pregnancy to postpartum lactation and eventually to the resump-tion of cycling can be squarely couched in terms of a life history trade-off between current and future reproduction. Women's time and resources can be most attuned to mating (begetting a future baby) when they are cycling, but from pregnancy to postpartum, women's efforts shift toward maternal investment (caring for the newly conceived offspring). A second reason for this focus on females is that the physiological changes women undergo in the peripartum period are dramatic, and they guide the intensive metabolic and behavioral adjustments women undertake. We attend to impacts on males too, but these entail less pronounced life history and physiological shifts compared with those in females, and they are often hitched to the peripartum fluctuations of their mates.

THE PHYSIOLOGICAL JOYS OF PREGNANCY

Suppose the marvels of conception have yielded that rare occurrence: a new life, growing, taking hold within a woman's uterus, the incarnation of an-other generation of human sexuality. What immediate impacts might this pregnancy have on a couple's sexual behavior?

One way to address this question is by focusing on the physiological shifts women undergo during pregnancy. Among other changes, women's ts enlarge, their blood volume increases, their pelvic ligaments loosen, their uterus expands (Levay and Baldwin 2009). These are changes that cilitate pregnancy and eventually birth. The expanded fat reserves, including in the breasts, provide a metabolic reservoir to draw on for lactation, which burns more calories than even late gestation. The enhanced blood volume can help in shuttling nutrients and wastes for two. The amazing transformation in the uterus provides shelter for the growing life.

Hormones orchestrate many of the physiological shifts women undergo during pregnancy (Gardner and Shoback 2007; Petraglia, Florio, and Torricelli 2006). The increases in estrogens, particularly estriol, facilitate fat deposition in the breasts and gluteofemoral regions. The rising levels of progesterone literally protect and promote gestation, including development of the uterus. The combination of rising estrogens and progesterone prepare the breasts for lactation; the withdrawal of progesterone at birth, combined with suckling, begins the flow of milk. While the gestating woman serves as the initial source of hormones regulating pregnancy, the expanding placenta, at the interface between mom and baby in the womb, takes on an important endocrine role, also releasing hormones, such as human placental lactogen (closely related to growth hormone and prolactin), that prepare the woman for continued maternal investment.

The physiological changes females undergo during pregnancy are geared toward the baby in the making, not toward making a new baby (Ellison 2001). This is an important consideration in terms of the impacts of pregnancy on a women's sexual desire and a couple's sexuality. The increases in progesterone and prolactin can blunt a woman's desire. At the same time, the rising levels of estrogens and even testosterone may have by-product effects enhancing some features of a woman's sexuality. The rising levels of sex steroids may increase sexual desire in some women and facilitate vaginal elasticity and lubrication. Other impacts may be less direct: changes in appearance, mood, fatigue, and nausea may variably alter female sexuality, with some of these effects most pronounced during the first and third trimesters of pregnancy.

Through their classic research on human sexual response, Masters and Johnson (1966) further elucidated some of the physiological shifts women undergo during pregnancy. During the first trimester, the development of the fetus creates vasocongestion in the pelvic region from increases in blood volume. This vasocongestion persists throughout gestation and can cause feelings of "sexual tension" due to the increased pressure in the pelvic region. As the breasts change, many women report breast tenderness during the first trimester and frequent nipple erections and breast engorgement throughout pregnancy. Some of these changes in breast sensitivity may impinge on breast stimulation during sexual behavior. Women frequently report increases in sexual desire during the second trimester, an effect that may trace to increased estrogen and testosterone levels, and also to reductions in nausea compared with the first trimester.

SEXUALITY DURING PREGNANCY

So how do those female physiological shifts impact actual patterns of sexual behavior? With physiological changes orienting a woman's body toward gestating and preparing for continued investment in a current offspring, we have seen how those changes may undercut or sometimes yield by-product benefits for female sexuality. Another major consideration is that a pregnant female does not yet have a nursling in hand (if this is her first child)—that change will arrive at birth. Gestation thus can offer, by comparison, a less encumbered time during which females can engage in sex, perhaps in ways serving adaptive social but nonprocreative functions. Here, we consider patterns of nonhuman primate sexual behavior during pregnancy, followed by patterns among people, observing both similarities (sex during pregnancy is more common, compared with periods of lactational amenorrhea) and differences (potential social functions of gestational sex differ in most other primates compared with people).

Various nonhuman female primates engage in gestational sex. The socially monogamous gibbon has sex relatively infrequently, but does remain sexually active during pregnancy (Barelli et al. 2008). Females of various other primate species, ranging from baboons to mangabeys to long-tailed macaques, also engage in sex during pregnancy (Ziegler 2007). Among white-faced capuchin monkeys in Costa Rica, males courted females and females solicited sexual advances from males as often during pregnancy as during times when females were cycling; however, the males engaging in gestational sex with females tended to be lower-ranking ones, with dominant males "allowing" this even within plain view, although they would not tolerate it during times when females were cycling (Perry and Manson 2008).

e all of this sex during pregnancy? At minimum, an answer
ize that other animals also engage in nonprocreative sex, but a
er must go beyond that observation. Some of the patterns may
by-products of the physiological changes noted above. Increases
oid levels may foster ongoing sexual desire and receptivity in fe-
ring pregnancy, even if the sex is clearly not going to lead to an-
nception. The main social function advanced for primate females
g in gestational sex is that it may confuse paternity in the case of
es living in multi-male, multi-female groups (Hrdy 1981). Suppose,
imple, that a female wants her group's males to be nice to her baby; by
g sex with the lot of males during gestation, she may make them
k" they could be the offspring's father, and thus they will treat her
nt better than they otherwise would.

Exactly this sort of interpretation—paternity confusion—has been ad-
nced to explain why, for example, langur monkeys and chimpanzees
ate during pregnancy (Stumpf and Boesch 2005). Further, in blue mon-
eys, females have sex with more males while pregnant during years when
nore males are in the group (Pazol 2003), and female baboons continue to
have sex during pregnancy to maintain "friendships" with male consorts who
may provide protection to them and their offspring (Palombit, Seyfarth,
and Cheney 1997). In the case of gorillas, which mate in polygynous groups,
female sexual behavior during pregnancy is generally rare, but when it does
occur it tends to be in response to the male's mating with another female in
the group. An interpretation of this pattern is female-female competition to
maintain adaptive relationships with a mate within a polygynous social
group (Doran-Sheehy, Fernandez, and Borries 2009).

In an interesting discussion of captive data on cotton-top tamarin and
common marmoset sexual behaviors, Snowdon and Ziegler (2007) note
that these South American monkeys tend to engage in frequent sex throughout the ovulatory cycle and pregnancy. In these species, females and males
tend to mate within long-term reproductive bonds that are generally monogamous, and males provide considerable parental investment. Interpreting
the patterns of sexual behavior among cotton-top tamarins, they write that
"frequent sexual behavior across the ovulatory cycle and during pregnancy
may be one way in which adults maintain the relationship. . . . [P]airs increase sexual behavior following the challenges of separation and following
presentation of odors from a novel female, especially when the novel female

is fertile. . . . Increased sexual arousal in males coupled with increased proceptivity by females led to increased sexual behavior that appears to maintain and reinforce pair bonds in the face of threats to the relationship" (Snowdon and Ziegler 2007: 46–47). We suggest that their interpretation of long-term bonded pregnancy sex among these monkeys has parallels in humans—that humans too can deploy gestational sex to maintain a relationship and paternal investment. So we will now jump into the patterns of human sexual behavior and pregnancy.

In their classic survey of human sexuality, Ford and Beach (1951) compiled cross-cultural data on sexuality from sixty small-scale societies of hunter-gatherers, horticulturalists, pastoralists, and farmers. Although they lacked quantitative data on sexual behavior, they observed patterns in attitudes toward sexuality across pregnancy. They found that 70 percent of societies permitted sexual intercourse during early pregnancy, with that percentage remaining relatively high through the middle of pregnancy and then dropping steadily. When women were in the seventh month of gestation, 50 percent of societies allowed intercourse, but only 25 allowed it did during the ninth month of pregnancy. There are several reasons why such taboos are in place. As one common rationale for sex taboos, in many societies there is a belief that sex during pregnancy can harm the growing fetus. Among Trobriand Islanders, for example, "As pregnancy progresses and the woman becomes big, sexual intercourse must be abandoned, for, as the natives say, 'the penis would kill the child.'" (Malinowski 1929: 193). Furthermore, the societies that tended to prohibit female sex during the earliest stages of pregnancy were also most likely to be polygynous societies, the implication being that a husband whose pregnant wife is tabooed from having sex may have other wives with whom he can have sex.

In a large, quantitative, international compilation of data specifying effects of pregnancy on female sexuality, Brewis and Meyer (2005b) integrated data on the recent sexual behavior of more than 90,000 women from nineteen countries, including Bolivia, Nepal, Peru, Dominican Republic, and Cameroon. They concluded that effects of pregnancy on women's sexual activity during the preceding four weeks were robust and patterned: "Pregnancy has a negative effect for the most part on sexual frequencies. . . . [G]enerally, coital frequencies are significantly lower in pregnancy compared with the non-pregnant state. . . . [O]verall . . . sexual frequencies decline in the first trimester, decrease further in the second, and still further in the third"

nd Meyer 2005b: 508). These data buttress the cross-cultural con-
resting on sexual attitudes, showing that rates of sexual behavior
across pregnancy. But there remain additional contributors to varia-
n female sociosexuality during pregnancy.

some societies, sex during pregnancy is thought to serve important
ctions. For example, among the Aka of Central Africa, semen is thought
help promote growth of the infant, and thus pregnant women and their
usbands have frequent sex during pregnancy (Hewlett and Hewlett,
2008). Some Aka men report feelings of exhaustion—that sex with a demanding pregnant wife is hard work—based on the need to have repeated
sex. Another interpretation of this pattern could be that frequent sex during
pregnancy enhances the mating bond and facilitates the husband's fidelity
and continued investment. Among many native Amazonian societies, all
men having intercourse with a pregnant woman are thought to contribute
to a fetus's conception and growth in a process that anthropologists have
labeled partible paternity. As an example, Bari women have sex with multiple males throughout pregnancy, and the men with whom she has sex are
considered secondary fathers (Beckerman and Valentine, 2002). It may be
no coincidence that most of these Amazonian societies characterized by
partible paternity are matrilocal, meaning that a mother tends to live with
her maternal kin, and her mating relationships with men tend to be more
fluid. In this context, an interpretation of pregnant females mating with
multiple men could be that the practice recruits the potential investment of
typically two men, an advantage over a single man in a mating context in
which a single mate may be unreliable or insufficient to provide resources
such as fish, meat, and other means of support (see Walker, Flinn, and Hill
2010).

In a narrower cultural scope, with almost all studies conducted in the
United States and Western Europe, von Sydow (1999) reviewed fifty-nine
sexuality studies addressing patterns of sexual behavior across the full
length of pregnancy. Coital frequency tended to remain high the first trimester, was quite variable the second trimester, and dropped considerably
the third trimester. These patterns are generally similar, then, to the international findings reported above. Positions during intercourse also changed
across gestation, with most third-trimester intercourse entailing positions
with the man behind the woman and with partners side by side, a contrast
to more frequent face-to-face positions previously employed.

THE BIRTH OF A BABY, AND THE DECLINE OF SEX

While human sexual behavior typically drops off across the course of pregnancy, there is also the occasional encouragement to have sex as birth nears. In some societies, including among the Aka, sex is encouraged in order to facilitate birth. Is there any physiological substance to such a belief? Delving back into the components of semen, which we discussed earlier, we saw that it contains many compounds designed to enhance the prospects of the sperm they accompany. The prostaglandins in semen that aid cervical ripening and smooth muscle contraction in the uterus may in fact assist birth outcomes. Indeed, in biomedical contexts, synthetic prostaglandins (along with synthetic versions of oxytocin) may be employed to facilitate births.

But say birth occurs, with or without assistance from prostaglandins—what does this entail for a couple's sex life? Additional aspects of the birth process may matter, not least of which is the social and biomedical context in which it occurs. Births entailing cesarean section (surgical removal of a baby through a woman's abdomen) or episiotomy (cutting the outer area by the posterior vagina to enlarge the birth outlet) are associated with delays in the resumption of postpartum sexual behavior (Abdool, Thakar, and Sultan 2009). Whether a birth occurs within a monogamous or a polygynous family, among other social factors, also affects the resumption of postpartum sex, as we shall show below. In a dramatic cultural practice concerning the intersection of birth and sex, among the Marquesans, inhabitants of an island in Polynesia, a woman fresh from birth was encouraged to have sex the same day with her husband, in a stream. According to the ethnographer who recorded this practice, "This was believed to be of therapeutic value in producing a good flow of blood and in aiding the internal organs of the mother to return to normal condition" (Suggs 1966: 25). However, this was a one-shot practice, followed by more typical postpartum patterns.

We now describe some of the physiological shifts females face during the postpartum period. The transition from pregnancy to the postpartum period is marked by rapid declines back to baseline levels of estrogens, testosterone, and progesterone in women. The declines in estrogen and testosterone can downregulate sexual desire and have a negative impact on other traits, such as vaginal lubrication, making sex both less desirable and maybe even painful. The postpartum phase is characterized in nonhuman primates and even most humans today by lactation, with metabolic and hormonal factors helping suppress a female's return to cycling (in this state of lactational amenorrhea). The nipple stimulation during lactation facilitates prolactin release and the continued production of milk, simultaneously serving to inhibit the female hypothalamic-pituitary-gonadal (HPG) axis—in others words, helping delay her return to cycling. As perhaps the archetypal symbol of maternal investment, lactation utilizes women's resources to sustain a newborn, and comes at the expense of preparing to have the next child (that is, it helps postpone a return to ovulatory cycling). With her time and attention focused on her new offspring, this may also come at some expense to her time and intimacy with a partner, not to mention any older children. All said, the physiological shifts postpartum women face help explain their further reductions in sexual behavior during this period compared with pregnancy and, especially, times when they are cycling.

The inhibitory effects of the postpartum period—marked by lactation and other intensive forms of maternal investment—on nonhuman primate female sexual behavior are dramatic. Among white-handed gibbons in Thailand, only one female of seven observed in a postpartum lactational state was seen mating, even though females were sexually active during pregnancy and while cycling (Barelli et al. 2008). In mountain gorillas, postpartum females might typically remain abstinent for more than three years after having a baby (Doran-Sheehy, Fernandez, and Borries 2009). Among white-capped capuchin monkeys, females were significantly less sexually active during the postpartum state in contrast to times when they were cycling or pregnant; further, the data indicate that postpartum females both sought mating less and were courted less by males compared with females in other reproductive states (Manson, Perry, and Parish 1997). Among rhesus monkeys, Maestripieri (2007) comments on the notable reductions in postpartum female sexual behavior compared with times when females are cycling or pregnant. Through our discussions with primatologist colleagues, we learned that the dramatic reductions in female postpartum sexual behavior are so obvious and taken for granted that our queries about this subject seemed quite silly (*of course* they are not having sex for prolonged periods).

If there is a nonhuman primate counterpart that represents more rapid resumption of female sexual behavior, the best candidate may well be several species of small South American callitrichid monkeys. Common marmosets and cotton-top tamarins have been argued as having a postpartum estrus, a phase arising quite quickly after birth when they are sexually receptive (see McNeilly 2006). However, that pattern has been best documented in well-fed, generally sedentary monkeys, and it is not clear if the same pattern holds among their wild brethren (Lottker et al. 2004). If it does, then this pattern would be yet another derived element of their unusual reproductive physiology and behavior, characterized by regular twinning (or "tripleting" in captivity), highly invested male partners in long-term bonds and in paternal care, and in an intensive feeding strategy relying on nutritionally dense gums, insects, and more. If a postpartum estrus does hold among some of these callitrichids, then it would mean that they have evolved mechanisms to ovulate shortly after birth while intensely lactating, bypassing some of the otherwise well-documented inhibitory effects of lactation on the HPG axis in humans and other primates.

The nonhuman primate patterns, perhaps excepting the callitrichids, underscore features of the human postpartum sexuality data. Various quantitative studies of human postpartum sexual behavior indicate that women also engage in lower coital frequency postpartum compared with pregnancy or while cycling; we are part of the wider primate fabric, after all. Yet women often resume sexual behavior much earlier than would be expected for an ape of our body size. Part of this latter pattern is that humans begin cycling earlier than might be expected (more on this below) and also have sex even when they are not cycling, perhaps to facilitate relationship maintenance. Put another way, postpartum women may have sex earlier than would be expected in order to foster an ongoing relationship with an investing partner, a human pattern that has few, if any, analogues in other primates.

A notable feature of female sociosexuality that varies cross-culturally is the duration of postpartum sex taboos. These are taboos that specify how long a new mother should refrain from intercourse. Cross-culturally, postpartum sex taboos are in place the longest in societies where males have sexual access to other women (polygynous societies). For example, women in certain Malaysian groups refrain from sexual activity for years after giving birth. Their partners, however, are free to have sex with the new mothers' younger sisters during the postpartum abstinence period (Dentan 1997). Postpartum sex taboos are longest (restraint from sexual activity for one or more years after giving birth) in African societies, where polygyny is more frequently practiced. For example, the Nso and Fulani of Cameroon are highly polygynous groups that have a two-year postpartum sex taboo. These groups believe that semen can poison the breast milk, afflicting the infant (Yovsi and Keller 2003). Sex taboos after birth are shortest in Eurasia and the Mediterranean region, lasting around one to five months (Frayser 1985). When men have alternative sexual outlets, there may not be as strong a push for women to resume sexual intercourse with their partner after giving birth. Practices among Trobriand Islanders illustrate this pattern: "Sexual intercourse between [husband and wife] is strictly taboo . . . until the child can walk. But the stricter rule is to abstain from intercourse until it is weaned—that is, some two years after its birth— and this stricter rule is said always to be observed by men in polygamous households" (Malinowski 1929: 197).

Providing international reports on behavior rather than taboo lengths, Brewis and Meyer (2005b) explored a sample drawn from nineteen countries in Africa, Asia, and the Americas to investigate various influences on married women's recent sexual behavior. The data were drawn from masses of quantitative Demographic and Health Surveys conducted in countries such as the Philippines, Kenya, and Bolivia, with health workers having asked married women with children about their most recent sexual behavior. Through statistical techniques, the authors modeled the fluctuations in women's sexual intercourse across their prime reproductive years. The fluctuations are, in one word, dramatic. As one example, women in the Dominican Republic who have been married around five years vacillate between having virtually no sex during the early postpartum period to a probability of daily intercourse of around 18 percent when they are cycling. To take another example, women in the Philippines have almost no sex early postpartum but a daily estimated probability of around 10 percent when cycling. While some of these frequency estimates seem low, perhaps due to methodological limitations, the overall pattern shows clear troughs in coital frequency during the early postpartum phase and far higher rates while women are cycling.

In a more geographically circumscribed view, drawing on a review of fifty-nine predominantly North American and Western European societies, all of which allow only monogamous marriage, reductions in sexual behavior postpartum are also typical but marked by relatively short durations of postpartum abstinence when viewed from a cross-cultural perspective or with respect to other primates. As the author of this review summarizes, "Compared with the prepregnancy period, coital frequency is reduced in most couples during the first year after birth" (von Sydow 1999: 36). Another recent scholarly paper on Western sexuality points to the magnitude of these effects in its tongue-in-cheek title: "Which is worse for your sex life: Starving, being depressed, or a new baby?" (Carter et al. 2007).

Yet, additional patterning indicates that marital quality is associated with earlier resumption of sexual behavior postpartum in Western samples (De Judicibus and McCabe, 2002). Better relationships may be reinforced by sexual behavior, perhaps also consistent with a relationship-maintenance function to sex that females employ to foster continued investment by a valued partner. As Meston and Buss (2007) put it: "Women often use sex in many different ways to protect their relationships." (111), adding that "women are motivated to have sex to mate guard because the costs of not doing so can be catastrophic. A woman who fails at mate guarding can lose the support provided by her mate, whether it be material or emotional" (113–114). The relationship-maintenance function of earlier and more frequent postpartum sex may help a woman address a partner's sexual desire and encourage his continued investment in her and their offspring.

LACTATION: GOOD FOR BABY, BAD FOR SEX

The physiology of lactation, briefly described above, helps account for its inhibitory influence on female sexuality. We can add other aspects of lactation, as well, to its impact on sexuality. The oxytocin increases that women experience during lactation help foster milk release. Oxytocin can also increase maternal attachment to her infant. In a study of mothers of young children in Israel, maternal blood oxytocin levels were positively correlated with several measures of maternal behavior toward young children (Feldman et al. 2007). The same attachment and investment a mother experiences with her infant may, however, come at the expense of her emotional investment with a romantic partner. Further, the physical contact women experience with lactation highlights the role of breasts as food providers rather than as sexual stimuli. While the benefits of breast-feeding—from nutritional to immunological to enhancing maternal-offspring relationships—are remarkable, one thing lactation is not good for is a couple's sex life. Lactating women tend to experience lower sexual desire than nonlactating women, and men whose partners are nursing report lower sexual satisfaction (Hyde et al. 1996).

Lactation and relationships are tied together in other ways. In a cross-cultural study investigating correlates of breast-feeding duration, several patterns emerged (Quinlan and Quinlan 2008). Hunter-gatherer mothers nursed longer than mothers in societies of agriculturalists, pastoralists, and other types of societies. Indeed, recently studied foragers tend to nurse offspring for around three years prior to weaning, with supplementary foods being introduced at around six to nine months (Marlowe 2005). Among other patterns, women nursed longer when divorce rates were lower and when they had access to other childcare providers, such as older siblings and others. These patterns make sense. Women living in societies with lower divorce rates can afford to prolong investment in their current offspring before shifting attention to a potential next of kin (another baby); this is another incarnation of the life history trade-off between current and future reproduction. Women who have access to other caregivers can afford to nurse longer, since they may be able to turn over the care of older children to others. Conversely, women without a partner's (that is, a father's) support, may wean earlier as part of a strategy to find a new, potentially investing mate.

In contemporary urban contexts, considerable debate concerns whether women's labor is compatible with breast-feeding. Some of the same factors are relevant: if women can rely on a consistent partner for support and resources, they may be able to prolong breast-feeding, including choosing forms of paid employment that might be more compatible with breast-feeding. However, formal education, including of older children, and sparse availability of female kin or other caregivers can push women toward a faster return to paid work, suggesting that reductions in alternative caregivers may also be associated with shorter lactation duration. The low for lactational duration, however, at least in the United States, occurred in the early 1970s, when use of the Pill had expanded, feminism was flourishing, milk formulas were marketed as progressive and positive, and many women opted out of lactation altogether.

In addition to the social context to lactation, energetic factors play important roles. In a "metabolic load" model, Peter Ellison (2001) has postulated how the energetic costs of lactation relative to maternal energetic condition (for example, fatness) and changes in energetic condition (changes in fatness) affect the duration of lactational amenorrhea. This is an important model because it looks beyond nursing frequency and intensity as a driver of lactational amenorrhea. As an example, even though !Kung foragers of southern Africa and the Toba of Argentina both nurse frequently on demand, Toba women have a much shorter duration of lactational amenor-rhea and interbirth interval. In large part, this is because their bodies have more fat on average and in turn are better able to return to an energetic condition making it possible for them to resume ovulatory cycling sooner.

Through research tracking C-peptide of insulin levels in the urine of postpartum Toba women, Peter Ellison and Claudia Valeggia (2003) found that this metabolic hormone rises around the time women resume their ovulatory cycle. In other words, this energetic signal indicates that women's relative metabolic condition permits a return to ovulatory cycling; they have the energetic resources to start cycling and perhaps have another baby. In a wider scholarly picture, this metabolic-load model is an important addition to Aristotle's ancient wisdom, pronounced in *Generation of Animals*: "While women are suckling children, menstruation does not occur according to nature, nor do they conceive; if they conceive, the milk dries up." For not just lactation duration but also women's relative energetic status help account for the pattern of lactational amenorrhea.

So what do these energetic factors have to say about postpartum sexual-ity? By yielding variation in the duration of lactational amenorrhea, these factors also alter the timing of when women may be physiologically primed to resume sexual behavior (LaMarre, Paterson, and Gorzalka 2003). Women who are not cycling may, as noted above, be more apt to report sexual pain due to vaginal dryness or reduced sexual desire. In a wider comparative primate context, ape females in the wild do not typically resume cycling for years, and at the same time, they engage in little sex while in lactational amenorrhea. Humans, even by hunter-gatherer standards, resume postpar-tum ovulatory cycling faster than would be expected. Several energetic and social factors may be central to this human-derived trait: human female foraging strategies enable them to acquire more energetically dense extractive foods (such as tubers obtained with a digging stick), and food processing techniques (from pounding to leaching to cooking) further amplify the energetically rich diet available to human forager females. Add to that other food providers, including grandmothers and dads, and women may enjoy further energetic benefits, enabling their more rapid resumption of postpar-tum ovulatory cycling (Reiches et al. 2009).

In a variety of subsistence populations, from the highland horticultural Gainj of New Guinea to *ribereinhos* of the Brazilian Amazon (Piperata and

Guatelli-Steinberg 2011), the duration of female postpartum amenorrhea ranges from around ten to twenty months. Among the Ariaal, pastoralists of northern Kenya, 80 percent of Ariaal women report resuming menses by eighteen months after giving birth (E. Miller 2011). These durations of lactational amenorrhea are short by comparative ape standards, but long in comparison with well-fed, sedentary, nonlactating, healthy women living in contemporary cities, where women may begin cycling within a couple of months postpartum.

Eventually, though, females do begin postpartum cycling again. Interestingly, females typically begin cycling even while breast-feeding, as the relative metabolic load tilts toward conditions favorable for preparing for a subsequent pregnancy. This is another of those simple yet profound observations concerning the evolutionary nature of human sexuality (Lancaster and Kaplan 2009). In ancestral or natural-fertility human populations, or among a typical nonhuman primate community in the wild, we would find many of the reproductive-age, cycling females winding down their lactation. This means that female choice in sexual partners takes place within a context of lactational hormones and with a weanling in hand, in addition to male sexual partner preferences mapping onto the realities of lactating females. Does this matter in any measurable way?

Among the Hadza of Tanzania, breast-feeding women preferred higher pitched, less masculine voices compared with non-breast-feeding women (Apicella and Feinberg 2009), one indication that the lactational status of women may be a relevant factor for mate choice. Michelle Escasa-Dorne recently investigated the effects of breast-feeding on women's sociosexuality and mate preferences in Manila, in the Philippines (unpublished data). She found, too, that lactating women expressed a preference for higher pitched male voices. However, breast-feeding women did not prefer less masculine male faces compared with nonlactating, cycling women. In an Internet-based study Peter conducted with Escasa-Dorne and Heidi Manlove, they also found that breast-feeding (or cycling status) did not affect postpartum women's preferences for male facial masculinity or femininity (unpublished data). Thus, several studies point toward differences in women's evaluations of male voices, but not male faces, depending on breast-feeding status.

Suppose a woman has a baby but perceives she will receive inadequate support from a partner. What impact might that have on decisions to continue investing in an offspring, or to shift attention to finding another partner? The regular occurrence of abortion and infanticide in the historical and cross-cultural record in the contexts of perceived inadequate partner support and young female ages indicates that women often opt out of current reproduction to wait for a later, perhaps better time (Hrdy 1999). In other cases, women may keep their progeny but seek an alternative mate. Interestingly, one of the most cross-culturally salient reasons women divorce their husbands is inadequate support (Betzig 1989), indicating that marital dynamics hinge in part on male investment. Through study of modal divorce patterns in some fifty-eight societies, Helen Fisher (1989) found that divorce was more common after a couple had been together around four years. She interpreted this pattern to mean that those couples remain together long enough to raise a dependent offspring to a point at which it could be weaned and survive without intensive biparental care.

If a woman perceives inadequate partner support, waiting several years to divorce might enable her to recruit his investment long enough to ensure her offspring will survive, before she shifts her romantic attentions elsewhere. More overt female solicitation of new mates in the peripartum period might detract from even inadequate perceived parenting effort provided by a child's father. Still, we know virtually nothing about how women with children go about separating from one partner and seeking another, an area where we will surely see more research in the future. We do know that, whether in the United States, among the Tsimane of Bolivia, or indeed in the wider set of societies studied by Helen Fisher, divorce is more common among younger couples, when juggling considerations of partner quality and investment from the standpoint of current versus future reproduction are especially salient.

ON MANLY PERIPARTUM PURSUITS

Envision a male fantasy. What if the manly beasts among us, ourselves included, could gestate a baby? What if we could nurse our offspring? These are, at first glance, preposterous suppositions, and yet some of the reaches of vertebrate biology encompass variation falling within these bounds. Take, among gestating males, syngnathid fish, or pipefish and sea horses. What these males do is akin to pregnancy: females deposit eggs in a male's pouch that he fertilizes and then tends in his pouch until they are ready for release into the cruel oceans of the world. More directly in the human realm, Malcolm Potts and Roger Short speculated on human male pregnancy: "In theory, however, a fertilized egg placed in an ectopic site in the male could last the whole of pregnancy and produce a viable offspring, just as it sometimes does in a woman" (Potts and Short 1999: 129). Even more, a recent story of a genetic female who underwent a sex change and then gave birth has its own website (www.malepregnancy.com).

As for male lactation, two species of mammals—Dayak fruit bats and masked flying foxes—have been found to produce small amounts of milk, though it is not clear if they nurse their young (the nipples of the male fruit bats did not look like they had been used for suckling, careful observations suggested) (Kunz and Hosken 2008). Some human males also possess an ability to produce milk and even breast-feed; this may occur during unusual physiological states, such as during refeeding after near-starvation and among overweight men (Bribiescas 2006). The shifts in sex steroids under such circumstances can permit these unusual human male states. Peter has been told by two students that they knew of men who had actually nursed their offspring, presumably providing small amounts of milk as a kind of supplement to the mother's primary source, and a YouTube search can yield the occasional man featured for his devoted lactating abilities.

In the remainder of this section, we focus on the impacts of pregnancy and the postpartum period on male sexuality. We leave aside the fantastical world of pregnant or lactating men to focus on the more typical male experiences during these times. How does male sexuality fare amid pregnant partners and nursing infants? Some of the material above indirectly alludes to impacts on males, but we will attempt to crystallize some patterns here.

The male libido is less dramatically affected by a partner's pregnancy and postpartum investment than is the female partner's libido. Men do not undergo the same dramatic physiological shifts during pregnancy and postpartum that, in simple terms, favor investment in the new progeny instead of in begetting another one. To be sure, a nascent literature indicates that male psychology, physiology, and health can be affected by a partner's pregnancy and postpartum investment (Gray and Anderson 2010). As an illustration, the most consistent hormonal correlate in males of fatherhood is having lower testosterone levels, a finding documented in places as diverse as Boston, Beijing, Kingston, and northern Tanzania (Gray and Campbell 2009). In the most robust of such studies, relying on a sample of around five hundred men in a longitudinal research design, set in Cebu City, Philippines, Lee Gettler and colleagues (2011) found that men with higher testosterone levels were initially more likely to marry, but experienced more dramatic declines in testosterone over time once they became fathers, compared with married nonfathers and unmarried men. A few other studies indicate increases in male prolactin levels in response to infant stimuli, among other effects in the handful of investigations into the hormonal and neural correlates of paternal care.

Even if males exhibit some physiological impacts of a partner's pregnancy and postpartum social life, the impacts on male libido are modest (Gray and Anderson 2010). One pattern manifest in von Sydow's (1999) review of predominantly North American and European samples is that libido in the male diminished somewhat during his partner's third trimester, and it remained lower in approximately 20 percent of men up to a year postpartum. The magnitude of these effects pales in comparison to those observed among females. As a behavioral assay of libido, several studies tracking masturbation rates in the United States and Germany found that, despite diminished coital frequency, masturbation frequency was little altered from pregnancy to the postpartum period. If we add to that equation likely lower coital frequency during those time frames, then constant masturbation rates may be indicative of slightly declining libido (not making up for lower coital frequency with higher masturbation frequency, in other words), and thus we have another line of evidence pointing toward modest impacts of a partner's peripartum period on male libido.

As part of a 2011 fatherhood study conducted in Jamaica, approximately 2,500 men whose partner had given birth in the previous day or two were asked questions about their recent sexual behavior and desire. This study is part of a larger longitudinal study of parents and children in Jamaica, and Peter is involved in the fatherhood side of it. Data entry, much less analysis, has only begun, yet a preliminary look at responses of approximately fifty fathers indicated that a notable fraction reported diminished libido in the recent weeks, in addition to engaging in masturbation or sex with an additional partner. Findings from this large-scale quantitative study will surely help expand our understanding of fatherhood and sexuality (for example, how relationship quality factors into male sexuality).

Many of the studies that have examined impacts on men's libido during the peripartum period are likely some of the same ones apt to document any such effects, if they exist at all. These European and North American samples involve men living in societies where only social monogamy is legally permitted, and where fathers tend to be relatively invested in a partner and in direct offspring care, by cross-cultural standards (Gray and Anderson 2010; Marlowe 2000). If, in these societies, the impacts of the peripartum period on male libido are modest, then they are likely even less pronounced in other societies where men have multiple available sexual partners (societies permitting polygynous marriage, for example) and where men devote little time to direct childcare. In this observation we hearken back to a pattern manifest in the earlier cross-cultural surveys of sexuality during pregnancy and postpartum: the role of polygyny. In polygynous societies, taboos against sexual activity were more pronounced, and they kicked in at earlier stages of pregnancy compared with taboos in socially monogamous societies. Further, postpartum sex taboos had longer durations among polygynous societies. The important point with polygyny is that a male has multiple, socially sanctioned potential sexual partners—his multiple wives. So even if one wife's reproductive state places her "off limits," he may have another partner whose reproductive state makes her eligible for sex. Or he may seek to marry yet another partner, another woman to marry and have sex with. Among the Turkana, pastoralists of northern Kenya, Pierre Lienard notes that some men seek another wife or other sexual partner during this exact time (personal communication).

In socially monogamous societies, like the United States, male and female conflicts on peripartum sexual desire may be more visible. In a polygynous society, by contrast, a wife in lactational amenorrhea may go about her business without the unnecessary indulgence or pestering of her spouse, since her co-wives may serve his sexual needs. But where socially monogamy prevails, then the sex differences in sexual desire reach their peak during the peripartum period. This is a notable feature of human sex differences in sexual desire that has not received the attention it is due. Yet even the renowned sex scholars Masters and Johnson cautioned long ago on the potential significance of the peripartum period for a couple's relationship dynamic, albeit in socially monogamous twentieth-century society in Saint Louis, Missouri: "Six weeks before and six weeks after delivery usually are proclaimed restricted periods by medical interdiction. Many male partners first break marital vows during this three-month period." (Masters and Johnson 1966: 168)

For socially monogamous males whose partners are in the throes of the peripartum period, what does this mean for their sexual behavior? If Masters and Johnson are correct, this may be a particularly vulnerable time for sexual fidelity in a couple's relationship, particularly if this is a young couple. We do not know of quantitative data specifying how often men cheat during this period. We also do not know of any data indicating whether men in these circumstances appeal more often to prostitutes, where accessible, as alternative sexual outlets. However, recent research emphasizing women's active role in having sex for relationship-maintenance functions suggests that women may be engaging in sexual behavior in the peripartum period more than otherwise expected in order to reduce the chance that a valued partner will stray. In the end, in a socially monogamous context, the fallout of competing male and female sexual desires is likely some kind of compromise. The sexual battlefield, or bedroom, or some other sexual setting (choose your preferred image), has a lengthy evolutionary history and ongoing social significance.

WHY CARE ABOUT PERIPARTUM SEXUALITY

As we wrap up this discussion of peripartum sexuality, we underscore some of the reasons why these patterns are important. One is that most of our ancestral social lives unfolded in a nexus of females who were pregnant or lactating during the bulk of their reproductive years. Most hominin conceptions occurred while women were breast-feeding, and the profile of adult mates was dominated by already-partnered prospects, most with offspring. The abundance of nulliparous, cycling women on a university campus or in some other corner of contemporary social life is quite derived and, frankly, weird when situated within this evolutionary perspective.

A second reason is that peripartum fluctuations in women's (and their partner's) sexual behavior are the evolutionarily "normative" pattern. Our ancestors lived and mated in a world characterized by rises and falls in libido and sexual behavior across the peripartum period. That knowledge is important to any would-be or actual parents faced with these realities. That reality is also relevant to therapists or clinicians who advise couples about what constitutes "normal" sexual behavior. A nascent clinical literature discusses peripartum shifts in sexuality, in some cases implicitly comparing peripartum patterns with a baseline of sexuality of the cycling woman (Abdool, Thakar, and Sultan 2009); that comparison may be one that the couple or the clinician aspires to, but it is not the evolutionary normative one. This point was driven home forcefully in an Internet-based survey of more than 200 postpartum women's sociosexual behavior, conducted by Peter, Michelle Escasa-Dorne and Heidi Manlove (unpublished data). Employing one of the gold-standard measures of female sexual function (the Female Sexual Function Inventory), the average score on it (21.5) was well below the supposed clinical cutoff (26.6) for defining "sexual dysfunction" (see Wiegel, Meston, and Rosen 2005). In other words, the typical postpartum women's sexual function in this survey classified her as sexually dysfunctional. Does it make any sense to view these low female sexual function scores among postpartum women as "dysfunctional"? Does it not make more sense to view those low sexual function scores as the normal fallout of fluctuating sexuality across the peripartum period?

Some very prominent sex research projects have overlooked the obvious and profound importance of reproductive status (such as the various trimesters of pregnancy, and postpartum lactational amenorrhea) in their compilation of sex data. In the only two U.S. attempts to document nationally representative probability samples of sexual behavior (Herbenick et al. 2010a; Laumann et al. 1994), neither explicitly incorporated pregnancy or postpartum variables, although plenty of space was given to other salient variables including sex, age, and socioeconomic status. For such reasons, research and practice based on an appreciation of peripartum shifts in human sexuality is likely to yield continued and fruitful insights.

CHAPTER ELEVEN

The Sands of Time

Aging and Sexuality

> Generally, the intensity of physiologic reaction and duration of anatomic response to effective sexual stimulation are reduced through all four phases of the sexual cycle with the advancing years. Senile involution of the target organs (breasts, labia, vagina, uterus) is evidence of postmenopausal sex-steroid starvation. Regardless of involutional changes in the reproductive organs, the aging human female is fully capable of sexual performance at orgasmic response levels, particularly if she is exposed to regularity of effective sexual stimulation.
>
> —William Masters and Virginia Johnson, *Human Sexual Response*

ALTHOUGH MASTERS AND JOHNSON'S prose could hardly be compared with poetry or a sultry romantic novel, their groundbreaking research on the mechanisms of human sexual response showed that changes occur with age. Women and men both exhibit diminished capacities for sexual response with advancing age. Some of those age-related changes reflect, at a proximate level, declines in steroid hormones, as alluded to in the passage above (sex-steroid starvation is a vivid phrase). However, there is considerable variation in the specific manifestations of age-related change in sexual function across populations and time, a consequence of human biological changes playing out in specific socioecological settings. Further, any human age-related patterns of sexual function need to be placed within a wider phylogenetic and functional account in the first place. Why do human libidos and sexual responses diminish at the ages they do?

THE FOUNTAIN OF YOUTH

Explorers such as Juan Ponce de León to nineteenth-century doctors like Charles-Édouard Brown-Séquard have long sought the fountain of youth. In de León's case, he imagined finding it in Florida, a haven today for sun-seeking retirees. In Brown-Séquard's case, he tried his luck with injected extracts of guinea pig and dog testes, thinking that something in them gave him a new vigor. These attempts aside, evolutionary scholars have much to say about who holds the fountain of youth, and they would point to some surprising candidates. Take tortoises, such as the ones Darwin encountered in the Galapagos (meaning tortoise in Spanish) Islands, some of which out-live the oldest people alive today (Ricklefs and Finch 1995). Or consider bowhead whales, one of which was found in 2007 with a harpoon in it embedded that dated to 1879, indicating that it had been struck over a cen-tury before and also revealing that some bowhead whales, too, can outlive the oldest humans (Haag 2007). Really, though, far more plants than ani-mals hold the secrets to longevity: individual creosote bushes can live hun-dreds of years in the Mojave Desert, with a cloned group (offshoots of a single plant) estimated to be a remarkable 11,700 years of age (Vasek 1980); bristlecone pines in parts of the Western U.S. have reached over 6,000 years of age, and some long-lived eucalyptus trees in Australia similarly hold their own for thousands of years. What do trees know about living long lives that the smartest doctor does not?

An evolutionary perspective on aging builds on the concept that species should invest in maintenance functions (from chemical defenses in trees to immune systems in mammals) if the circumstances pay. So if members of a species, like tortoises or bristlecone pines, for example, continue to repro-duce and are subject to low-enough extrinsic mortality (mortality that is difficult to avoid, such as predation), then they can afford to maintain them-selves, investing in their own capacities for continued survival and in turn living long lives. This is basic life history theory making sense of species differences in senescence, or loss of function with advancing age. But is there any empirical support for these theoretical accounts?

As Steven Austad (1999) showed, opossums adhere to expectations. Through a contrast of mainland opossums living in the U.S. South with opossums living on a nearby island, where they had been isolated from predators for thousands of years, Austad found support for the hypothesis that the island opossums would senesce at slower rates than their mainland cousins. Under the idea that the islanders faced lower extrinsic mortality due to lower rates of predation, he found that the island opossums died at slower rates with advancing age (meaning the average islander lived a longer life) and also scored lower on a physiological measure of stress (wear and tear to their tail tendon fibers). The mainlanders, compared with the islanders, lived fast and died young. These opossums are hardly outdone, though.

Bats tend to outlive their similarly sized mammalian cousins, such as squirrels (Austad 1999). Why? Consistent with the theory, bats can more readily escape would-be predators by flying away, enabling them to invest more energy in maintenance functions. Many birds, such as parrots, also live long lives, in part because lower predation rates enable them, too, to plan for the future by investing in themselves and their offspring. Perhaps to the chagrin of some parrots, their long lives mean that they may outlive the people who buy and keep them as pets. In one other example, among social insects such as termites, the better-protected, egg-laying queen tends to senesce at a lower rate than other hive members engaged in more risky activities, such as foraging and colony defense. All told, such examples illustrate how age-specific fertility and mortality schedules shape the rate of senescence among organisms.

MONKEYING AND APING AROUND

What does all of this discussion about evolution and senescence mean for humans? Against a phylogenetic backdrop of other primates, many Old World monkeys, such as rhesus of India or vervets in Africa, may live into their twenties in the wild but longer in captivity. As an illustration, the oldest baboons in the wilds of East Africa are thought to reach their early thirties (ages of long-lived species in longitudinal studies may be estimates, since births may not have been observed, and animals may come and go; Bronikowski et al. 2011). Other primates, from lemurs to squirrel monkeys to muriquis (woolly spider monkeys), may reach maximum estimated life spans from the twenties to around forty, in the case of muriquis. All of these primates senesce at considerably faster rates than humans.

Apes tend to senesce at slower rates than lemurs as well as Old and New World monkeys. Tallying wild chimpanzee mortality data from five field sites in Africa, virtually every chimpanzee was dead by age fifty, with only a few stragglers at one field site still hanging on at that age (Hill et al. 2001). Still, the overall pattern of senescence and average life span among these wild chimpanzees surpass those of nonapes. Among wild orangutans, some may live into their fifties too (Wich et al. 2009). Comparing wild and captive orangutans, more recent studies of captive orangutans indicate that they live longer lives, on average, than their wild counterparts; however, that was not true during the mid-twentieth century, when captive environments (including medical care, diets, social deprivation) were less conducive to survival and captive orangutans lived shorter lives on average (Wich et al. 2009). Gorillas senesce at faster rates than either chimpanzees or orangutans, consistent with their faster rate overall of juvenile development and earlier age of puberty. Captive gorillas may survive into their forties, longer than they would typically live in the wild (Margulis et al. 2007).

Females tend to live longer than their male counterparts in most primate species, a sex difference in mortality thought to reflect in good part the greater sexual selection pressures operating on males to undertake risks for potential mating opportunities (see Geary 2010; Hrdy 1981). Consider a baboon male, who emigrates around the time of puberty in the hope that he will find a willing mate in some other group, putting himself in harm's way (from predators and other baboons) on his journey, and then, if he successfully enters a group, having to deal with many other baboon males who may not want him around as a competitor. The exceptions to this pattern wherein males and females exhibit similar mortality profiles with advancing age—titi monkeys, gibbons, lemurs—are also socially monogamous primates, meaning the reproductive stakes are more similar between females and males (Allman et al. 1998). Indeed, across various studies of mammals and birds in the wild, polygynous species typically display sex differences in mortality (males having shorter lives, on average), whereas among monogamous species male and female mortality profiles are similar (Clutton-Brock and Isvaran 2007).

The species-specific, sex-specific, and context-specific (captive versus wild) nature of senescence is important to understanding the evolution of human senescence, and eventually to making sense of changing sexuality with advancing age. Humans appear to senesce at slower rates than apes living in even the best of captive environments. This is nicely shown by contrasting recent studies of human hunter-gatherers and horticulturalists, such as the

!Kung, Hadza, Ache, and Tsimane, which reveals that they have longer lives than apes living either in captivity or in the wild (Gurven and Kaplan 2007). Of course, recent human hunter-gatherers also have technological and cognitive shields (controlled use of fire, projectile weapons, knowledgeable elders) to environmental insults that their more distant foraging ancestors lacked (Marlowe 2005). How far back mortality patterns from recent foragers can be projected to more distant foragers is an open question. The available bio-archaeological samples (such as estimated age profiles of past hominin species) contain few old individuals, although older individuals become more common more recently in the genus *Homo* (Caspari and Lee 2004). When combining this line of evidence with other paleoanthropological data on rates of development, brain size estimates, and archaeological evidence of tool use and subsistence, along with the comparative ape and contemporary hunter-gatherer, the best guess would be that the slower rate of human senescence took root in the genus *Homo* at some point within the past two million years (see Lieberman 2011). A more specific educated guess would be that only with modern humans ourselves, within the past 150,000 years or so, have people been living long enough to even contemplate a postreproductive existence (Crews and Stewart 2010).

Here is where postmenopausal grandmothers enter the discussion. Considerable debate has arisen concerning why human females live well beyond their reproductive capacities (Hawkes et al. 1998; Williams 1957). The current formulations of the grandmother hypothesis recognize that human female reproductive senescence is virtually identical to that of chimpanzees (for example, rates of follicular atresia are remarkably similar between chimpanzees and humans of similar age; Jones et al. 2007), with an extended life span the derived feature that gives rise to postreproductive female existence. While some argue that the cause of extended human life spans is grandmaternal care (that is, a postreproductive female caring for her grandchildren generating indirect fitness benefits favoring her continued survival), others view grandmothering as a by-product of life spans that have been extended for other reasons. We side with the by-product view of grandmothering for several reasons, including the fact that selection acting late in life tends to be weak. Additionally, the comparative nonhuman evidence convincingly demonstrates that postreproductive female lives are hardly unique to humans, although humans, even among forager groups, appear to spend a significantly longer proportion of their lives in postreproductive states than is the case in postreproductive captive animals. Among captive Japanese macaques, for example, almost all (twelve of fourteen) that survived to reach age twenty-five had ceased reproduction (Fedigan and Pavelka 2011). Among older gorilla females living in U.S. zoos, five of twenty-two had ceased cycling altogether (were postmenopausal, in other words), and another 32 percent of older females displayed irregular cycles akin to a perimenopausal state. Interestingly, the peri- and postmenopausal females no longer maintained the same social affiliations with silverbacks, though whether that was because the females were not interested in doing so or the males were not so interested in the females (or both), we do not know (Margulis et al. 2007). As Finch and Holmes put it forcefully, after reviewing the evidence that in many captive mammalian groups, from mice to opossums to golden lion tamarins: "A gradual pattern of female reproductive aging, with substantial postreproductive periods consistent with expectations for mammals with fixed oocyte pools, appears to be a conservative mammalian trait." (Finch and Holmes 2010: 89). Put another way, the overall theoretical and empirical evidence is most consistent with the view that human senescence has slowed recently in hominin evolution (perhaps only with modern humans), has a number of parallels with captive nonhuman animals, and has given rise to common and extended postreproductive lives that can be spent having sex and caring for descendant kin, among other activities.

A few other evolutionary angles on human life spans remain. While some diehards do not think anything about human senescence requires special explanation (including Austad himself), we still believe that the extended postreproductive stage among recently studied foragers (compared with, say, coddled lab rats, or captive apes) requires explanation, even if only to ask how that might differ from ancestors a mere 200,000 years ago in Africa. We suggest, as have others, that modern humans themselves have been responsible for extending their lives, in a process biologists refer to as niche construction (Odling-Smee 2003). By living in larger groups (less predation), employing an increasingly specialized and varied technological arsenal (enabling access to new plant and animal resources, among other things), and enhancing capacities for social information exchange (learning and telling others where to find water, for example), human social and cultural practices undertaken during development and adulthood could have made the difference. The focus of selection, in such a scenario, is on developmentally plastic traits, expressed earlier in life, with later consequences for extended life spans. Natural selection, as two scholars put it recently, "may never have acted directly to extend human life span or postreproductive survival. Rather, life spans extending well beyond those of other large-bodied apes appear to be an epiphenomenon of natural selection acting on aspects of early life history, altricial offspring, slow growth, and an extended growth phase that stretched out the period for attaining maximal reproductive potential" (Crews and Stewart 2010: 545). Such a view highlights the role that we humans may have played in extending human life spans recently in evolution, giving rise to conundrums that become increasingly relevant, such as how a postmenopausal woman should spend her days and nights. Of course, changing global demography means that such dilemmas, once rare in human evolution, are now quite common around the world, as more and more people are living to advanced ages (Vaupel 2010).

After all those thoughts on grandmothers and longevity, what do older males have to say about themselves, about their own existence? Though grandfathers have hidden in scholarly shadows, they might want to share a few words too. They might agree with the points about extended human life spans but would want to speak up on the subject of age-specific fertility (see Bribiescas 2006; Gray and Anderson 2010; Marlowe 2000). If considering the available estimates of age-specific fertility in forager societies such as the !Kung, Hadza, Agta, and Ache, these would illustrate that the trajectory of male age-specific fertility has a similar shape to that of females in all of these populations, in large part because most human reproduction occurs within long-term bonds (intertwining a couple's fertility) (see Marlowe 2010; Tuljapurkar, Puleston, and Gurven 2007). Still, male age-specific fertility would be shifted toward slightly older ages than females' fertility patterns in these same populations, in part because males tend to be older at marriage and may be more likely to reproduce after divorce. Yet male reproductive senescence does dwindle downward, with age-specific fertility quite low by the fifties and onward. This could suggest an evolutionary rationale for why male reproductive equipment may be fully maintained through the twenties and thirties, start to show some downward movement in the forties, and eventually pick up steam and decline rapidly by around the fifties to sixties. The fact that male age-specific fertility is spread across a greater span of the adult years, including at older ages, is likely more of a derived feature of human life histories than the age spans at which human females reproduce. In other words, human males continue to father offspring at older ages than other apes, whereas the age-related profile of human female reproductive senescence (in thirties and forties) looks more similar to that of female chimpanzees. While estimates of age-specific fertility in wild primates such as apes are challenging to obtain, in part because of the difficulties of paternity assignment, male chimpanzee, bonobo, gorilla, and orangutan fertility is likely confined to a narrower age spread during times when males are able to successfully compete for mates (see Strier 2011). With human males, who stake the bulk of their reproductive success in long-term partnerships and repeated reproductive bouts, this provides a rationale (maintained age-specific fertility) for males to stick around rather than keel over, and may help account for less pronounced sex differences in senescence in humans compared with, say, orangutans or gorillas.

THE LIVES OF OLDER LADIES

Among !Kung women after age forty, when they have ceased reproduction, "An older woman may take a younger man as a lover and do this more openly. . . . There seems to be less danger attached to it, since husbands are often away for long periods or perhaps are less jealous. In a number of recorded cases, after divorce or widowhood a woman has married a man younger than herself, in some instances ten or twenty years younger" (Lee 1992: 41). This portrait of a hunter-gatherer population illustrates some of the changes in older women's sexuality. No longer fertile, a woman's sexual liaisons are less dangerous since they cannot yield an offspring that a cuckolded husband will be asked to invest in. A female may be better positioned to flaunt her sexuality in order to entice attention and resources from younger men who perhaps lack the social might to attract a woman of reproductive age. To make just a few observations on the changing nature of female sexuality with advancing age, let us consider some of the major life history transitions structuring a changing nature of womanhood.

The most marked aspect of female aging is menopause, or the cessation of ovulatory cycling. While menopause is often defined as the lack of menstruation for one year, and typically occurs around age fifty in most populations of women, female reproductive function declines long before menopause actually takes hold (Sievert 2006). Indeed, the peak of human fertility occurs between the ages of twenty-five and thirty-five, with diminished chances of conception and delivery thereafter (Ellison 2001; Vitzthum 2009). Among the !Kung, whose limited diet, high activity load, and frequent nursing constrained female energetic resources, their typical age of last reproduction was in the thirties, whereas for women in other forager populations, such as the Hadza, it might be around forty instead. All that said, female cycles become less regular, less fertile, and eventually cease altogether with menopause, heralding an important transition in the lives of women in many societies. Indeed, we might expect the menopausal transition to entail downregulated sexual function (Dennerstein, Alexander, and Kotz 2003), alongside changes in other feminine domains.

Put in terms of life history theory, a woman leaving behind her fertile years faces other fruited plains. These can include a release of social constraints on her sexuality, alongside an expanded potential for political power and enhanced investment in descendant kin. With her physiology shifting away from support for mating, she can rechannel her time and investments to enhance the survival and well-being of her children and grandchildren, perhaps also maintaining an ongoing relationship with a partner (such as her spouse). A study investigating women's preferences for male faces found that postmenopausal women reported lower preferences for masculine faces, one indication of women's shifting priorities with age (Little et al. 2010). This discussion invokes yet again the grandmother hypothesis, with the view that while a postmenopausal woman is alive, she may have beneficial effects on her descendants' survival and fertility.

Cross-cultural research indicates that women do often experience an uptick in their power in the family and more widely with menopause. As one review of menopause in many small-scale societies indicated, upon reaching menopause, women are released from some of the feminine ideals, meaning that they are subject to fewer restrictions (see Frayser 1985). That may mean a postmenopausal woman in Saudi Arabia no longer requires a family chaperone when going about her business. It may also mean, as it does among the !Kung, that older women need not exert such care to conceal their genitals (their pubic aprons and postures need not be so carefully guarded). Any society with menstrual taboos, calling for women to avoid certain foods or certain activities, or maybe to find their way to a menstrual hut, are out the window— they no longer apply. With fewer restrictions on their behavior, postmenopausal women are better able to expand the scope of their influence.

Grown children who might themselves be entering the reproductive arena may be another sphere in which postmenopausal women exert greater control (Kerns and Brown 1992). Older women can play instrumental roles in marital transactions, particularly when these are arranged. In societies with patrilocal residence, where a woman as a young bride had moved into her husband's household under his mother's watchful eye, that woman is now the watchful eyes; no longer so subject to the pressures of others, she is more able to exert similar pressures herself. Having to devote less time to intensive maternal care of young children, postmenopausal grandmothers have greater opportunity to guide those older children and other younger members of the community—helping to mobilize others for cooperative labor, for religious groups, for other political ends. Putting flesh to these generalizations, Kerns and Brown (1992) compile ethnographic accounts of middle-aged women during this transitional phase. In one account, of the Bakgalagadi of Botswana, we learn that "as a woman approaches middle age her status improves. She is no longer under the tutelage of her mother-in-law and is normally in charge of her compound. She has adult children to support and assist her. As head of her household, she can delegate chores and responsibilities to daughters and to daughters-in-law. . . . [A] middle-aged woman, by virtue of her control over agricultural harvests, participates more substantially in public exchanges through both gift giving and sales of agricultural produce" (Solway 1992: 52).

The beneficial effects that grandmothers have on their descendants' survival and reproductive success have been quantified in a number of recent studies, in societies ranging from the Gambia in West Africa to historical Swedish agriculturalists to the Khasi in India. In a review of such work, Rebecca Sear and Ruth Mace (2008) note that in eight of ten studies the presence of a grandmother is positively associated with grandchildren's survival. In other words, a grandmother might be able to enhance the survival of her grandchildren, perhaps by helping care for the child or providing resources to the mother. As another illustration, among Trobriand Islanders, grandmothers sometimes adopt a family member, and they are more apt to adopt the firstborn child of a daughter than a later-born child. As Wulf Schiefenhovel and Andreas Grabolle (2005) explain it, "Their custom could represent a sensible adaptive choice: as young, unmarried people have the freedom to have love affairs . . . it is possible that some firstborn children may not have been fathered by the man who is the legal husband at the time of their birth" (187).

In a less explored aspect of postmenopausal women's lives, it could be that the pull of descendant kin is strong enough to move them away from their ongoing romantic partnerships. If a postmenopausal woman can exert her political influence without need for a spouse, and if she has no need for his resources, such as land or salary, then her time and investments might be more efficiently spent on her descendant kin. Such a view would suggest that, among postmenopausal women, diminished interest in sex and in ongoing romantic partnerships are sometimes adaptive. Indeed, through research Peter has conducted with Maureen Samms-Vaughan on the physiology and context of grandmaternal care in greater Kingston, Jamaica, some women expressed such sentiments (Gray and Samms-Vaughan 2010). Very few of the highly investing grandmothers in that study (such as grandmothers who lived with a grandchild) were actively partnered, whereas many of the less centrally involved grandmothers (such as women who lived with a spouse elsewhere in the city and spent some time with their grandchildren) were partnered, consistent with a trade-off between investment in a grandchild versus ongoing partnership. Further, from interviews, the typical sentiment expressed by the unpartnered, involved grandmothers was that they were not interested in finding a new partner—that their lives were now instead oriented toward a grandchild's well-being. As one woman put it in describing romantic partnerships, "Finished with that years ago," whereas when commenting on her grandchildren she noted, "They take up all of my time."

GENTLEMEN, TURN DOWN YOUR ENGINES

There has been little research on the changing lives of men from middle aged onward, thus yielding a very short section here, despite an interesting story that is waiting to be told. As noted by anthropologist Polly Wiessner, older men can play socially stabilizing influences in society. Less prone to quick trigger fingers, and with descendant kin of concern, older men may have an eye out toward constraining behaviors, by younger men, in particular, that could harm the public good. Among the Mae Enga of highland

New Guinea, for example, elder men were called on to help keep the peace after eruptions of violence among younger men (Wiessner and Pupu 2012).

Other lines of evidence indicate age-related changes in men's behavior. As Martin Daly and Margo Wilson (1988) pointed out long ago, most violence is committed by young men, with the flip side of that coin being that older men are less likely to kill each other over perceived slights to personal honor or the sexual jealousies that so often erupt into violence among their younger compatriots. One of the best predictors of male involvement in violent crime is simply age. Yet against the general background that male investment in "mating effort" (male-male competition, mate choice, and sexual coercion) declines with advancing age, there is variation in the specific reproductive behaviors demanded of men in different social contexts. If older men are allowed to marry polygynously, they may be more prone to engage in higher rates of male-male competition and be less involved in direct care of children, all else being equal. Indeed, cross-cultural work indicates that in societies with polygynous marriage men do tend to be less invested in direct childcare, but often quite invested in moral training and transmitting resources such as land or livestock to their progeny (Gray and Anderson 2010). Further, international data on remarriage rates indicate that middle-aged men, whether widowed or divorced, are far more likely to remarry than are similarly aged women, although the likelihood of men remarrying declines regularly with advancing age (Chamie and Nsuly 1981).

Some of the emotional and behavioral changes men undergo with age likely track their partners' own changes in reproductive effort. Since most sex occurs within long-term bonds—a theme of this book—a man married to a postmenopausal wife may find himself less interested sexually in her, and her in him, meaning that he may also be less concerned with attempting to oversee and defend her sexuality. As observed among the Hadza of Tanzania, men spend less time interacting with a wife, the older she is (Marlowe 2010). With their children older as well, an aging man may find his time and efforts shifting toward male-male political relationships and concern for the well-being of his grandchildren. However, that pattern can be attenuated by the local marital system: if polygynous marriage is allowed, an older man may be torn between seeking an additional wife for himself versus helping his children and grandchildren. Or a man may be divorced, perhaps if he is unable to continue providing resources during middle age, a pattern widespread in the wake of the fall of communism in Russia and many Eastern

European countries, when many middle-aged men lost their jobs and struggled to adapt (Francouer and Noonon 2006).

SEX FOR THE AGES

Although reconstructing the details is sketchy, Kublai Khan, the grandson of Genghis Khan, is thought to have continued taking wives and consorts as he aged, even in shaky health (he is thought to have suffered from gout, in part due to living a life of alcoholic excess) (Weatherford 2004). The genetic finding that one of every eight Y-chromosome variants in Central Asia is identical, and can be traced to a common ancestor living around 800 years ago, points toward the Khan legacy (Zerjal et al. 2003), but the historical material suggests that it was more likely Kublai than Genghis who was responsible for the sexual and reproductive profligacy of that era. Yet that degree of sexual activity late in life is hardly the norm, whether then or today. Among contemporary examples of the age range for sexual activity, the aging sex icon Hugh Hefner looks his years but projects an image of polygynous sexual license with the young women in his Playboy mansion. In parts of India, a couple is supposed to stop engaging in sex once their eldest son marries, a sexual change in recognition of shifting family priorities. In the nineteenth-century Oneida commune of upstate New York, young men were sexually kept by postmenopausal women, thus avoiding pregnancies while also occupying the young men so that political leaders, such as John Noyes, could have less encumbered sexual access to the group's young women.

In light of these anecdotes, what about broader patterns of sexual behavior? What are the sex lives of aging women and men like? Here, we present the results of several large, recent, quantitative U.S. studies before also contemplating patterns drawn from several international studies.

Stacy Lindau and colleagues (2007) recently conducted a nationally representative U.S. survey of sexual behavior that was based on interviews with around three thousand adults age fifty-seven to eighty-four. In answer to whether study participants had had sex in the previous twelve months, the proportion of men and women who were sexually active declined with age, and was lower at all ages among women. More specifically, 84 percent of men age fifty-seven to sixty-four reported sex in the past twelve months, compared with 67 percent among sixty-five- to seventy-four-year-olds and 39 percent of men age seventy-five to eighty-four. Among women, 62 percent of participants age fifty-seven to sixty-four, 40 percent of those sixty-five to seventy-four, and 17 percent of those age seventy-five to eighty-four reported sexual activity in the previous twelve months. The falloff in sex with advancing age is, thus, a robust feature of this study.

As for sexual practices, the vast majority reported that sex entailed vaginal intercourse. For example, 87 percent of women and 91 percent of men age fifty-seven to sixty-four reported usually having vaginal intercourse. Of other sexual practices, rates of both oral sex and masturbation dropped off steeply with advancing age. For oral sex, 53 percent of women age fifty-seven to sixty-four reported engaging in this practice in the previous twelve months, whereas 35 percent of women age seventy-five to eighty-four reported this; for males, the parallel reports are 62 percent among participants age fifty-seven to sixty-four and 28 percent of those age seventy-five to eighty-four. For masturbation, the rates drop from 32 of women age fifty-seven to sixty-four reporting this in the previous twelve months, to 22 percent among those sixty-five to seventy-four, to 16 percent among women age seventy-five to eighty-four. Parallel rates in men are 63 percent of those age fifty-seven to sixty-four, 53 percent of sixty-five- to seventy-fouryear-olds, and 28 of seventy-five- to eighty-four-year-olds. The drop in masturbation, which does not have the same negotiations with a partner as oral sex or intercourse, suggests that other factors, such as declining sexual desire, are wielding measurable effects on age-related changes in sexual behavior.

How should we interpret the sex differences in age-related patterns of sexuality? One view is that these may be indicative of sex differences in sexual desire that persist at advancing ages; that interpretation could be most sensible with respect to sex differences in masturbation. However, another major cause of these patterns is that the demographic landscape changes with advancing age: with men often older than their sexual partners, men dying at younger ages, and women being less likely to remarry (more on that later), the demographics at older ages increasingly tilt toward living females. The consequence is that "the impact of age on the availability of a spouse or other intimate partner is particularly marked among women. A total of 78% of men 75–84 years of age, as compared with 40% of women in this age group, reported having a spousal or other intimate relationship" (Lindau et al. 2007: 772). Since most sex occurs within long-term relationships such as marriage, older women simply have fewer sexual options available to them. An interesting question would be how women cope when seeking a sexual partner in such a female-heavy demographic: does this situation elicit more female-female indirect aggression, or even more same-sex female sexual behavior? Or at these advanced ages, without direct reproductive stakes, perhaps such concerns fade away, unlike at earlier ages when reproductive-age females may resort to more aggressive tactics (co-wife poisonings and harsh words serving as good examples; Jankowiak, Sudakov, and Wilreker 2005). Although researchers were not able to address these questions directly, other variables associated with sexual behavior did arise that apply to females and males alike: self-reports of poor health, for example, were associated with lower rates of sexual behavior.

Lest we get ahead of ourselves, this same study chronicles the most commonly reported sex problems. These problems differ between men and women, and sometimes by age group too. For men, 28 percent age fifty-seven to sixty-four reported a lack of interest in sex, with 44 percent of women in the same age group also reporting a lack of interest in sex, although these percentages did not change substantially with increasing age. For men, their other primary sex problems were erectile problems (31 percent of men fifty-seven to sixty-four, increasing to 44 percent of older men), climaxing too quickly (30 percent of men fifty-seven to sixty-four, though that decreased to 21 percent among men age seventy-five to eighty-four), and an inability to climax (15 percent of men age fifty-seven to sixty-four, increasing to 33 percent of men seventy-five to eighty-four), with only a small percentage of men reporting either pain during intercourse or sex not being pleasurable. For women, apart from a lack of interest in sex, their other reported sex problems were difficulty with lubrication (36 percent of women age fifty-seven to sixty-four, increasing to 44 percent among women age seventy-five to eighty-four), inability to climax (between 33 percent and 38 percent of women across age groups), pain during intercourse (between 12 percent and 18 percent of women across age groups), and sex not being pleasurable (approximately 25 percent of women across age groups). The interpretation of these sex differences in sex problems would recognize the relevance of age-related physiological changes (such as declines in estrogen postmenopausally, which diminish female libido and vaginal lubrication: Masters and Johnson 1966; Meston 1997) as well as functional considerations. (With few ancestral men and women sexually active, much less reproducing, at these ages, why maintain the sex-specific machinery needed for sex and reproduction?)

An even more recent U.S. survey yields generally similar patterns of age-related changes in sexual behavior, although relying on a wider age spread (participants were fourteen to ninety-four years of age) and different methods (data collection over the Internet, for example, rather than in interviews) (Herbenick et al. 2010a). As an example, during the past twelve months 51 percent of women age fifty to fifty-nine, 42 percent of women age sixty to sixty- nine, and 22 percent of women age seventy and older reported having engaged in vaginal intercourse, whereas among men of these same groups 58 percent, 54 percent, and 43 percent, respectively, reported engaging in vaginal intercourse in the past year. Further, 54 percent of women in their fifties, 47 percent of women in their sixties, and 33 percent of women in their seventies reported masturbating alone during the past twelve months, with comparable rates among men of 72 percent, 61 percent, and 46 percent. This survey also presented data on other time frames (the past month, and during one's life) and for a greater variety of categories of sexual behavior (distinguishing giving from receiving oral sex, for example), but the core age-related declines and lower rates of sexual behaviors among females appeared in this study, as in the earlier one. Also of note, among men in the age groups fifty to fifty-nine and older, a higher percentage reporting giving and receiving oral sex from a male compared with rates of females giving and receiving oral sex, an indication that at advanced ages, rates of same-sex male sexual behavior were higher than rates of same-sex female sexual behavior.

One of the best-studied aspects of aging and human sexuality has been across the menopausal transition. While aging alone may take its toll on sex, a related question concerns effects specifically traced to the physiological impact of menopause itself. Reviewing approximately fifteen studies, both cross-sectional and longitudinal designs, Lorraine Dennerstein, Alexander, and Kotz (2003) found that the best of these (studies of longer duration, those measuring hormones) consistently documented an influence of the menopausal transition on measures of sexuality such as elevated dyspareunia (pain during intercourse) and reduced sexual desire, with effects linked to declining estrogen. As a gold standard of U.S. longitudinal studies of this type, the SWAN study (Study of Women's Health across the Nation), which involved more than 3,000 women age forty-two to fifty-two, observed similar effects of the menopausal transition on dyspareunia and sexual desire, but without effects on intercourse frequency (Avis et al. 2009).

Taking aging and sexuality global, Ed Laumann and colleagues (2005) investigated sexual attitudes and behavior in twenty-nine countries, including South Africa, Japan, Australia, and Mexico, surveying 27,500 sexually active men and women age forty to eighty. By all accounts this was an impressive effort. The data were presented differently, and thus the survey does not allow for direct comparisons by age or sex with rates of sexual behavior in the U.S. data; instead investigators focused on clustering the global patterns of aging and sexuality by attitudinal and relationship factors. As a central finding, there was a positive relationship across countries in the percentage of participants who reported satisfaction with their sexual function and relationship satisfaction. At the low end, around 60 percent of Japanese and Taiwanese participants reported satisfaction with sexual function, with around 15 percent of Japanese and 20 percent of Taiwanese respondents reporting relational satisfaction. At the high end, approximately 90 percent of Australians reported satisfaction with sexual function and around 70 percent reported relational satisfaction. Important messages, then, from this international study are that sexual and relationship satisfaction are highly related (no surprise, given the primacy of long-term relationships to human sexuality), and that considerable variation in sexual and relational satisfaction exists among aged populations. A related analysis from this study showed that a host of sexual problems increased with advancing age: low sexual desire and problems with vaginal lubrication among women, and erectile and ejaculatory problems among men (Laumann et al. 2005).

Another important finding from this international study concerns sex differences in reported sexual well-being. Men reported higher levels of physical pleasure, emotional pleasure, satisfaction with sexual function, and importance of sex than did women. However, the magnitude of these sex differences varied by "clusters" (in which cluster 1 consisted of countries, such as Australia, with high sexual and relationship satisfaction, and cluster 3 consisted of countries such as Japan and Taiwan, with low rates of both). In this vein, the findings "support the notion that the ideal of companionate relationships tends to value positively sexual competencies, interests, and performance between intimate sex partners. In other words, sex in companionate relationships serves not only reproductive purposes, but also expresses the quality of the relationship" (Laumann et al. 2005: 158).

In an attempt to systematically investigate aging and sexuality in a sample of small-scale societies of hunter-gatherers, agriculturalists, pastoralists, and the like, Rhonda Winn and Niles Newton (1982) dug into the Human Relations Area Files for data referring to older men's and women's sexuality. The effort was marked as much by gaps in data as by rare insights. Still, their effort stands as one of the few attempts to uncover patterns of aging and sexuality from a cross-cultural perspective. So what did they find?

Older males were sexually active in twenty of twenty-eight societies, and older females sexually active in twenty-two of twenty-six societies from which they could obtain data. Thus, at least with the admittedly slim data available here, it appears quite common that older women and men remain sexually active with advancing age. As an example, among the Marquesans, "Some sexual relations are carried on by individuals in the sixties and seventies. Males of this age are often capable of normal copulation, but with greatly reduced frequency" (Suggs 1966: 122). Beyond this coarse-grained observation, many ethnographic details point toward similarities and differences cross-culturally in both male and female sexuality with advancing age.

One means by which older men remain sexually active in some societies is by marrying younger wives, sometimes in polygynous unions. The desire to have more children is sometimes articulated as a reason for these men to remain sexually active with their younger wives, as among the Kuwaiti Bedouin and Siberian Chukchee. However, a refrain in at least five societies, including among prospective young brides in the Trobriands and among the Lovedu of South Africa, is that old men are not physically attractive. In some other cases, such as Taiwan's Hokkien, the thought that older men and women remain sexually active is found distasteful, but for a different reason: that they already have grandchildren.

One of the noteworthy patterns of female sexuality is that reports from eighteen of twenty-two societies indicated that women had sex with younger men, sometimes considerably younger. In some cases, it appears as if their younger male partners were desirable, with some of them being married men. In the majority of cases, however, this seems to reflect a context in which older women are not viewed as attractive but are taken as partners because they are the only ones available to some men. As Winn and Newton (1982) put it, "Older women were resorted to as sexual partners only by young men who were able to find no other outlets, as in the case of the

Marquesa Islanders of French Polynesia, the aboriginal Easter Islanders, the Inca of Peru, the Nambicuarar of Brazil, and the Trumai of Mato Grosso. Older women were also utilized by young boys as sexual initiators and sexual instructors in several of the societies reported, as in the Tupinamba of northeast Brazil" (293). Given marital customs whereby a young husband might marry a much older woman (as, for example, in a levirate marriage, when he "inherits" her as a spouse after her older husband [his brother] dies), he might find himself sexually active with a considerably older wife. However, in these cases it appears that young husbands frequently find their much older wives unattractive, as among the Ngonde of Tanzania, where "men who inherited wives were required by ritual laws to have intercourse with all of them, but if the wife were 'old,' the husband would never lie with her again, being repulsed by her age and 'ugliness'; [moreover], in the case of the Nambicuara and Trumai cultures of South America, young men who married older women because of the unavailability of younger women were reported to decline sexual activity with their wives, expressing lack of interest in these women" (Winn and Newton 1982: 298).

Another noteworthy pattern in female sexuality was that older women frequently experienced a loosening of constraints to their sexual expression. With their reproductive capacities past, older women might make ribald jokes and display an enhanced sexual aggressiveness, like among the Taiwanese Hokkien and the Muria of India. "These behaviors included exposure of the genitals, suggestive dancing, imitation of copulatory movements in public, clutching the genitals of men in public, and requesting sexual intercourse from strangers," according to Winn and Newton (1982: 295). Older women, then, are subject to fewer constraints and inhibitions, sometimes carrying their sexuality on their sleeve, and it is also worth highlighting that men do not appear to undertake this same degree of expanded latitude with age. Aging women altogether avoid outlets for their sexuality, too: among the Tallensi of Ghana and Kanuri of Nigeria, older women might shun or be shunned by an aged husband, and move in with an adult son instead.

A few additional ethnographic reports among foragers indicate age-related changes in sexuality, also arguing against any fantasy that our aged hunter-gatherer forebears lived in a sexual golden era. Among the Aka foragers of Central African Republic, Hewlett and Hewlett (2010) report: "Frequency of sex over age 45 was not calculated because most of the older women were single due to the loss of their husband, divorced or stopped having sex with their husbands. There is general belief and practice that when one's own children start to have children of their own it is time to reduce sexual activity to help with the grandchildren and make sure existing dependent children survive" (111). And among Arnhem Land Australian aborigines, "Old people who are less active sexually tend to exaggerate their conquests; and men insist that their wife or wives are completely satisfied with their untiring attention. Usually a man will remain sexually active longer than will a woman; there are many cases of men sixty to seventy years who still indulge in coitus. Women after the menopause usually continue to have coitus, but this becomes less as they grow older. . . . Copulation with a young girl is said to rejuvenate the elderly man, by strengthening the erectness of his penis; while copulation with a young man is said to make an old woman more attractive." (Berndt and Berndt 1951: 104).

PROBLEMS WITH AGING

During interviews with Ariaal agropastoralist men in Kenya concerning aging and quality of life, one of the topics explored was erectile function. While an unusual subject to touch on, it nonetheless elicited considerable discussion, particularly when it came to light that a blue pill existed (Viagra) that could aid in treating erectile difficulties. The study indicated that men age sixty and older reported lower rates of erectile function, a noteworthy finding given that the Ariaal men have lean body mass, do not have metabolic diseases such as diabetes and hypertension, and do not take the piles of prescription medications that so many aged men now take, some of which are known to compromise erectile function (Gray and Campbell 2005). Various other studies, conducted in targeted clinical samples as well as well-designed epidemiological designs, indicate that older men frequently report erectile problems, whether in Malaysia, England, or the United States. Thus, a decline in erectile function (or, put another way, an increase in erectile dysfunction) is a robust feature of male aging, particularly from around age sixty onward, and is but one indication of the potential concerns over reproductive function that may arise with advancing age. The fact that Viagra and its pharmaceutical kin (such as Cialis) have been so marvelously successful also speaks volumes to the relevance of erectile dysfunction among older men, not to mention the occasional anthropological anecdote about treatments for "potency" (Aka hunter-gatherer men, for example, are pur-ported to use two different plant-based treatments for male potency, though it is not clear whether they work).

Apart from erections, other functions go down with advancing male age (Knobil and Neill 2006; Masters and Johnson 1966; Meston 1997). Male testosterone levels tend to decline with age, with those declines being grad-ual from around age thirty onward, but also varying in trajectory across populations (Bribiescas 2006). In the United States, for example, where men's testosterone levels are higher early in adulthood compared with, say, the Ache of Paraguay or even the Ariaal of Kenya, U.S. men's testosterone has more room to fall with age, yielding a steeper rate of decline compared with some other subsistence societies (Ellison et al. 2002). At any rate, the declines in male testosterone with age may be both indicative of diminished reproductive investment (in libido, muscle, hematocrit—the kinds of things that testosterone promotes and that aid male success in competitive repro-ductive arenas) and reflective of changing health status (overweight men, for example, especially those suffering from metabolic diseases such as type II diabetes, tend to have lower testosterone levels). Less immediately obvi-ous, age-related decreases in male testosterone may have impacts on erectile function itself (through reduced maintenance of muscle at the base of the penis, as well as vasculature).

The recognition that men's testosterone decreases with age, and that the administration of testosterone to hypogonadal men (those with low testos-terone levels) can positively influence mood, libido, muscle mass, and other outcomes, has generated considerable interest in the use of testosterone as an intervention to enhance aging male well-being (Bhasin et al. 2006). From an evolutionary perspective, we are on novel terrain here, since we do not think many ancestors of ours even 150,000 years ago lived long post-reproductive lives, nor of course would they have had access to such inter-ventions. Is this simply a new grafting of rooster testes into men's blood-streams, or are there legitimate benefits that can outweigh the risks? (Risks are typically couched in terms of prostate cancer progression or increased vulnerability to strokes or heart attacks, but with potential impacts on social relationships that warrant attention too.) During Peter's postdoctoral years, working under Shalender Bhasin, a physician and endocrinologist, ques-tions concerning the risks and benefits of testosterone treatments were front and center, with Peter conducting psychological testing on men involved in trials designed to help address that side too. More recently, Bhasin and other key stakeholders have successfully initiated the first large-scale clinical trial in older men that will quantify risks and benefits of testosterone treatment. In the meantime, Peter occasionally glimpses signs and advertisements featuring Jeffry Life, a buff-looking seventy-year-old man working at Las Vegas–based Cenegenics, the largest training facility in the United States for doctors who practice so-called anti-aging medicine. Aging men, all this says, attempt to seek solace in a chemical fountain of youth—testosterone—largely for issues couched in well-being or health rather than reproduction.

In less dramatic fashion (no visually stunning advertisements), other features of male reproductive anatomy and physiology wane with age. Sperm concentrations, sperm motility, and the volume of semen released at ejaculation all decrease with advancing age, while older fathers have increased risk of transmitting heritable disorders (such as schizophrenia) to their offspring (Knobil and Neill 2006). A straightforward way to view these age-related changes is that they reflect male reproductive senescence; that we have few recent ancestral males who reproduced beyond their fifties means that there had been diminished selection on maintaining these other aspects of male fertility. This might be more troublesome today, should an older man wish to have children with a fertile partner, yet we do occasionally hear of a ninety-four-year-old man overcoming such obstacles (Seymour et al. 1935).

If men have their age-related reproductive concerns, so do women (Knobil and Neill 2006; Masters and Johnson 1966). Indeed, some of the central age-related concerns raised by women are hormone-related too: including the more sudden declines in ovarian steroids, such as estrogen, preceding and at menopause, and the question of whether or not to take exogenous hormones (synthetic estrogen and progesterone) to ameliorate some of the effects of "steroid starvation" (as Masters and Johnson had put it in the quote at the start of this chapter).

As menopause approaches, FSH and LH levels increase as the ovaries become less responsive to the push to release steroids, leading to diminished release of estrogen and progesterone. The reduced estrogen can affect various traits under its influence, traits that share an agenda to aid fertility (from positive effects on libido to vaginal lubrication to the maintenance of fat reserves useful during reproduction). Some of the most consistently documented effects of reduced estrogen at menopause are hot flashes, night sweats, and diminished vaginal elasticity and lubrication. The development of synthetic estrogen, combined with Robert Wilson's (1966) book *Feminine Forever,* seemed to offer a formula for rescue and a rallying cry for holding off some of the less savory symptoms of advancing age, including those experienced at menopause. The later realization that while replacement estrogen does stave off key symptoms, such as hot flashes, but at the cost of an increased risk of endometrial cancer, led to development of progestin interventions too, in order to allay the negative uterine effects.

In the new millennium, results from the first of several large-scale, well-designed clinical trials designed to quantify the risks and benefits of several types of female hormone intervention at advancing age, including menopause, became available. The Women's Health Initiative study, which enrolling more than 16,000 women, raised concerns over hormone treatments. Among early findings, researchers reported that women on a combined estrogen and progestin therapy exhibited increased risk of coronary heart disease compared with women taking a placebo (Rossouw et al. 2002). And since that initial report, follow-up analyses from the same study as well as others have painted an unclear picture, finding some risks to combined treatments, but with the effects also contingent on other factors, such as whether a woman initiated treatment at menopause or at a later age. Just like for men, such discussions of hormone replacement therapy put us in new territory since few females would likely have had postmenopausal lives until our species evolved, and obviously there was no access to the novel compounds giving rise to the debate over hormone replacement in the first place. Hormone-replacement therapy may have tangible benefits to an individual's well-being, with many benefits also realized in terms of ongoing sexual and romantic partnerships (rather than fertility, although a complementary discussion centers around interventions to enable women to reproduce at ages when they otherwise cannot), but such benefits are traded off against modest yet quantified health risks.

Would women everywhere express a concern over the menopausal symptoms highlighted above (such as hot flashes), much less seek some sort of hormone-based therapy to ameliorate them? The answer is clearly no, as existing cross-cultural research on menopause reveals. For example, in a review of relevant studies concerning culture and menopause conducted between 2000 and 2004, Melissa Melby, Lock, and Kaufert (2005) showed that the frequency of hot flash reporting among postmenopausal women varied considerably; while 16 percent among Han Chinese women in Taiwan reported recent hot flashes, 47 percent of postmenopausal Moroccan women did. A variety of factors, from smoking to perceived stress to higher body masses, were associated in specific studies with increased hot-flash reporting, but there also seem to be differences in symptom reporting requiring explanation in their own light. As an extreme example noted in this review, rural Mayan women in southern Mexico did not report hot flashes or vasomotor symptoms (Melby, Lock, and Kaufert 2005). The specifics of a woman's physiology (body mass, diet, pregnancy history) and cultural milieu (norms of communication, including whether it is acceptable or not to report symptoms such as hot flashes) can underscore what Margaret Lock (1993) described as a "local biology." Lock's (1993) data showed that few Japanese women reported hot flashes during the menopausal transition compared with Western samples. However, more recent work among Japanese Americans, which combined reports with physiological measures, suggests reality may be more complicated: Japanese Americans reported fewer hot flashes than European Americans, but these differences were not reflected in skin conductance measures of hot flashes, which did not differ by ethnic group (Brown et al. 2009). The upshot from this research is that "local biologies" can be real (based on differences in body mass, reproductive history, and such), but that reported experiences of the menopausal transition may represent a mix of cultural norms regarding communication and physiological underpinnings. Furthermore, whether women themselves will seek a pharmaceutical answer to their menopausal concerns varies too; availability, expense, and knowledge matter, to be sure, yet in findings stemming from research with postmenopausal Jamaican women, Peter noted that none reported being on hormone-replacement therapy, even though many were aware of such treatments.

With these observations on hormone replacement therapy, we wrap up our discussion of aging and human sexual behavior. Across the studies of aging and sexual behavior, several patterns stand out. One is that the frequency and diversity of sexual behavior tends to decline with age, which is consistent with expectations of evolutionary theory and age-related changes in mechanisms of sexual response. Another pattern is that sexual behavior is typically expressed within long-term partnerships, marking their importance from a recently derived evolutionary standpoint, and also with regard to demographic and cultural aspects of partner availability. Finally, the specifics of aging and sexual behavior are clearly highly socially contextualized, highlighting the importance of an integrative approach that combines evolutionary, mechanistic, and social accounts to best make sense of age-related patterns in human sexual behavior.

CHAPTER TWELVE

Sexual Revolutions

Contemporary Human Sexual Practices

> Humans would probably have lived, as already stated, as polygamists or temporarily as monogamists. Their intercourse, judging from analogy, would not then have been promiscuous. They would, no doubt, have defended their females to the best of their power from enemies of all kinds, and would probably have hunted for their subsistence, as well as for that of their offspring. The most powerful and able males would have succeeded best in the struggle for life and in obtaining attractive females.
>
> —Charles Darwin, *The Descent of Man and Selection in Relation to Sex*

IN HIS CLASSIC book, which has served as a foundation for evolutionarily minded scholars of human behavior and sexuality ever since, Darwin outlined his views of human mating. He left out active female mating strategies from this summary, and there is no hint of a developmental account. Yet he captured a few elements of human sexuality that remain on center stage: the evolutionary context of monogamy and polygyny (but not promiscuity); the involvement of men in family life; the relationship between status and reproductive success.

With *Descent of Man* (1871), Darwin broke with some earlier characterizations of the natural history of human sexuality. By some popular accounts, humans lived and bred in "primal hordes," collectives of men and women who engaged in promiscuous sex. The basis for those accounts has often been linked to Lewis Henry Morgan and reflects his misreading of the native North American society (the Iroquois) that he studied and wrote about. The Iroquois had a system of matrilineal inheritance, but they were far from promiscuous, and their mating relationships sometimes entailed a woman with a man who would come and go (perhaps because of long-distance warfare). However, Morgan's view of "primal hordes" struck an intellectual chord, and influenced other key thinkers, such as Friedrich Engels and Karl Marx, facilitating misleading accounts of the foundations of human sexual communion. Darwin basically disagreed, though, finding in his comparative and cross-cultural investigations support for monogamy or polygyny as the recent human ancestral condition, not a promiscuous horde.

These nineteenth-century historical accounts show how depictions of human sexuality in its evolutionary scope have been subject to debate. While some views still hold water, others no longer do. In a recent popular book on the evolution of human sexuality, Ryan and Jetha (2010) have effectively attempted to resurrect the view that recent humans evolved in primal hordes. Through distorted data and an emphasis on male reproductive strategies, they contend that people today need to take a break from the difficulties of fidelity in long-term bonds in favor of actualizing the promiscuity they see as being built into us. We hope that the mounds of data, both cross-cultural and biological, we have presented in this volume make it clear that promiscuity is a poor characterization of recent human sexual evolution. In light of comparative data, especially from primates, greater promiscuity may have held among our more distant hominin ancestors, roughly six million years ago. But a sexual revolution occurred among our more recent hominin ancestors. This sexual revolution pushed us toward slight polygyny, likely within the genus *Homo,* in several steps that unfolded in the past two million years.

In this final chapter, we present some of the "take-home" points of the book. We then turn to several topics in which human sexuality has front-page significance. We consider HIV and AIDS, drawing on its nonhuman primate origins and variable sexual transmission across human societies. We touch on human sexuality in the Internet age, finding that our evolved ways rear their heads again in the new technological world. Then we wrap up with a nod to some possible ways in which human sexuality will continue unfolding in the near future (and even dare to consider sex in outer space).

TAKE-HOME POINTS

Sexuality fuels much of the passionate world, from movies about desirable vampires to gifts of precious gems excavated from the earth. Engagement

in, and desire for, sexual behavior and romantic partnerships is a remarkably consistent human experience. The cause of such proclivities, sexual reproduction, is a force of nature. While there remains some debate about the
evolutionary costs and benefits of sexual reproduction, the generation of
novelty remains one of the primary outcomes, helping to yield new ways to
combat the environmental challenges faced by descendants. As one example,
sexual reproduction yields offspring that may have greater success on the
disease battlefield, against constantly attacking pathogens that seek to maximize their own evolutionary best interest. Sexual reproduction is an adaptation in its own right.

The origins of sexual reproduction set the stage for each of the sexes to
evolve. One being (female) possessed the larger, more sessile gametes (eggs)
and the other being (male) harbored the smaller, more mobile gametes (sperm).
Those beings evolved sex-specific strategies to see to the success of their respective gametes. Starting with those differences in gametes, Bateman (1948)
argued that females typically invest more energy in gametes and are thus the
limiting resource. Later, Trivers (1972) showed that in mammals the sex differences in parental investment guide sex differences in traits such as mate
choice (in which females tends to be more discriminating). More broadly, the
reproductively limiting sex (often, but not always females) is the one practicing more careful mate choice, and the other sex is engaged in more same-sex
competition and courtship. This view resonates well with what Darwin
wrote so long ago in his classic work on sexual selection.

The field of life history theory expands on the theoretical basis of sex differences in reproductive strategies. Life history theory specifies that organisms allocate limited time and resources to competing agendas of growth,
maintenance, and reproduction. This body of scholarship does at least three
important things for our understanding of human sexuality. It gives greater
weight to direct reproduction as a process that can favor the evolution of sex
differentiated traits—it is not just that males fight and court more that
makes them different from females; rather, the sexes are also different because female mammals have evolved marvelous adaptations, such as mam-
mary glands and uteruses, that foster direct reproductive success. Life history theory also gets us to think in terms of adaptation and a life-course
perspective on human sexuality, representing developmentally shifting priorities in sex and reproduction as sensible adjustments in priorities enhancing lifetime reproductive success. So much of the evolution-oriented work

on human sexuality focuses on emerging adult (that is, college age) mating strategies. However, as this book has highlighted, sex differentiation begins before birth, children are developing sociosexual animals, puberty and adolescence turn up the volume on sexuality, sexual priorities can wax and wane during the reproductive years based on female cyclical processes (ovulatory cycles; pregnancy and lactation), and sexual function slides downhill with advancing age. Finally, life history theory also recognizes the context-specific attunement of an individual's and a population's sexual and reproductive efforts. What makes sense in one place may not make sense in another (for example, it may have made more sense for our grandparents to have had more kids than it does for many adults today, in light of changes in formal education, job markets, and so forth).

Considering these broad theoretical views, how does human sexuality show when placed under an evolutionary spotlight? Here, a number of central patterns are visible. For one, sex differences exist, and they have evolved, adaptive underpinnings, just as in other sexually reproducing creatures. Males have higher sexual desire than females, for example, although the magnitude of that difference represents an average and fluctuates by context and across development (being at its greatest during the reproductive years in the immediate postpartum period). As another theme, human sociosexual behavior typically unfolds within the context of long-term relationships. People regularly form long-term partnerships (to the puzzlement of rhesus monkeys and bonobos, among most mammals) in which they focus their sexual energies, and in which they have and raise offspring. Evolutionary models, as we have pointed out, suggest that these long-term bonds have been derived recently, perhaps in the past two million years of hominin evolution, and thus they speak to a truly profound sexual revolution in our recent ancestors' ways. That said, another theme of this book is that these long-term bonds seem to fit best within a "slight polygyny" or "mostly monogamy" context, in which the most socially dominant or prestigious men have slightly more sexual access and reproductive success (perhaps having two wives rather than one in a foraging context). Further, in this context females may be partnered with a long-term mate at the same time they are attentive to potentially more desirable mates, and sometimes may accede to sharing a mate (as co-wives, or in affairs) if their own circumstances (such as mate value) or their partner's characteristics warrant it (for example, a male able to provide resources to her and her offspring).

Another theme is that these sex-specific and species-specific selection pressures have shaped our anatomy, physiology, and behavior in synergistic ways. The mating psychology and behavior that perhaps is the focus of an evolutionary psychologist's research has a functionality to it that resonates with the functionality evidenced by reproductive anatomy and physiology. The effects of hormones at different stages of the life course help orchestrate a suite of characteristics (motivational processes in the brain, secondary sexual characteristics such as female fatness and male muscle at puberty) that are geared toward working together to enhance an individual's lifetime reproductive success. There is no mind/brain dualism at play; there is instead orchestration, coordination, and communication throughout the central and peripheral areas of our being.

The evolutionary process has led individuals to be conceived with a sex- and species-specific agenda. Put another way, females and males possess sex-specific genes that guide and respond to sex-specific physiological mechanisms (sex hormones) that in turn set the stage for the sexes to experience the world in different ways across the life span. Further, individuals of one species have inherently different agendas than members of other species (such as a human having some greater disposition toward long-term sociosexual bonding than a chimpanzee). After all, no matter how hard one tries (and as we have seen, experimenters have actually tried), a chimpanzee develops differently from a human or a bonobo, even when raised in a similar environment.

Yet—and this is an important yet—the developmental process is a dynamic one, entailing interaction between physiology and socioecological context. An individual is not a preprogrammed sexual robot; rather, an individual is a "plastic" creature, an ongoing production of her or his biological inheritance interfacing with social, nutritional, disease, and other features of her or his environment throughout life. The nature of this plasticity gives rise to the atypical patterns of sex differentiation highlighted earlier (such as complete androgen insensitivity syndrome, where an XY individual looks and lives almost entirely as a female, owing to a nonfunctioning androgen receptor). The nature of this plasticity enables the ways in which one is raised to shape her or his physiology, including neural substrates underlying attitudes and behaviors. A teenage girl raised among the Canela in the Amazon might have found it sensible to engage in sex with lots of men, sometimes one after another, in the forest; whereas that same behavior would seem outlandish in many other societies. The plastic developmental process and the physiological mechanisms that facilitate it allow individuals to practice a "local sexuality" that makes sense in light of the context in which she or he has been raised. Simplistic views that genes, hormones, neurotransmitters, or just "biology" writ large are deterministic are simply unfounded.

Another take-home point concerns the nature of human sociosexual variation. If we take as a starting point our ancestors living in Africa around 70,000 years ago, archaeological and genetic data suggest they were relatively sparsely distributed on the landscape, living in small groups of hunter-gatherers. Educated guesswork, in part fueled by studies of more recent foragers, suggests that hunter-gatherer sexuality at that time would likely have entailed slightly polygynous long-term sociosexual unions, and groups would have included children who probably engaged in sex play during normal development, and adult couples who engaged in less intercourse in late pregnancy or postpartum. But as humans expanded through more of Africa and indeed throughout most of the world, the array of sexual practices also expanded.

Cultural practices arose to control individuals' libido and fidelity. As an example, practices such as female genital modification likely emerged in concert with other practices in ancient Egypt. Extremes of reproductive skew (such as political leaders having many concubines) arose in hierarchical societies around the globe, for example, among the Inca of South America and in imperial China. Variation in many characteristics simply expanded, such as sexual positions; sexual partners (including domesticated animals); rules governing acceptable extrapair sex (from outright bans to conditional acceptance of nonromantic sex to full "swinging" at sex clubs); and regulations and behaviors concerning adolescent sex (from honor killings to sex education to help prevent disease and teen pregnancy). This variation can be viewed in terms of cultural evolutionary processes, including ways in which features of sexuality appear adaptive to individuals within specific socioecological contexts. There is interesting history, genuine contingency, and lots of variation. Yet there are often patterns underlying human sexual behavior that can be unpacked best from an evolutionary perspective, as we have attempted to do throughout the book. Seemingly exotic behaviors, such as infanticide or having sex with a spouse's sibling, often have a coherent and understandable cultural or evolutionary logic when put in context. Further, within a given sociosexual context there are often politics at play.

Parent-offspring conflict indicates why parents and their offspring may not always see eye to eye (such as over an appropriate marital partner for a son or daughter); sexual conflict reminds us that the agendas of females and males may differ (romantic comedy or action adventure?); and same-sex conflict has helped shape sexual dimorphisms such as enhanced male musculature (fighting begets traits that abet success in fighting). Nevertheless, the social and sexual dynamics within a given context also have political fault lines along which there is no simple harmony.

In articulating these take-home points, we are emphasizing some of the central theoretical and empirical aspects of human sexual behavior within an integrative evolutionary perspective. For some readers, this perspective is already second nature; for others, it may differ from ways in which they would typically think of human sexual behavior. If this is a new view, the effect of fully internalizing this outlook on the world can be profound. An evolutionary perspective on human sexuality helps make sense of many disparate things in a coherent fashion. It provides a way to look at new phenomena, with an informed view consistent with the wider sciences. Once you see the world in this light, there may be no turning back.

HIV/AIDS

One mark of a successful theory is how it copes with novelty. In this case, what does an integrative evolutionary perspective have to say about the origins and expansion of HIV and AIDS in humans? A central example of a sexually transmitted disease, the human immunodeficiency virus (HIV)— and its associated disease, acquired immunodeficiency syndrome (AIDS)— has primate origins and has run a variable course among humans around the world, in part due to variable transmission opportunities. When it comes to human sexual behavior, one of the primary expenses attendant to it is the risk of a sexually transmitted disease (STD) traveling along for the ride. STDs are also referred to as venereal diseases and sexually transmitted infections (STIs), the latter term often being preferred because it can speak to the presence of infection without specific disease symptomology.

In humans, HIV takes either of two primary types, with both traced to nonhuman primate roots (Cohen 2010). HIV-1, the primary type worldwide, was acquired from chimpanzees in west-central Africa, while HIV-2 stems from sooty mangabeys, a West African monkey. Apparently, both of these strains jumped from their primate hosts into humans during the twentieth century, likely due to human consumption of primate meat (the evidence for this seems more likely than bestiality as a mode of transmission, which some urban legends have proposed). As part of a bushmeat trade, these primates (and others) have been hunted for food; during the course of butchery and consumption the virus would have been allowed to "jump ship" into humans.

One puzzle has been why HIV-1, in particular, can be so deleterious to humans when it seemed to be operating under the radar in other primates. Indeed, various animals, including cats (who can have feline immunodeficiency virus, or FIV, as cat owners are likely aware), harbor immunodeficiency viruses in the same family as HIV. However, recent data from chimpanzees in Gombe, Tanzania, reveal that some chimpanzees are infected with their equivalent (SIV, simian immunodeficiency virus) while others are not, and that those infected with the virus live shorter lives, on average (Keele et al. 2009). Thus there may be deleterious effects of the virus in at least some other creatures too.

As to the transmission dynamics of HIV-like viruses in other primates, we do not know much. As sexually promiscuous as chimpanzees are, it seems surprising that all adults would not be infected with the virus if it were transmitted during intercourse. In a comparative primate survey, Nunn and Altizer (2006) found that promiscuous species had higher blood leukocyte counts than monogamous primate species, which is consistent with the view that sexual promiscuity facilitates greater transmission of pathogens, as those pathogens elicit immune responses that can be indexed by leukocyte counts. Importantly, in this data set, humans cluster with the monogamous primate species, another line of evidence consistent with the view that we have mostly monogamy or slight polygyny, but not promiscuity, bred into our physiology. Interestingly, across eight species of Old World monkeys, females have a higher prevalence of STDs, such as SIV, than males (Nunn and Altizer 2006). This sex difference could be due to several factors: multi-male, multi-female mating patterns in which typical females have many mates but some males have less sexual access than other males, and the fact that for a given act of sexual intercourse females are at greater risk of STD transmission than males partially because they may experience greater internal abrasions during intercourse.

Ever since HIV made the leap into humans, it has been one of the most dreaded of human STDs ever discovered (Pisani 2008; World Health Organization). Since its origins in west-central Africa in humans, it has spread around the world, in part through sexual contact, including famously through "patient 0," a Canadian male flight attendant, in the early 1980s.

The highest prevalence of HIV/AIDS in the world occurs in southern and East African countries, with approximately one-fourth of adults in Botswana infected with the virus, for example. In these regions, HIV is primarily spread through heterosexual intercourse. A truck driver visiting a brothel may acquire HIV from a prostitute. A young woman having sex with an older male "sugar daddy" may acquire the virus, only to pass it along to a mate of more similar age. A married man cheating on his wife may acquire the virus from a girlfriend or prostitute and transmit it to his wife. His wife, in turn, may transmit HIV to her offspring during pregnancy, birth, or lactation. These are some of the most common dynamics of HIV transmission in African countries and elsewhere. The nature of these dynamics has also generated heated discussions about the shape that these sexual networks take, and the number of complex social, economic, political, and physiological factors that are at play. Formal polygynous marriage, now in decline, has its correlate in multiple ongoing partnerships; the colonial and postcolonial breakdown of formal marriage in Africa has fueled greater partner fluidity (sexual fidelity in a polygynous marriage has lower STD risk than multiple-partner networking in less stable relationships). Moreover, the co-occurrence of other STDs, such as gonorrhea and syphilis, facilitates further STD transmission. Conversely, while working in a Muslim community in Kenya in 2001, Peter noted that seemingly fewer people in the community reported relatives or friends dying of HIV/AIDS, compared with people in other parts of East Africa, a pattern that turns out to hold: African Muslim communities have been less hard hit by HIV outbreaks (Gray 2004). The reasons for this are complicated, but they likely reflect more stable sexual networks in the Muslim communities, which in part may reflect less social disruption from colonial or other Western influences.

Outside of Africa, most sexual transmission of HIV occurs via anal intercourse between males and during heterosexual sex with prostitutes. Jails serve as important breeding grounds for HIV transmission, given the relative unavailability of female partners and the (often coercive) male-male sex that may serve as a substitute. Female and male recipients of anal sex (those who are penetrated) are at greater risk of STD transmission during a given sexual act, for reasons akin to those noted above with respect to sex differences in primate STD prevalence: the friction and abrasions that can occur increase the likelihood of STD transmission.

Within the United States, HIV rates have also been increasing among the elderly. As we have discussed earlier in this book, and elsewhere (see Gray and Garcia 2012), sexual activity continues well into post-reproductive years. In the United States, the two age demographics hit hardest by HIV are young adults and the elderly. Of particular concern are rising rates of HIV in senior citizen homes. Because those men and women do not worry about pregnancy, they are likely to forgo the use of condoms. During the HIV/AIDS scare of the 1980s, many of today's senior citizens were already settled with families and long-term partnerships (way into their thirties if not older), and they missed much of the crucial internalizing of sexual risk. As an undergraduate, Justin volunteered with an HIV/AIDS nonprofit organization, and the story of one woman who spoke to a public audience still resonates with him. She had been married to one man, her only sexual partner, for most of her life. After divorcing later in life, she started dating another man when she was well into her sixties. She did not use condoms with her new partner, as she had been postreproductive for well over a decade. She contracted HIV late in life from her second lifetime sexual partner.

The social nature of STDs like HIV rears up regularly, affording some of the most weighty evidence for why we must attend to the realities of human sexual behavior without undue blushing and with honesty (Wellings et al. 2006). As human sociosexual dynamics have departed from the small-scale foraging past, now enabling a degree of sexual networking with prostitutes and with many same-sex partners within short time frames, for example, there are evolutionarily novel opportunities for STDs to spread. In the United States, the 1960s era of the Pill and free love ushered in glory days for STDs, with genital herpes thriving and perhaps later allowing HIV greater opportunity to spread. Sexual networks of adolescents today, in their world of pre-child-rearing hooking up, continue to lend themselves to STD risk unless risk is reduced through protections like condoms. Military forces stationed around the world lend themselves to STD risk, when male soldiers stationed far from home, and often single, seek local sexual outlets. Long-distance travel and trade of other sorts facilitate the spread of STDs, as observed in the infertility affecting some Australian aborigine communities in the wake of STDs introduced by outsiders. So far as we know, there have been no links drawn explicitly between humans and the animals with whom they have had sex, but earlier discussions of bestiality make that a plausible scenario, and we know that many human pathogens were otherwise acquired through close contact with domesticated animals.

SEX ON THE INTERNET

Two of the earliest economic successes on the Internet were websites offering genealogical services and websites offering pornography. Apparently, early web users were keen both to find out who their ancestors were, and to find sexual stimuli that their deeper ancestors equipped them to seek. As it has matured, the Internet has drastically changed our ability to consume sexual stimuli, and to connect with others for what may become sexual relationships. In this technologically new world, how do evolved human sexual tastes find their expression? Sex differences in sexual desire have their counterparts in Internet use patterns—one sign that some things never change, even when the world around us does. Furthermore, a growing number of romantic relationships are initiated through Internet means. While in previous times and places, one's parents might have arranged romantic and marital relationships, today a company on the web can fulfill a similar role.

As a means of gaining traction in the Internet search world, we draw from a recently published book that surveyed exactly this terrain. Ogi Ogas and Sai Gaddam's (2011) *A Billion Wicked Thoughts* covers the results of more than 400 million web searches from about 2 million users, primarily from the United States but also from India, Nigeria, Canada, and the United Kingdom. The searches were conducted between July 2009 and July 2010, with 13 percent of them containing erotic content. This focal web study was supplemented by an AOL search database on over 650,000 users, many other online content searches, and online conversations with sporadic users. The book provides an important landmark in the sexual survey of the Internet, and identifies a number of patterns that speak to concerns over evolutionary aspects of human sexuality placed within this new world.

Summing up mounds of work, Ogas and Gaddam (2011) write, "On the Web, men prefer images. Women prefer stories. Men prefer graphic sex.

Women prefer relationships and romance. This is also reflected in the divergent responses of men and women when asked about what sexual activities they perform on the Internet" (19). Further, of those preferred sexual activities on the Internet, 37 percent of men listed viewing erotic pictures and movies (versus 6 percent of women), and 21 percent of women listed staying in contact with love or sex partners (versus 8 percent of men). To interpret these sex differences in web use, the authors draw on Don Symons's (1979) *Evolution of Human Sexuality,* finding that, despite having been written long before such web content could be fathomed, Symons articulated sex differences in interest in erotic stimuli that conform quite well to contemporary web patterns.

In examining males' Internet sexuality content, other patterns arise. One is that young women (ages sixteen to eighteen) draw the most interest, which is consistent with human preferences for long-term mates of high reproductive value. Women of other ages (for example, MILFs—google that one if you do not recognize the term) also draw attention. The MILF or "cougar" phenomenon (older sexually agentic women) typically represents a young single male fantasizing about a mature woman taking the sexual lead, with little long-term commitment involved. The preferred physiques of women are not thin but of medium weight or even fat, a pattern that other scholars have recognized in magazines and other media (thinness in females is driven more by female-female competition than by appealing to males). Males often focus on or enter search terms related to male genitalia, particularly penises. While in day-to-day reality few men see others' penises, that is not true of primates in the wild, nor was it true of our ancestors until they adopted clothing, suggesting that males may find signaling value in checking out each others' testimony (a word, incidentally, referring to the former practice of men holding their testicles while speaking under oath). Among gay men, sexually explicit images are often favored, typically of masculine men, even if they are heterosexual (robust cowboys), and often young men; gay men employ search patterns akin to those of heterosexual men, except with a focus on men rather than women. Further, gay male online porn features viewers' fantasies more often as "bottoms" during penetrative anal sex, indicating the important role of male-male performative power dynamics at play.

As for female-oriented web use, a variety of other patterns hold. Women's searches for male celebrities online (and paralleling magazine offerings) may appeal to the desire to evaluate high-status and desirable men, including uncovering the gossip behind their glossy images. This resonates with the assessment of prospective high-status long-term mates, drawing on both visual cues and richer stories and contexts that may illustrate a man's character. In the growing niche of electronic books, women's erotic and romance genres have been very successful. This parallels the wider book-publishing industry, in which romance novels have been depicted by evolutionary scholars, such as Catherine Salmon, as "female pornography." The keys of that genre, whether in new digital or more traditional formats, are female protagonists with whom the female reader identifies; a high-status, desirable male (often a doctor, king, boss, or some other figure who wields power); and central female character who elicits such a powerful desire in the male protagonist that he overcomes his base behaviors and falls passionately in love with her (but with no other woman). In her work on Harlequin romance novels, Maryanne Fisher has demonstrated that many evolutionary models are at play, such as a male hero, unlike the heroine, who often undergoes a personal transformation from sexy-cad figure to loving-dad figure (Fisher and Cox 2010). In some of the more controversial varieties of this genre, less common today than in the past, the female protagonist may so excite the male that he sexually coerces her (representing a forceful-submission fantasy), but key interpretations of this view are that it signals the power of a woman's sexual allure to entice a desirable mate, and it notably does not include fantasizing about low-status, undesirable men (Hawley and Hensley 2009).

In one of the very few online porn sites that has been profitable targeting heterosexual women—Sssh.com—the content clearly differs from that served up to typical male viewers. In *A Billion Wicked Thoughts,* the founder of Sssh.com is quoted as saying: "Women want to see foreplay, a lot of kissing, a lot of talking before the action gets going. They like to see women with a little more weight on them, a little older, not skinny young girls. The guys have to be clean, well-dressed, and well-kept. They hate men that are sloppily dressed" (Ogas and Gaddam 2011: 161). This pattern reflects a slower pace to female sexual arousal, entails greater context, and involves a decent man. This fits with earlier work we considered in Chapter 8 on the human sexual response, and the sex differences in it.

Several websites have also emerged to help individuals find a sexual or romantic partner. In the case of sex partners, a variety of services have become available in the past few years. Some websites help one find prostitutes (often calling themselves something else, such as "escorts") who offer everything from a thirty-minute intercourse-only encounter to the all-night "girlfriend/boyfriend experience" in which one may have dinner, a movie, and a night of sexual activity. Other websites are designed specifically for those looking for extrarelational affairs—yes, there is a site devoted to those looking for extrapair copulations. "Bulletin board" websites such as craigslist.org offer sections for those looking for a "casual encounter," with every imaginable combination available (individuals looking for men, women, transgender dates, couples, and so on). However, in 2010 craigslist.org removed the section where one could easily find local prostitutes, following a series of criminal investigations.

Specialty sites have also emerged, allowing individuals to better find exactly what they are looking for in an ever expanding mating market—whether it be someone of a particular ethnic group, a particular body type, and even a similar medical condition. The Internet allows people to expand their search radius and, for the most part, find whatever sexual outlet fits their fancy.

While sex on the Internet has gained in popularity, so has the pursuit of romantic relationships. Today there are a variety of websites that help one find a romantic partner—from short-term dating to long-term dating to marriage. Some websites allow users to browse for others in their area, some use personality profiles to pair singles, and others use "guided communication" questionnaires to facilitate initial interactions with a potential partner. A decade ago, there was greater stigma surrounding online dating than there is today (it was seen as a sign of weakness in the dating game). But by now many individuals, young and old, around the globe, have turned to online dating as a way to have fun and meet a preferred partner. A 2009 study commissioned by the international online dating site Match.com suggests that one in five singles in the United States has at some point dated someone he or she met online, and one in six people who married within the past three years met his or her partner through the Internet. In a sample of 7,000 recently married adults in the United States, the number of people who met their new spouse through online dating was more than twice the number of those who met at bars, clubs, and other social events.

Starting in 2010, Justin began working with Match.com, as part of a study on single Americans. In collaboration with anthropologist Helen

Fisher, this ongoing study seeks to better understand who single Americans are, and how, as a demographic, they are shaping American culture. At any given time, roughly one-third of the U.S. adult population is estimated to be single. This is vastly different from the recent historical record, where as recently as the mid-1900s single adults were often thought to be aberrant and treated as inferior to those who were married (Coontz 2005). The findings from this study on single Americans are still to come, but it raises many interesting prospects about the future of human close relationships. As more interactions turn to the Internet, including dating and the pursuit of love and sex, we are left to wonder how patterns of human sexual behavior will continue to change. Thus far, many of these newly emerging trends are consistent with a broad evolutionary account that takes into consideration life history and environmental context.

FUTURES OF HUMAN SEXUALITY

Shining the spotlight of human sexuality into the near future, what do we see? Just a few days before first writing this passage, NASA's Mars rover *Curiosity* traveled over eight months and 350 million miles to attempt to find evidence of life on Mars, even past microbial life. Could, some day, such an outer-space mission reveal other forms of sexual reproduction, and other sexual forms, in the universe? If so, that would be a sexual revelation. In a more immediately plausible scenario, what would happen if people were to travel to Mars? Already astronauts spend months aboard a space station, though travel to Mars would take longer and has raised the question of even making a one-way journey. In a recent book on life in space, the popular science writer Mary Roach (2010) devoted a chapter to the subject of sex in space. She pointed to some of the challenges: gravity is a big one. The mechanics of human sexual behavior unfolded in Earth's gravity, so subtracting that from the equation could make intercourse quite tricky. Ejaculation in zero gravity would pose many of the same problems as urinating or drinking in space, which require special receptacles to prevent fluid from scattering inside the living area. Were conception to occur, it's also not clear how or whether gestation would progress, since maternal circulation, among many things, would be greatly affected. Interestingly, while recent biomedical intervention often places women on their back during birth (as, for example, in U.S. hospitals), the typical birth position for humans cross-culturally, and even for monkeys and apes, entailed some sort of vertical position, in part to utilize the benefits of gravity to facilitate delivery. That could change too.

Returning to Earth, what are some other near-future sexual projections? We share a few off the top of our heads, though readers may carry out their own thought experiments.

One of the ongoing and projected trends concerns declining human fertility, something we also featured in a previous chapter. More and more people in more and more parts of the world are having fewer and fewer children. About one-third of the world's countries are having fewer children than replacement (in other words, their populations would be or are declining without immigration), and that fraction will likely keep growing. Continued declines in fertility have various impacts on sexuality; put simply, the world will keep tilting toward one in which adult sexuality gains political traction at the expense to adult reproduction (child rearing).

With lower fertility, women will spend even more of their reproductive years cycling (rather than pregnant or in a state of lactational amenorrhea), meaning they will be subject less to the peripartum downregulations in libido and sexuality. Women's health demands and expenditures will tilt more toward preventing children (through greater access to fertility control) than having children. The political meaning of female breasts will favor sexual over lactation functions. A higher fraction of people's economic expenditures will be dedicated to "mating effort" over "reproductive effort," meaning that women will spend more on the cosmetics and other consumer products that advertise their sexual allure, and men will spend more on courtship rituals (dates) and male-male efforts (wearing a favorite team sport's colors) rather than on diapers, baby books, or children's sport equipment. Las Vegas, where Peter lives, seems to be ahead of this curve already: the city's 1990s experiment to attract families has been cast aside in favor of the more economically profitable niche of adult entertainment. Hotel-casinos designed with kid-friendly themes have scrapped those schemes in favor of adult-oriented marketing, as any parent of young children watching some of the more sexually charged shows will attest. In the puberty and adolescence chapter we discussed material displays in light of mating effort, and the projected further declines in fertility will mean that this material-display function will occupy more adult hours too.

Pushing the lower fertility projection into more tenuous areas, we imagine it will contribute to continued reductions in sexual coercion and relaxation of sexual constraints. This reasoning begins with the fact that there are more and more women in public spheres who are not as heavily affected by work/family trade-offs (they are not carrying, nursing, and primarily raising young children), and are thus able to be more involved in shaping political and legal discourse. As an example, Steven Pinker (2011) describes declines in wife beating and other forms of sexual coercion (such as wife killing), attributing it in part to shifts in attitudes and policing, and we would additionally note that those shifts may have been politically and legally facilitated in part through declines in fertility.

In the United States and other parts of the world, there have been dramatic attitudinal changes in the past generation toward same-sex sexuality: fewer and fewer people (though certainly not zero, as any individual who has been discriminated against will note) think engaging in same-sex sexual behaviors is a concern. Moreover, attitudes and prejudice toward same-sex families in the United States have also been improving (Massey, Merriwether, and Garcia 2013). When we teach in university settings (admittedly, a biased sample), students overwhelmingly seem unconcerned whether a friend or classmate has sex with someone of the same sex, and we imagine that attitudes will keep moving in the same direction. Of course, this has not always been so. It was only in 1974 that the *Diagnostic and Statistical Manual of Mental Disorders,* the Bible of clinical diagnosis, gave up referring to homosexuality as a disorder. We project that popular and legal constraints on sexual practice (including same-sex sexual behavior, but also prostitution and pornography, among other things) will continue to loosen, in part because (to put it bluntly) there are fewer parents of young children to object, and in part because the more inclusive political process today further relaxes sexual binds that may have had stronger hold in pronatalist, patrilocal social contexts. In such a time, the power of romantic love, whether opposite-sex or same-sex, transcends the reproductive motive.

Another projected trend is the globalization of sexual culture. The more interconnected the world becomes, the more the segmented cultural variation in sexuality of one place encounters that of another. This is already happening all around us through daily Internet reports, and international travel and immigration, but the pattern will expand (for example, as demographics of U.S. immigration continue orienting toward Latin America and Asia). In

France, there are legal battles going on over whether women and girls are allowed to wear the *hijab* in school. In the United States, clitoridectomy has been outlawed, though some immigrants from Sudan arrive having already been through the procedure. Cultural attitudes toward affairs, using prostitutes (for example, for sexual initiations of an adolescent son), and masturbation, among other aspects of human sexuality, vary among those who will continue to encounter each other in the night clubs of the world or on the Internet. Doctors will need to weigh new standards, the legal system will need to find new solutions, and intimate couples will need to sort through the realities of an increasingly globalized sexual culture.

The health and medical aspects of sexuality will grow in various ways in the near future. For one thing, as the size of humanity's belts expands, so does the prevalence of sexual dysfunction and infertility (see Knobil and Neill 2006; Power and Schulkin 2010). The expanding global obesity epidemic carries with it obesity-related sexual and reproductive challenges. In women, obesity is often associated with difficulties conceiving. In the case of polycystic ovarian syndrome (PCOS), for which obesity is a risk factor, endocrine dysregulation includes elevated androgen levels that contribute to menstrual irregularities and difficulties conceiving, and it may also have ancillary psychological effects, such as mood alterations. In the case of obesity in men, the extra fat can also alter steroid hormone levels, tilting toward lower androgen-estrogen ratios, and in turn contributing to male breast development, lower libido, and erectile dysfunction. Metabolic diseases such as Type II diabetes, riding along with the growing obesity epidemic, have additional effects on sexual function. All that said, today's obese and diabetic children, not to mention adults, will have a near future in which sexual and reproductive problems await, and where medical intervention will likely be required.

As even this brief account of hormones, obesity, and sexual function indicates, biomedical science has uncovered some of the physiological mechanisms underpinning human sexual function. In earlier chapters, we visited the hormones, neurotransmitters, and brain areas involved in love and sexual response, among other features of human sexuality. Masters and Johnson (1966) helped rev up those engines decades ago, and they have not stopped. In 2011, Barry Komisaruk and colleagues employed functional imaging to demonstrate the effects of female masturbatory orgasm on brain activity, finding that genital and nipple stimulation have an overlapping geography in the brain. The burgeoning field of epigenetics is helping show how early-life social and nutritional variables influence an organism's gene expression across her or his life, including processes of sex differentiation. Building on this expanding knowledge of the mechanisms of human love and sex, what will it mean for the future?

The more scientists learn about the mechanisms of sexual function, the greater our capacity for intervention will be—consider our discussions of aging and hormone supplementation in Chapter 11. This expanded knowledge will probably give rise to interventions that legitimately improve the well-being of users (as Viagra did for many), but will also likely beget more of the products that speak to insecurities and profitable bottom lines. Penis extension, anyone? Or how about an oxytocin spray to increase your "trustworthiness" in a dating exchange? The scientific evidence for most of these products is minimal, and inconclusive at best. A large and growing pharmaceutical arsenal also has plenty of unintended sexual side effects. The use of selective serotonin reuptake inhibitors (SSRIs, such as Prozac) may have impacts on the dynamics of courtship and relationship maintenance (Fisher and Thomson 2006), while the use of cholesterol-lowering statins may increase erectile dysfunction (Rizvi, Hampson, and Harvey 2002). The expanded knowledge of mechanism and sexual behavior will give rise to new questions: during an interview on CNN discussing a paper on dopamine-receptor variation and promiscuity that Justin coauthored, the reporter asked whether one should collect hair from a lover to have it genotyped, to predict whether that partner will cheat. Justin's answer: Well, no, because there are all kinds of reasons someone might cheat, and the likelihood that it could be attributed to genetic makeup at a single locus alone is quite small.

It is important to remember that our investigations into the nature of human sexuality do not seek to justify one behavior over another. By understanding the influence of the evolutionary process, the role of biology in behavioral proclivities, and the cross-cultural patterns, we can better understand sexuality and the human condition. These lines of research do not seek to justify or condemn sexual behavior. Humans are not prisoners of our natural history. We have evolved remarkably complex (and big) brains for the explicit purpose of making decisions. We have evolved intricate cultural practices to modify the mind. It is our hope that with this book we have demonstrated the incredible diversity that exists in human sexuality, and imparted an appreciation for why this diversity exists in a particular socio-ecological context.

From an intellectual standpoint, the greater knowledge base will continue adding layers to a given sociosexual phenomenon (love is a chemical experience), perhaps, to some, seemingly diminishing that reality (if love is only chemistry, then I feel sad because it seemed like something more). However, the romantics and sexually sparked bloggers have nothing to fear from the intrusions of new biomedical data: these additions complement rather than replace the immediate feelings, moods, and behaviors surrounding the joys and pitfalls of sexual experience. As Helen Fisher has often been quoted as saying in response to studying love and sex: Knowing the recipe to a fantastic dessert does not make it taste any less delicious. There are more layers to reality, but our immediate reality remains as rich as ever.

As a final wrap-up, we might speculate on things that just will not go away, whether in the near or the more distant future. One is sex itself. Sex is a driving engine of humanity's creative efforts, and that will not stop. Even in a world where sex is no longer required to have children (just mix some gametes in a test tube, and maybe recruit a surrogate), would-be parents typically choose sex over the in vitro method. Further, if we are to make sense of human sexual behavior in the future, whether relying on test tubes or not, whether on Earth or in outer space, then we can rest assured that our ability to do so will trace to none other than Charles Darwin and his evolutionary contemporaries. His insights on evolution and sexual selection, and their refinements in generations after him, are fundamental to our understanding of human sexuality. Darwin's intellectual contributions will continue to shape an understanding of human sexuality in laboratories and bedrooms of past, present, and future.

Acknowledgments

We have many people to thank for helping make this book possible. We thank Michael Fisher at Harvard University Press for his support of this project from its inception. We thank those who professionally guided the book through the editorial and production processes—Michael Fisher, Lauren Esdaile, Bryan Cholfin, Lisa Roberts, and Anne Zarella of Harvard University Press; Edward Wade of Westchester Publishing Services; and Julie Hagen. We thank reviewers for feedback that has enhanced the readability and connections made in the book. We are indebted to the many scholars who have preceded us and to those who continue to engage in explorations of human sexuality. We are constantly drawn back to Charles Darwin and his intellectual breadth and rigor. The work of pioneers such as Alfred Kinsey, Clellan Ford and Frank Beach, William Masters and Virginia Johnson, Sarah Hrdy, and others shines throughout the book, and that is mentioning only a few of our intellectual heroes. We also take inspiration from the scholars who conduct research and teach in these areas; we are grateful to those scholars who form the very bedrock of our understanding of evolution and human sexual behavior from biological and cross-cultural perspectives. We are also inspired by those who attempt to reach wider audiences in their books and blogs, a difficult but important undertaking. Darwin's own books remain such useful guides, even in the present, for that same reason; they were written for an intelligent reader, whether a scientist or a thoughtful layperson.

We thank our peers for lots of wonderful moments and reflections on the fossil, genetic, neurobiological, animal and nonhuman primate, forager, psychological, and other evidence helping to make sense of who we are and how we became this way. We have both taught a variety of university courses that have profoundly influenced the scope of this book—its breadth and tone—in what we hope are positive ways. We thank our students for the good questions, and for the connections made between ideas and their passions. We are humbled by our friends and colleagues in the field who provided detailed thoughts, advice, suggestions, and commentary on various aspects of this book, particularly Haley Moss Dillon, Alice Andrews, Bill Jankowiak, and Helen Fisher. And a special thank you to Megan Gray, who tirelessly provided us with feedback on chapter drafts.

Peter has a number of intellectual debts that he wishes to acknowledge. The roots of this book, without his knowing it, were laid in graduate school at Harvard University. In addition to the tremendous academic environment there overall, he internalized an approach to investigating human behavior from an evolutionary perspective. He thanks his mentors Peter Ellison, Frank Marlowe, Richard Wrangham, and Cheryl Knott. He thanks his postdoc mentor, Shalender Bhasin, for giving him an opportunity to live for two years in clinical endocrinology, and for sharpening Peter's thinking on hormones and the relevance of evolution to biomedical and clinical contexts. At the University of Nevada, Las Vegas, where Peter began teaching in 2005, he thanks his wonderful colleagues; they provide a warm, supportive, and integrative environment that can make such a book viable. Peter thanks his own students—Michelle Escasa-Dorne, Dana Foster, Heidi Manlove, Tom Steiner, Tiffany Alvarez—and members of a weekly, informal reading group for their feedback on earlier versions of the chapters. Above all, Peter thanks his family—his extended family, his parents, David and Kay Gray (who either shudder or chuckle at various passages in this book), his sister, Diana, and most of all Megan, Sophie, and Stella. Evolution has made the reproductive process immensely rewarding, a reason why few joys compare with those of a beloved partner and kids.

Justin also acknowledges those who have supported him and his work. He is indebted to his friends and colleagues at Binghamton University, which provided an environment that took evolutionary studies seriously and encouraged broad interdisciplinary training, allowing him to pursue a research agenda outside the box. For their guidance, support, and collaboration, he thanks his mentors David Sloan Wilson, J. Koji Lum, Chris Reiber, Ann Merriwether, and Anne Clark. He is also grateful to his colleagues at The Kinsey Institute and Indiana University for opening their doors to him, for their incredible knowledge, and for their unwavering support of his intellectual interests—Julia Heiman, Erick Janssen, Stephanie Sanders, Virginia Vitzthum, Liana Zhou, Debby Herbenick, and Jenny Bass. A special thank you goes to Helen Fisher, who has been an extraordinary mentor, colleague, and friend. Most of all, Justin thanks his family, his kin, for their unbelievable support for his unusual career path, their willingness to talk about anything, anytime, and for maintaining a tribe filled with love and laughter—his parents, Helen and Mike, and cousins Jen, Dave, Emily, and Lexi. Last, he thanks his extended family and close friends—there are too many to name—who have maintained a constant flow of support and humor.

References

Abbott, E. 2010. *Mistresses: A History of the Other Woman.* London: Duckworth Overlook.

Abdool, Z., R. Thakar, and A. H. Sultan. 2009. "Postpartum female sexual function." *European Journal of Obstetrics & Gynecology and Reproductive Biology* 145: 133–137.

Abma, J. C., G. M. Martinez, and C. E. Copen. 2010. "Teenagers in the United States: Sexual activity, contraceptive use, and childbearing: National Survey of Family Growth 2006–2008, National Center for Health Statistics." *Vital and Health Statistics* 23.

Ackerman, J. M., V. Griskevicius, and N. P. Li. 2011. "Let's get serious: Communicating commitment in romantic relationships." *Journal of Personality and Social Psychology* 100: 1079–1094.

Adkins-Regan, E. 2005. *Hormones and Animal Social Behavior.* Princeton, NJ: Princeton University Press.

Adovasio, J. M., O. Soffer, and J. Page. 2007. *The Invisible Sex: Uncovering the True Roles of Women in Prehistory.* New York: HarperCollins.

Aldrich, R., ed. 2006. *Gay Life and Culture: A World History.* New York: Universe.

Alexander, G. M., and M. Hines. 2002. "Sex differences in response to children's toys in a non-human primate *(Cercopithecus aethiops sabaeus)."* *Evolution and Human Behavior* 23: 467–479.

Allman, J., A. Rosin, R. Kumar, and A. Hasenstaub. 1998. "Parenting and survival in anthropoid primates: Caregivers live longer." *Proceedings of the National Academy of Sciences* 95: 6866–6889.

Alvergne, A., C. Faurie, and M. Raymond. 2009. "Father-offspring resemblance predicts paternal investment in humans." *Animal Behaviour* 78: 61–69.

Alvergne, A., and V. Lummaa. 2010. "Does the contraceptive pill alter mate choice in humans?" *Trends in Ecology and Evolution* 25: 171–179.

Anderson, K. G. 2006. "How well does paternity confidence match actual paternity?" *Current Anthropology* 47: 513–520.

Andersson, M. 1994. *Sexual Selection.* Princeton, NJ: Princeton University Press.

Apicella, C. L., and D. R. Feinberg. 2009. "Voice pitch alters mate-choice-relevant perception in hunter-gatherers." *Proceedings of the Royal Society B* 276: 1077–1082.

Apicella, C. L., D. R. Feinberg, and F. W. Marlowe. 2007. "Voice pitch predicts reproductive success in male hunter-gatherers." *Biology Letters* 3: 682–684.

Archer, J. 2006. "Testosterone and human aggression: An evaluation of the challenge hypothesis." *Neuroscience and Biobehavioral Reviews* 30: 319–345.

Armstrong, E. A., P. England, and A. C. K. Fogarty. 2009. "Orgasm in college hookups and relationships." *Families as They Really Are,* ed. B. J. Risman, 362–377. New York: W. W. Norton.

Armstrong, E. A., P. England, and A. C. K. Fogarty. 2012. "Accounting for women's orgasm and sexual enjoyment in college hookups and relationships." *American Sociological Review* 77: 435–462.

Armstrong, N. R, and J. D. Wilson. 2006. "Did the 'Brazilian' kill the pubic louse?" *Sexually Transmitted Infections* 82: 265–266.

Austad, S. N. 1999. *Why We Age.* Hoboken, NJ: Wiley.

Avis, N. E., S. Brockwell, J. F. Randolph Jr., S. Shen, V. S. Cain, M. Ory, and G. A. Greendale. 2009. "Longitudinal changes in sexual functioning as women transition through menopause: Results from the study of women's health across the nation (SWAN)." *Menopause* 16: 442–452.

Bancroft, J. 2009. *Human Sexuality and Its Problems,* 3rd ed. Edinburgh: Churchill Livingstone.

Barber, N. 2001. "Mustache fashion covaries with a good marriage market for women." *Journal of Nonverbal Behavior* 25: 261–272.

Barelli, C., M. Heistermann, C. Boesch, and U. H. Reichard. 2008. "Mating patterns and sexual swellings in pair-living and multimale groups of wild white-handed gibbons, *Hylobates lar.*" *Animal Behavior* 75: 991–1001.

Bartels, A., and S. Zeki. 2000. "The neural basis of romantic love." *NeuroReport* 11: 3829–3834.

———. 2004. "The neural correlates of maternal and romantic love." *NeuroImage* 21: 1155–1166.

Bartlett, T. 2011. "The Hylobatidae." In *Primates in Perspective,* 2nd ed., ed. C. J. Campbell, A. Fuentes, K. C. MacKinnon, S. K. Bearder, and R. Stumpf. New York: Oxford University Press.

Bateman, A. J. 1948. "Intra-sexual selection in drosophila." *Heredity* 2: 349–368.

Baumeister, R. F. 2000. "Gender differences in erotic plasticity: The female sex drive as socially flexible and responsive." *Psychological Bulletin* 126: 347–374.

Baumeister, R. F., K. R. Catanese, and K. D. Vohs. 2001. "Is there a gender difference in strength of sex drive? Theoretical views, conceptual distinctions, and a review of relevant evidence." *Personality and Social Psychology Reviews* 5: 242–273.

Beach, F. A. 1976. "Sexual attractivity, proceptivity, and receptivity." *Hormones and Behavior* 7: 105–138.

Beckerman, S., and P. Valentine, eds. 2002. *Cultures of Multiple Fathers: The Theory and Practice of Partible Paternity in South America.* Gainesville: University of Florida Press.

Beetz, A. M., and A. L. Podberscek, eds. 2005. *Bestiality and Zoophilia.* Lafayette, IN: Purdue University Press.

Beksinska, M. E., H. V. Rees, I. Kleinschmidt, and J. McIntyre. 1999. "The practice and prevalence of dry sex among men and women in South Africa: A risk factor for sexually transmitted infections?" *Sexually Transmitted Infections* 75: 178–180.

Bentley, G. R. 1999. "Aping our ancestors: Comparative aspects of reproductive ecology." *Evolutionary Anthropology* 7: 175–185.

Bentley, G. R., and C. G. N. Mascie-Taylor, eds. 2000. *Infertility in the Modern World.* Cambridge: Cambridge University Press.

Berndt, R., and C. Berndt. 1951. *Sexual Behavior in Arnhem Land.* New York: Viking Fund.

Betzig, L. 1986. *Despotism and Differential Reproduction: A Darwinian View of History.* New York: Aldine.

Betzig, L. 1989. "Causes of conjugal dissolution: A cross-cultural study." *Current Anthropology* 30: 654–676.

Bevc, I., and I. Silverman. 2000. "Early separation and sibling incest: A test of the revised Westermarck theory." *Evolution and Human Behavior* 21: 151–161.

Bhasin, S., G. R. Cunningham, F. J. Hayes, A. M. Matsumoto, P. J. Synder, R. S. Swerdloff, and V. M. Montori. 2006. "Testosterone therapy in adult men with androgen deficiency syndromes: An Endocrine Society clinical practice guideline." *Journal of Clinical Endocrinology and Metabolism* 91: 1995–2010.

Birkhead, T. 2000. *Promiscuity.* Cambridge, MA: Harvard University Press.

Birkhead, T., and A. P. Moller, eds. 1998. *Sperm Competition and Sexual Selection.* New York: Academic Press.

Blackwood, E., and S. E. Wieringa, eds. 1999. *Female Desires: Same-Sex Relations and Transgender Practices across Cultures.* New York: Columbia University Press.

Blaustein, J. D., and M. S. Erskine. 2002. "Feminine sexual behavior: Cellular integration of hormonal and afferent information in the rodent forebrain." In *Hormones, Brain and Behavior,* ed. D. W. Pfaff. New York: Academic Press.

Block, J. 2009. "And then we were poly." *In One Big Happy Family: 18 writers talk about polyamory, open adoption, mixed marriage, househusbandry, single motherhood, and other realities of truly modern love,* ed. R. Walker. New York: Riverhead Books.

Bogin, B. 1999. *Patterns of Human Growth,* 2nd ed. New York: Cambridge University Press.

Bolin, A., and P. Whelehan. 2009. *Human Sexuality: Biological, Psychological, and Cultural Perspectives.* New York: Routledge.

Borgerhoff Mulder, M. 1988. "Kipsigis bridewealth payments." In *Human Reproductive Behaviour: A Darwinian Perspective,* ed. L. Betzig, M. Borgerhoff Mulder, and P. Turke, 65–82. New York: Cambridge University Press.

——. 2009. "Tradeoffs and sexual conflict over women's fertility preferences in Mpimbwe." *American Journal of Human Biology* 21: 478–487.

Boyd, R., and P. J. Richerson. 1985. *Culture and the Evolutionary Process.* Chicago: University of Chicago Press.

Brennan, P. L. R., R. O. Prum, K. G. McCracken, M. D. Sorenson, R. E. Wilson, and T. R. Birkhead. 2007. "Coevolution of male and female genital morphology in waterfowl." *PLoS ONE* 2: e418.

Brewer, G., and C. A. Hendrie. 2011. "Evidence to suggest that copulatory vocalizations in women are not a reflexive consequence of orgasm." *Archives of Sexual Behavior* 40: 559–564.

Brewis, A., and M. Meyer. 2005a. "Demographic evidence that human ovulation is undetectable (at least in pair bonds)." *Current Anthropology* 46: 465–471.

Brewis, A., and M. Meyer. 2005b. "Marital coitus across the life course." *Journal of Biosocial Science* 37: 499–518.

Bribiescas, R. G. 2006. *Men: Evolutionary and Life History.* Cambridge, MA: Harvard University Press.

Bribiescas, R. G., P. T. Ellison, and P. B. Gray. Forthcoming. "Reproductive effort and paternal investment plasticity in men: Implications for the evolution of *Homo.*" *Current Anthropology.*

Bronikowski, A. M., J. Altmann, D. K. Brockman, M. Cords, L. M. Fedigan, A. Pusey, T. Stoinski, W. F. Morris, K. B. Strier, and S. C. Alberts. 2011. "Aging in the natural world: comparative data reveal similar mortality patterns across primates." *Science* 331: 1325–1328.

Broude, G. J., and S. J. Greene. 1976. "Cross-cultural codes on twenty sexual attitudes and practices." *Ethnology* 15: 409–429.

Brown, D. E., L. L. Sievert, L. A. Morrison, A. M. Reza, and P. S. Mills. 2009. "Do Japanese American women really have fewer hot flashes than European Americans? The Hilo Women's Health Study." *Menopause* 18: 870–876.

Brown, P. J. 1991. "Culture and the evolution of obesity." *Human Nature* 2: 31–57.

Brown, W. M., M. Hines, B. A. Fane, and S. M. Breedlove. 2002. "Masculinized finger length patterns in human males and females with congenital adrenal hyperplasia." *Hormones and Behavior* 42: 380–386.

Buss, D. M. 1989. "Sex differences in human mate preferences: Evolutionary hypotheses tested in thirty-seven cultures." *Behavioral and Brain Sciences* 12: 1–49.

———. 2003. *The Evolution of Desire: Strategies of Human Mating.* New York: Basic Books.

Call, V., S. Sprecher, and P. Schwartz. 1995. "The incidence and frequency of marital sex in a national sample." *Journal of Marriage and Family* 57: 639–652.

Cameron, N. M. 2011. Maternal programming of reproductive function and behavior in the female rat. *Frontiers in Evolutionary Neuroscience* 3: 1–10.

Campbell, B. C. 2006. "Adrenarche and the evolution of human life history." *American Journal of Human Biology* 18: 569–589.

Campbell, C. J., A. Fuentes, K. C. MacKinnon, M., S. K. Bearder, and R. Stumpf, eds. 2011. *Primates in Perspective,* 2nd ed. New York: Oxford University Press.

Carnahan, S. J., and M. I. Jensen-Seaman. 2008. "Hominoid seminal protein evolution and ancestral mating behavior." *American Journal of Primatology* 70: 939–948.

Carter, C. S., L. Ahnert, K. E. Grossmann, S. B. Hrdy, M. E. Lamb, S. W. Porges, and N. Sachser, eds. 2005. *Attachment and Bonding: A New Synthesis.* Cambridge, MA: The MIT Press.

Carter, F. A., J. D. Carter, S. E. Luty, J. Jordan, V. V. W. McIntosh, A. F. Bartram, R. T. Mulder, J. M. McKenzie, and C. M. Bulik. 2007. "What is worse for your sex life: Starving, being depressed, or a new baby?" *International Journal of Eating Disorders* 40: 664–667.

Caspari, R., and S.-H. Lee. 2004. "Older age becomes common late in human evolution." *Proceedings of the National Academy of Sciences* 101: 10895–10900.

Chamie, J., and S. Nsuly. 1981. "Sex differences in remarriage and spouse selection." *Demography* 18: 335–348.

Chandra, V., M. Szklo, R. Goldberg, and J. Tonascia. 1983. "The impact of marital status on survival after an acute myocardial infarction: A population-based study." *American Journal of Epidemiology* 117: 320–325.

Chang, R. S., and J. R. Garcia. 2010 "Loverese: Bonding through intimate baby talk." Paper presented at NorthEastern Evolutionary Psychology Society conference.

Charnov, E. L., and D. Berrigan. 1993. "Why do female primates have such long lifespans and so few babies? Or, life in the slow lane." *Evolutionary Anthropology* 1: 191–194.

Chau, M. J., A. I. Stone, S. P. Mendoza, and K. L. Bales. 2008. "Is play behavior sexually dimorphic in monogamous species?" *Ethology* 114: 989–998.

Chivers, M. L., and J. M. Bailey. 2005. "A sex difference in features that elicit genital response." *Biological Psychology* 70: 115–120.

Chivers, M. L., M. C. Seto, M. L. Lalumeire, E. Laan, and T. Grimbos. 2010. "Agreement of self-reported and genital measures of sexual arousal in men and women: A meta-analysis." *Archives of Sexual Behavior* 39: 5–56.

Chumlea, W. C., C. M. Schubert, A. F. Roche, H. E. Kulin, P. A. Lee, J. H. Himes, and S. S. Sun. 2003. Age of menarche and racial composition in U.S. girls. *Pediatrics* 111: 110–113.

Clutton-Brock, T. H. 1991. *The Evolution of Parental Care.* Princeton, NJ: Princeton University Press.

Clutton-Brock, T. H., and K. Isvaran. 2007. "Sex differences in ageing in natural populations of vertebrates." *Proceedings of the Royal Society of London B* 274: 3097–3104.

Clutton-Brock, T. H., and K. McAuliffe. 2009. "Female mate choice in mammals." *Quarterly Review of Biology* 84: 3–27.

Clutton-Brock, T. H., and A. C. J. Vincent. 1991. "Sexual selection and the potential reproductive rates of males and females." *Science* 351: 58–60.

Cochran, G., and H. Harpending. 2010. *The 10,000 Year Explosion: How Civilization Accelerated Human Evolution.* New York: Basic Books.

Cohen, J. 2010. *Almost Chimpanzee.* New York: Times Books.

Cohen-Bendahan, C. C., C. van de Beek, and S. A. Berenbaum. 2005. "Prenatal sex hormone effects on child and adult sex-typed behavior: Methods and findings." *Neuroscience and Biobehavioral Reviews* 29: 353–384.

Coontz, S. 2005. *Marriage, a History: From Obedience to Intimacy, or How Love Conquered Marriage.* New York: Viking Press.

Copeland, S. R., M. Sponheimer, D. J. de Ruiter, J. A. Lee-Thorp, D. Codron, P. J. le Roux, V. Grimes, and M. P. Richards. 2011. "Strontium isotope evidence for landscape use by early hominins." *Nature* 474: 76–78.

Copen, C. E., A. Chandra, and G. Martinez. 2012. "Prevalence and timing of oral sex with opposite-sex partners among females and males aged 15–24 years: United States, 2007–2010." *National Health Statistics Reports, no. 56.* Hyattsville, MD: National Center for Health Statistics.

Crawford, J. C., M. Boulet, and C. M. Drea. 2010. "Smelling wrong: Hormonal contraception in lemurs alters critical female odour cues." *Proceedings of the Royal Society of London B* 278: 122–130.

Crews, D. E., and J. A. Stewart. 2010. "Human longevity and senescence." In *Human Evolutionary Biology*, ed. M. Muehlenbein, 528–550. New York: Cambridge University Press.

Crocker, W. H., and J. G. Crocker. 2004. *The Canela: Kinship, Ritual, and Sex in an Amazonian Tribe*. Belmont, CA: Thomson Wadsworth.

Daly, M., and M. Wilson. 1982. "Whom are newborn babies said to resemble?" *Ethology and Sociobiology* 3: 69–78.

———. 1988. *Homicide*. New York: Aldine.

Darwin, C. 1859. *Origin of Species*. Princeton, NJ: Princeton University Press.

———. 1871. *The Descent of Man, and Selection in Relation to Sex*. Princeton, NJ: Princeton University Press.

Davis, D. L., and R. G. Whitten. 1987. "The cross-cultural study of human sexuality." *Annual Review of Anthropology* 16: 69–98.

De Judicibus, M. A., and M. P. McCabe. 2002. "Psychological factors and the sexuality of pregnant and postpartum women." *Journal of Sex Research* 39: 94–103.

De Medeiros, C. B., S. L. Rees, M. Llinas, A. S. Fleming, and D. Crews. 2010. "Deconstructing early life experiences: Distinguishing the contributions of prenatal and postnatal factors to adult male sexual behavior in the rat." *Psychological Science* 21: 1494–1501.

Dennerstein, L., J. L. Alexander, and K. Kotz. 2003. "The menopause and sexual functioning: A review of the population-based studies." *Annual Review of Sex Research* 14: 64–82.

Dentan, R. K. 1997. *Malaysia and the "Original People": A Case Study of the Impact of Development on Indigenous Peoples*. Boston: Allyn and Bacon.

Dettwyler, K. 1995a. "A time to wean." In *Breastfeeding: Biocultural Perspectives*, ed. P. Stuart-Macadem and K. A. Dettwyler. New York: Aldine.

Dettwyler, K. 1995b. "Beauty and the breast: The cultural context of feeding in the United States." In *Breastfeeding: Biocultural Perspectives*, eds. P. Stuart-Macadem and K. A. Dettwyler. New York: Aldine.

De Vries, G. J., E. F. Rissman, R. B. Simerly, L. Y. Yang, E. M. Scordalakes, C. J. Auger, A. Swain, R. Lovell-Badge, P. S. Burgoyne, and A. P. Arnold. 2002. "A model system for study of sex chromosome effects on sexually dimorphic neural and behavioral traits." *Journal of Neuroscience* 22: 9005–9014.

DeWaal, F. 2006. *Our Inner Ape*. New York: Riverhead.

Diamond, J. 1997. *Why Is Sex Fun?* New York: Basic Books.

Diamond, L. M. 2008. *Sexual Fluidity*. Cambridge, MA: Harvard University Press.

Ditzen, B., M. Schaer, B. Gabriel, G. Bodenmann, U. Ehlert, and M. Heinrichs. 2009. "Intranasal oxytocin increases positive communication and reduces cortisol levels during couple conflict." *Biological Psychiatry* 65: 728–731.

Dixson, A. F. 1998. *Primate Sexuality.* New York: Oxford University Press.

———. 2009. *Sexual Selection and the Origins of Human Mating Systems.* New York: Oxford University Press.

Doran-Sheehy, D. M., D. Fernandez, and C. Borries. 2009. "The strategic use of sex in wild female western gorillas." *American Journal of Primatology* 71: 1011–1020.

Duffy, J. 1963. "Masturbation and clitoridectomy: A nineteenth-century view." *Journal of the American Medical Association* 3: 166–168.

Durante, K. M., N. P. Li, and M. G. Haselton. 2008. "Changes in women's choice of dress across the ovulatory cycle: Naturalistic and laboratory task-based evidence. *Personality and Social Psychology Bulletin* 34: 1451–1460.

Eaton, S. B., M. C. Pike, R. V. Short, N. C. Lee et al. 1994. "Women's reproductive cancers in evolutionary context." *Quarterly Review of Biology* 69: 353–367.

Eberhard, W. G. 1985. *Sexual Selection and Animal Genitalia.* Cambridge, MA: Harvard University Press.

———. 1996. *Female Control: Sexual Selection by Cryptic Female Choice.* Princeton, NJ: Princeton University Press.

Eckert, W. G., S. Katchis, and P. Dotson. 1991. "The unusual death of a pregnant woman by sexual foreplay." *American Journal of Forensic Medicine and Pathology* 12: 247–249.

Eens, M., and R. Pinxten. 2000. "Sex-role reversal in vertebrates: Behavioural and endocrinological accounts." *Behavioural Processes* 51: 1135–147.

Eerkens, J. W., A. G. Berget, and E. J. Bartelink. 2011. "Estimating weaning and early childhood diet from serial micro-samples of dentin collagen." *Journal of Archaeological Science* 38: 3101–3111.

Eggert, A., and S. K. Sakaluk. 1995. "Female-coerced monogamy in burying beetles." *Behavioral Ecology and Sociobiology* 37: 147–153.

Ellefsen, J. O. 1974. "A natural history of white-handed gibbons in the Malayan Peninsula." In *Gibbon and Siamang,* vol. 3, ed. D. M. Rumbaugh, 1–136. Basel: Karger.

Ellis, B. J. 2004. "Timing of pubertal maturation in girls: An integrated life history approach." *Psychological Bulletin* 130: 920–958.

Ellison, P. T. 2001. *On Fertile Ground.* Cambridge, MA: Harvard University Press.

Ellison, P. T., R. G. Bribiescas, G. R. Bentley, B. C. Campbell, S. F. Lipson, C. Panter-Brick, and K. Hill. 2002. "Population variation in age-related decline in male salivary testosterone." *Human Reproduction* 17: 3251–3253.

Ellison, P. T., and P. B. Gray, eds. 2009. *Endocrinology of Social Relationships.* Cambridge, MA: Harvard University Press.

Ellison, P. T., and C. Valeggia. 2003. "C-peptide levels and the duration of lactational amenorrhea." *Fertility and Sterility* 80: 1279–1280.

Engelman, R. 2008. *More: Population, Nature, and What Women Want.* Washington, DC: Island Press.

Escasa, M. J., J. Casey, and P. G. Gray. 2011. Salivary testosterone levels in men at a U.S. sex club. *Archives of Sexual Behavior* 40: 921–926.

Exton, M. S., A. Bindert, T. Krüger, F. Scheller, U. Hartmann, and M. Schedlowski. 1999. "Cardiovascular and endocrine alterations after masturbation-induced orgasm in women." *Psychosomatic Medicine* 61: 280–289.

Fagan, R. 1981. *Animal Play Behavior.* New York: Oxford University Press.

Fedigan, L. M., and M. S. M. Pavelka. 2011. "Reproductive cessation in female primates: Comparisons of Japanese macaques and humans." In *Primates in Perspective,* 2nd ed., ed. C. J. Campbell, A. Fuentes, K. C. MacKinnon, M. Panger, and S. K. Bearder. New York: Oxford University Press.

Feldman, R., A. Weller, O. Zagoory-Sharon, and A. Levine. 2007. "Evidence for a neuroendocrinological foundation of human affiliation." *Psychological Science* 18: 965–970.

Fernandez-Duque, E. 2007. "The Aotinae: Social monogamy in the only nocturnal haplorhines." In *Primates in Perspective,* ed. C. J. Campbell, A. Fuentes, K. C. MacKinnon, M. Panger, and S. K. Bearder. New York: Oxford University Press.

Fessler, D. 2007. "Neglected natural experiments germane to the Westermarck hypothesis: The Karo Batak and the Oneida Community." *Human Nature* 18: 355–364.

Finch, C. E., and D. J. Holmes. 2010. "Ovarian aging in developmental and evolutionary contexts." *Annals of the New York Academy of Sciences* 1204: 82–94.

Finer, L. B. 2007. "Trends in premarital sex in the United States, 1954–2003." *Public Health Reports* 122: 73–78.

Fisher, H. E. 1982. *The Sex Contract: The Evolution of Human Behavior.* New York: Quill.

———. 1989. "Evolution of human serial pairbonding." *American Journal of Physical Anthropology* 78: 331–354.

———. 1992. *Anatomy of Love: The Natural History of Monogamy, Adultery, and Divorce.* New York: Norton and Company.

———. 1998. "Lust, attraction, and attachment in mammalian reproduction." *Human Nature* 9: 23–52.

———. 2004. *Why We Love: The Nature and Chemistry of Romantic Love.* New York: Henry Holt.

———. 2009. *Why Him? Why Her? Finding real love by understanding your personality type.* New York: Henry Holt and Company.

———. 2011. "Serial monogamy and clandestine adultery: Evolution and consequences of the dual human reproductive strategy." In *Applied Evolutionary Psychology,* ed. S. C. Roberts. New York: Oxford University Press.

Fisher, H. E., A. Aron, and L. L. Brown. 2006. "Romantic love: A mammalian brain system for mate choice." *Philosophical Transactions of the Royal Society B* 361: 2173–2186.

Fisher, H. E., L. L. Brown, A. Aron, G. Strong, and D. Mashek. 2010. "Reward, addiction, and emotion regulation systems associated with rejection in love." *Journal of Neurophysiology* 104: 51–60.

Fisher, H. E., and J. A. Thomson, Jr. 2006. "Lust, romance, attachment: Do the side effects of serotonin-enhancing antidepressants jeopardize romantic love, marriage, and fertility?" In *Evolutionary Cognitive Neuroscience*, ed. S. M. Platek, J. P. Keenan, and T. K. Shackelford, 245–283. Cambridge, MA: The MIT Press.

Fisher, H. I. 1976. "Some dynamics of a breeding colony of Laysan albatrosses." *The Wilson Bulletin* 88: 121–142.

Fisher, M. L., and A. Cox. 2010. "Man change thyself: Hero versus heroine development in Harlequin romance novels." *Journal of Social, Evolutionary, and Cultural Psychology* 4: 305–316.

Fisher, M. L., J. R. Garcia, and R. S.Chang, eds. 2013. *Evolution's Empress: Darwinian Perspectives on the Nature of Women.* New York: Oxford University Press.

Foley, R., and C. Gamble. 2009. "The ecology of social transitions in human evolution." *Philosophical Transactions of the Royal Society B* 364: 3267–3279.

Ford, C. S., and F. A. Beach 1951. *Patterns of Sexual Behavior.* New York: Ace Books.

Fox, R. 1980. *The Red Lamp of Incest.* New York: Dutton.

Francis, R. C. 2011. *Epigenetics: The Ultimate Mystery of Inheritance.* New York: W. W. Norton.

Francoeur, R. T., and R. J. Noonan, eds. 2006. *The Continuum Complete International Encyclopedia of Sexuality.* New York: Continuum.

Frayser, S. 1985. *Varieties of Sexual Experience.* New Haven, CT: HRAF Press.

Freud, S. 1920. *The Complete Psychological Works of Sigmund Freud,* vol. 7: *Three Essays on the Theory of Sexuality.* London: Hogarth.

Frisch, R. E. 1984. "Body fat, puberty, and fertility." *Biological Reviews* 59: 161–188.

Gallup, G. G., and D. A. Frederick. 2010. "The science of sex appeal: An evolutionary perspective." *Review of General Psychology* 14: 240–250.

Gallup, G. G., R. L. Burch, and S. M. Platek. 2002. "Does semen have antidepressant properties?" *Archives of Sexual Behavior* 31: 289–293.

Gangestad, S. W., and R. Thornhill. 2008. "Human oestrus." *Proceedings of the Royal Society B* 275: 991–1000.

Garcia, J. R., M. J. Escasa-Dorne, and P. B. Gray. In review. "Women's salivary testosterone and estradiol levels following sexual activity in a non-laboratory setting."

Garcia, J. R., S. G. Massey, A. M. Merriwether, and S. M. Seibold-Simpson. In review. "Orgasm experience among emerging adult men and women: Relationship context and attitudes toward casual sex.

Garcia, J. R., and D. J. Kruger. 2010. "Unbuckling in the Bible Belt: Conservative sexual norms lower age at marriage." *Journal of Social, Evolutionary, and Cultural Psychology* 4: 206–214.

Garcia, J. R., J. MacKillop, E. L. Aller, A. M. Merriwether, D. S. Wilson, and J. K. Lum. 2010. "Associations between dopamine D4 receptor gene variation with both infidelity and sexual promiscuity." *PLoS ONE* 5: e14162.

Garcia, J. R., and C. Reiber. 2008. "Hook-up behavior: A biopsychosocial perspective." *Journal of Social, Evolutionary, and Cultural Psychology* 2: 192–208.

Garcia, J. R., C. Reiber, S. G. Massey, and A. M. Merriwether. 2012. "Sexual hookup culture: A review." *Review of General Psychology* 16: 161–176.

Gardner, D., and S. Shoback. 2007. *Greenspan's Basic and Clinical Endocrinology*, 8th ed. New York: McGraw-Hill.

Geary, D. C. 2010. *Male, Female: The Evolution of Human Sex Differences*, 2nd ed. Washington, DC: American Psychological Association.

Geher, G. 2009. Accuracy and oversexualization in cross-sex mind-reading: An adaptationist approach. *Evolutionary Psychology* 7: 331–347.

Geher, G., and S. B. Kauffman. 2013. *Mating Intelligence Unleashed: The Role of the Mind in Sex, Dating, and Love*. New York: Oxford University Pess.

Gentry, G. A., M. Lowe, G. Alford, and R. Nevins. 1988. "Sequence analyses of herpesviral enzymes suggest an ancient origin for human sexual behavior." *Proceedings of the National Academy of Sciences* 85: 2658–2661.

Georgiadis, J. R., R. Kortekaas, R. Kuipers, A. Nieuwenburg, J. Prium, A. A. T. Simone Reinders, and G. Holstege. 2006. "Regional cerebral blood flow changes associated with clitorally induced orgasm in healthy women." *European Journal of Neuroscience* 24: 3305–3316.

Gettler, L. T., T. W. McDade, A. B. Feranil, and C. W. Kuzawa. 2011. "Longitudinal evidence that fatherhood decreases testosterone in human males." *Proceedings of the National Academy of Sciences* 108: 16194–16199.

Giles-Sim, J. 1998. "Current knowledge about child abuse in stepfamilies." *Marriage and Family Review* 26: 215–230.

Gilmore, D. 1990. *Manhood in the Making.* Philadelphia: Temple University Press.

Glickman, S. E., G. R. Cunha, C. M. Drea, A. J. Conley, and N. J. Place. 2006. "Mammalian sexual differentiation: Lessons from the spotted hyena." *Trends in Endocrinology and Metabolism* 17: 349–356.

Goodale, J. 1971. *Tiwi Wives: A Study of the Women of Melville Island, North Australia.* Seattle: University of Washington Press.

Goodall, J. 1986. *The Chimpanzees of Gombe: Patterns of Behavior.* Cambridge, MA: Harvard University Press.

Gorer, G. 1938. *Himalayan Village: An Account of the Lepchas of Sikkim.* London: Michael Joseph.

Goy, R. W., F. B. Bercovitch, and M. C. McBrain. 1998. "Behavioral masculinization is independent of genital masculinization in prenatally androgenized female rhesus macaques." *Hormones and Behavior* 22: 552–571.

Graves, J. A. M., and S. Shetty. 2001. "Sex from W to Z: Evolution of vertebrate sex chromosomes and sex determining genes." *Journal of Experimental Zoology* 290: 449–462.

Gray, P. B. 2003. "Marriage, parenting, and testosterone variation among Kenyan Swahili men." *American Journal of Physical Anthropology* 122: 279–286.

———. 2004. "HIV and Islam: Is HIV prevalence lower among Muslims?" *Social Science and Medicine* 58: 1751–1756.

———. 2010. "Evolution and endocrinology of human behavior: A focus on sex differences and reproduction." In *Human Evolutionary Biology,* ed. M. Muehlenbein. New York: Cambridge University Press.

Gray, P. B., and K. G. Anderson. 2010. *Fatherhood: Evolution and Human Paternal Behavior.* Cambridge, MA: Harvard University Press.

Gray, P., and B. Campbell. 2005. "Erectile dysfunction and its correlates among the Ariaal of northern Kenya." *International Journal of Impotence Research* 17: 445–449.

———. 2009. "Human male testosterone, pair bonding, and fatherhood." In *Endocrinology of Social Relationships,* ed. P. T. Ellison and P. B. Gray. Cambridge, MA: Harvard University Press.

Gray, P. B., J. Flynn Chapman, M. H. McIntyre, T. C. Burnham, S. F. Lipson, and P. T. Ellison. 2004. "Human male pair bonding and testosterone." *Human Nature* 15: 119–131.

Gray, P. B., and D. Frederick. 2012. "Body image and body type preferences in St. Kitts, Caribbean: A cross-cultural comparison with U.S. university and Internet samples." *Evolutionary Psychology* 10: 631–655.

Gray, P. B., and J. R. Garcia. 2012. "Aging and human sexual behavior: Biocultural perspectives—A mini-review." *Gerontology* 58: 446–452.

Gray, P. B., and M. Samms-Vaughan. 2010. "Investigating potential hormonal associations of grandmaternal care in Jamaica." *Internet Journal of Biological Anthropology* 4, no. 1.

Gray, P. B., and S. M. Young. 2011. "Human-pet dynamics in cross-cultural perspective." *Anthrozoos* 24: 1–17.

Greenwood, P. J. 1980. "Mating systems, philopatry and dispersal in birds and mammals." *Animal Behaviour* 28: 1140–1162.

Gregersen, E. 1994. *The World of Human Sexuality: Behaviors, Customs, and Beliefs.* New York: Irvington.

Grimbos, T., K. Dawood, R. P. Burriss, K. J. Zucker, and D. A. Puts. 2010. "Sexual orientation and the second to fourth finger length ratio: A meta-analysis in men and women." *Behavioral Neuroscience* 124: 278–287.

Gueguen, N. 2007. "Women's bust size and men's courtship solicitation." *Body Image* 4: 386–390.

———. 2009a. "Menstrual cycle phases and female receptivity to a courtship solicitation: An evaluation in a nightclub." *Evolution and Human Behavior* 30: 351–355.

———. 2009b. "The receptivity of women to courtship solicitation across the menstrual cycle: A field experiment." *Biological Psychology* 80: 321–324.

Gurven, M., and H. Kaplan. 2007. "Hunter-gatherer longevity: Cross-cultural perspectives." *Population and Development Review* 33: 321–365.

Haag, A. L. 2007. "Patented harpoon pins down whale age." *NatureNews,* June 19.

Haig, D. 2000. "Genomic imprinting, sex-biased dispersal, and social behavior." *Annals of the New York Academy of Sciences* 907: 149–163.

Hall, L. S., and C. T. Love. 2003. "Finger-length ratios in female monozygotic twins discordant for sexual orientation." *Archives of Sexual Behavior* 32: 23–28.

Hamilton, B. E. and S. J. Ventura. 2012. "Birth rates for U.S. teenagers reach historic lows for all age and ethnic groups." NCHS Data Brief, no. 89. Hyattsville, MD: National Center for Health Statistics.

Hamilton, W. D., and M. Zuk. 1982. "Heritable true fitness and bright birds: A role for parasites?" *Science* 213: 384–387.

Han, C. S., and P. G. Jablonski. 2010. "Female genitalia concealment promotes intimate male courtship in a water strider. *PLoS One* 4: e5793.

Harlow, H. F. 1971. *Learning to Love.* San Francisco: Albion.

Harpending, H. and G. Cochran. 2002. "In our genes." *Proceedings of the National Academy of Science of the U.S.A.* 99: 10–12.

Harris, H. 1995. "Rethinking heterosexual relationships in Polynesia: A case study of Mangaia, Cook Island." In *Romantic Passion: A Universal Experience?* ed. W. R. Jankowiak. New York: Columbia University Press.

Hart, C. W. M., A. R. Pilling, and J. C. Goodale. 1988. *The Tiwi of North Australia.* New York: Holt, Rinehart and Winston.

Hatfield, E., and R. L. Rapson. 1993. *Love, Sex, and Intimacy: Their Psychology, Biology, and History.* New York: HarperCollins.

———. 2005. *Love and Sex: Cross-cultural Perspectives.* Lanham: University Press of America.

Hawkes, K., J. F. O'Connell, N. G. Blurton Jones, E. L. Charnov, and H. Alvarez. 1998. "Grandmothering, menopause, and the evolution of human life histories." *Proceedings of the National Academy of Sciences* 95: 1336–1339.

Hawkes, K., and R. Paine, eds. 2006. *The Evolution of Human Life History.* Santa Fe, NM: School of American Research.

Hawley, P. H., and W. A. Hensley. 2009. "Social dominance and forceful submission fantasies: Feminine pathology or power?" *Journal of Sex Research* 46: 568–585.

Heiman, J. R. 1977. "A psychophysiological exploration of sexual arousal patterns in females and males. *Psychophysiology* 14: 266–274.

Heiman, J. R., and J. LoPiccolo. 1988. *Becoming Orgasmic: A Sexual and Personal Growth Program for Women.* New York: Prentice Hall.

Heiman, J. R., D. L. Rowland, J. P. Hatch, and B. A. Gladue. 1991. "Psychophysiological and endocrine responses to sexual arousal in women." *Archives of Sexual Behavior* 20: 171–186.

Henderson, J. J. A., and J. M. Anglin. 2003. "Facial attractiveness predicts longevity." *Evolution and Human Behavior* 24: 351–356.

Herbenick, D. M., and J. D. Fortenberry. 2012. "Exercise-induced orgasm and pleasure among women." *Sexual and Relationship Therapy* 26: 373–388.

Herbenick, D., M. Reece, V. Schick, S. A. Sanders, B. Dodge, and J. D. Fortenberry. 2010a. "Sexual behavior in the United States: Results from a national probability sample of men and women ages 14–94." *Journal of Sexual Medicine* 7: 255–265.

Herbenick, D., V. Schick, M. Reece, S. A. Sanders, and J. D. Fortenberry. 2010b. "Pubic hair removal among women in the United States: Prevalence, methods, and characteristics." *Journal of Sexual Medicine* 7: 3322–3330.

Herdt, G. H. 2006. *The Sambia: Ritual, Sexuality, and Change in Papua New Guinea.* Belmont, CA: Thomson Wadsworth.

———. 1981. *Guardians of the Flutes: Idioms of Masculinity.* New York: McGraw-Hill.

Herdt, G. H., and J. Davidson. 1988. "The Sambia "turnim-man": Sociocultural and clinical aspects of gender formation in male pseudohermaphrodites with 5alpha-reductase deficiency in Papua New Guinea." *Archives of Sexual Behavior* 17: 33–56.

Herdt, G., and M. McClintock. 2000. "The magical age of 10." *Archives of Sexual Behavior* 29: 587–606.

Hess, E. 2008. *Nim Chimpsky: The Chimp Who Would Be Human.* New York: Bantam.

Hewlett, B. L., and B. S. Hewlett. 2008. "A biocultural approach to sex, love and intimacy in central African foragers and farmers." In *Intimacies: Love and Sex Across Cultures,* ed. W. R. Jankowiak. New York: Columbia University Press.

Hewlett, B. S. 1991. "Demography and childcare in preindustrial societies." *Journal of Anthropological Research* 47: 1–37.

Hewlett, B. S., and B. L. Hewlett. 2010. "Sex and searching for children among Aka foragers and Ngandu farmers of Central Africa." *African Study Monographs* 31: 107–125.

Hewlett, B. S., and M. E. Lamb, eds. 2005. *Hunter-Gatherer Childhoods: Evolutionary, Developmental and Cultural Perspectives.* New York: Transaction/Aldine de Gruyter.

Heyer, E., R. Chaix, S. Pavard, and F. Austerlitz. 2012. "Sex-specific demographic behaviours that shape human genomic variation." *Molecular Ecology* 21: 597–612.

Higham, J. P., C. Ross, Y. Warren, M. Heistermann, and A. MacLarnon. 2007. "Reduced reproductive function in wild baboons *(Papio hamadryas Anubis)* related to natural consumption of the African black plum *(Vitex doniana)*." *Hormones and Behavior* 52: 384–390.

Hill, K., C. Boesch, J. Goodall, A. Pusey, J. Williams, and R. Wrangham. 2001. "Mortality rates among wild chimpanzees." *Journal of Human Evolution* 40: 437–450.

Hill, K., and A. M. Hurtado. 1996. *Ache Life History.* New York: Aldine.

Hines, M. 2004. *Brain Gender.* New York: Oxford University Press.

Hobbs, D. R., and G. G. Gallup. 2011. "Songs as medium for embedded reproductive messages." *Evolutionary Psychology* 9: 390–416.

Hoff, C. C., and S. C. Beougher. 2010. "Sexual agreements among gay male couples." *Archives of Sexual Behavior* 39: 774–787.

Holmberg, A. 1946. "The Siriono." unpublished PhD diss. Yale University.

Howell, N. 2010. *Life Histories of the Dobe !Kung.* Berkeley: University of California Press.

Hrdy, S. B. 1981. *The Woman That Never Evolved.* Cambridge, MA: Harvard University Press.

———. 1999. *Mother Nature: A Natural History of Mothers, Infants, and Natural Selection.* New York: Pantheon.

———. 2009. *Mothers and Others: The Evolutionary Origins of Mutual Understanding.* Cambridge, MA: Belknap Press of Harvard University Press.

Huck, M., and E. Fernandez-Duque. 2012. "Building babies when dads help: Infant development of owl monkeys and other primates with allo-maternal care." In *Building Babies: Primate Developmental Trajectories in Proximate and Ultimate Perspectives,* ed. K. Clancy, K. Hinde, and J. Rutherford. New York: Springer Verlag.

Huen, K. F., S. S. Leung, J. T. Lau, A. Y. Cheung, N. K. Leung, and M. C. Chiu. 1997. "Secular trend in the sexual maturation of southern Chinese girls." *Acta Paediatrica* 86: 1121–1124.

Hughes, J. F., H. Skaletsky, T. Pyntikova, P. J. Minx, T. Graves, S. Rozen, R. K. Wilson, and D. C. Page. 2005. "Conservation of Y-linked genes during human evolution revealed by comparative sequencing in chimpanzee." *Nature* 437: 100–103.

Hughes, S. M., F. Dispenza, and G. G. Gallup. 2004. "Ratings of voice attractiveness predict sexual behavior and body configuration." Evolution and Human Behavior 25: 295–304.

Hughes, S. M., M. A. Harrison, and G. G. Gallup. 2007. "Sex differences in romantic kissing among college students: An evolutionary perspective." *Evolutionary Psychology* 5: 612–631.

Hyde, J. S., J. D. DeLamater, E. A. Plant, and J. M. Byrd. 1996. "Sexuality during pregnancy and the year postpartum." *Journal of Sex Research* 33: 143–151.

Inhorn, M., and D. Birenbaum-Carmeli. 2008. "Assisted reproductive technologies and culture change." *Annual Reviews of Anthropology* 37: 177–196.

Inhorn, M., and F. van Balen. 2002. *Infertility around the Globe.* Berkeley: University of California Press.

Insel, T. R. 2006. "Translational research in the decade of discovery." *Hormones and Behavior* 50: 504–505.

Jankowiak, W. R., ed. 1995. *Romantic Passion: A Universal Experience?* New York: Columbia University Press.

———, ed. 2008. *Intimacies: Love and Sex across Cultures.* New York: Columbia University Press.

Jankowiak, W. R., and E. F. Fischer. 1992. A cross-cultural perspective on romantic love. *Ethnology* 31: 149–155.

Jankowiak, W. R., and M. D. Hardgrave. 2007. "Individual and societal responses to sexual betrayal: A view from around the world." *Electronic Journal of Human Sexuality* 10.

Jankowiak, W. R., and T. Paladino. 2008. "Desiring sex, longing for love: A tripartite conundrum." In *Intimacies: Love and Sex across Cultures,* ed. W. R. Jankowiak. New York: Columbia University Press.

Jankowiak, W. R., M. Sudakov, and B. C. Wilreker. 2005. "Co-wife conflict and co-operation." *Ethnology* 44: 81–98.

Janssen, E., ed. 2007. *The Psychophysiology of Sex.* Bloomington: Indiana University Press.

Janssen, E., and J. Bancroft. 2007. "The dual control model: The role of sexual inhibition and excitation in sexual arousal and behavior." In *The Psychophysiology of Sex,* ed. E. Janssen. Bloomington: Indiana University Press.

Janssen, E., N. Prause, and J. Geer. 2007. "The sexual response." In *Handbook of Psychophysiology,* 3rd ed., ed. J. T. Cacioppo, L. G. Tassinary, and G. G. Berntson. New York: Cambridge University Press.

Jasienska, G. 2003. "Energy metabolism and the evolution of reproductive suppression in the human female." *Acta Biotheoretica* 51: 1–18.

Jasienska, G., and P. T. Ellison. 1996. "Physical work causes suppression of ovarian function in women." *Proceedings of the Royal Society of London B* 265: 1847–1851.

Joffee, T. H. 1997. "Social pressures have selected for an extended juvenile period in primates." *Journal of Human Evolution* 32: 593–605.

Johnston, R. E. 1986. "Effects of female odors on the sexual behavior of male hamsters." *Behavioral and Neural Biology* 46: 165–188.

Jokela, M. 2009. "Physical attractiveness and reproductive success in humans: Evidence from the late 20th century United States." *Evolution and Human Behavior* 30: 342–350.

Jones, J. H. 2011. "Primates and the evolution of long, slow life histories." *Current Biology* 21: R708–R717.

Jones, K. P., L. C. Walker, D. Anderson, A. Lecreuse, S. L. Robson, and K. Hawkes. 2007. "Depletion of follicles with age in chimpanzees: Similarities to humans." *Biology of Reproduction* 77: 247–251.

Judson, O. 2002. *Dr. Tatiana's Sex Advice to All Creation.* New York: Henry Holt.

Kahlenberg, S. M., and R. W. Wrangham. 2010. "Sex differences in chimpanzees' use of sticks as play objects resemble those of children." *Current Biology* 20: R1067–R1068.

Kaighobadi, F., T. K. Shackelford, and V. A. Weekes-Shackelford. 2011. "Do women pretend orgasm to retain a mate?" *Archives of Sexual Behavior.*

Kaplan, H. S. 1979. *Disorders of Sexual Desire.* New York: Brunner/Mazel.

Kaplan, H., K. Hill, J. Lancaster, and A. M. Hurtado. 2000. "A theory of human life history evolution: Diet, intelligence, and longevity." *Evolutionary Anthropology* 9: 156–185.

Karapanou, O., and A. Papadimitriou. 2010. "Determinants of menarche." *Reproductive Biology and Endocrinology* 8: 115.

Karney, B. R., and T. N. Bradbury. 1995. "The longitudinal course of marital quality and stability: A review of theory, method, and research." *Psychological Bulletin* 119: 3–34.

Karremans, J. C., W. E. Frankenhuis, and S. Arons. 2010. "Blind men prefer a low waist-to-hip ratio." *Evolution and Human Behavior* 31: 182–186.

Kauth, M. R., ed. 2006. *Handbook of the Evolution of Human Sexuality.* Binghamton, New York: The Haworth Press.

Keele, B. F., J. H. Jones, K. A. Terio, J. D. Estes, et al. 2009. "Increased mortality and AIDS-like immunopathology in wild chimpanzees infected with SIVcpz. *Nature* 460: 515–519.

Kellogg, W. N., and L. A. Kellogg. 1933. *The Ape and the Child: A Comparative Study of the Environmental Influence upon Early Behavior.* New York: Hafner.

Kelly, R. 1995. *The Foraging Spectrum.* Washington, D.C.: Smithsonian Press.

Kempadoo, K. 2004. *Sexing the Caribbean: Gender, Race, and Sexual Labor.* New York: Routledge.

Kerns, V., and J. K. Brown, eds. 1992. *In Her Prime: New Views of Middle-Aged Women,* 2nd ed. Urbana: University of Illinois Press.

Kiecolt-Glaser, J. K., J. P. Gouin, and L. V. Hantsoo. 2010. "Close relationships, inflammation, and health." *Neuroscience and Biobehavioral Reviews* 35: 33–38.

Killick, S. R., C. Leary, J. Trussell, and K. A. Guthrie. 2011. "Sperm content of pre-ejaculatory fluid." *Human Fertility* 14: 48–52.

Kinsey, A., W. B. Pomeroy, and C. E. Martin. 1948. *Sexual Behavior in the Human Male.* Philadelphia: Saunders.

Kinsey, A., W. Pomeroy, C. Martin, and P. Gebhard. 1953. *Sexual Behavior in the Human Female.* Philadelphia: Saunders.

Kirkpatrick, R. C. 2000. "The evolution of human homosexual behavior." *Current Anthropology* 41: 385–413.

Klein, R. 2009. *The Human Career,* 3rd ed. Chicago: University of Chicago.

Klotz, L. 2005. "How (not) to communicate new scientific information: A memoir of the famous Brindley lecture." *British Journal of Urology International* 96: 956–957.

Knobil, E., and J. D. Neill, eds. 2006. *Knobil and Neill's Physiology of Reproduction,* 3rd ed. St. Louis, MO: Elsevier Academic Press.

Knott, C. D., and S. M. Kahlenberg. 2011. "Orangutans." In *Primates in Perspective,* 2nd ed., ed. C. J. Campbell, A. Fuentes, K. C. MacKinnon, S. K. Bearder, and R. Stumpf. New York: Oxford University Press.

Kohl, J. V., and R. T. Francoeur. 2002. *The Scent of Eros: Mysteries of Odor in Human Sexuality,* 2nd ed. Lincoln, NE: iUniverse Press.

Kokko, H., and M. D. Jennions. 2008. "Sexual conflict: Battle of the sexes reversed." *Current Biology* 18: R121–R123.

Komisaruk, B. R., and B. Whipple. 2005. "Functional MRI of the brain during orgasm in women." *Annual Review of Sex Research* 16: 62–86.

Komisaruk, B. R., N. Wise, E. Frangos, W.-C. Liu, K. Allen, and S. Brody. 2011. "Women's clitoris, vagina, and cervix mapped on the sensory cortex: fMRI evidence." *Journal of Sexual Medicine* 8: 2822–2830.

Konner, M. 2010. *The Evolution of Childhood: Relationships. Emotion, Mind.* Cambridge, MA: Harvard University Press.

Kramer, A. C. 2011. "Penile fracture seems more likely during sex under stressful situations." *Journal of Sexual Medicine* 8: 3414–3417.

Kramer, K. 2008. "Early sexual maturity among Pume foragers of Venezuela: Fitness implications of teen motherhood." *American Journal of Physical Anthropology* 136: 338–350.

Kruger, D. J., and R. M. Nesse. 2006. "An evolutionary life-history framework for understanding sex differences in human mortality rates." *Human Nature* 17: 74–97.

Kunz, T. H., and D. J. Hosken. 2008. "Male lactation: Why, why not, and is it care?" *Trends in Ecology and Evolution* 24: 80–85.

Kuzawa, C. 1998. "Adipose tissue in human infancy and childhood: An evolutionary perspective." *Yearbook of Physical Anthropology* 41: 177–209.

Labuda, D., J.-F. Lefebre, P. Nadeau, and M.-H. Roy-Gagnon. 2010. "Female-to-male breeding ratio in modern humans: An analysis based on historical recombinations." *American Journal of Human Genetics* 86: 353–363.

Lalueza-Fox, C., A. Rosas, A. Estalrrich, E. Gigli, P. F. Campos et al. 2010. "Genetic evidence for patrilocal mating behavior among Neandertal groups." *Proceedings of the National Academy of Sciences* 108: 250–253.

LaMarre, A. K., L. Q. Paterson, and B. B. Gorzalka. 2003. "Breastfeeding and postpartum maternal sexual functioning. A review." *The Canadian Journal of Human Sexuality* 12: 151–168.

Lancaster, J. B., and H. S. Kaplan. 2009. The endocrinology of the human adaptive complex. In P. T. Ellison and P. B. Gray, eds., *Endocrinology of Social Relationships*. Cambridge, MA: Harvard University Press.

Lancaster, M. B., and C. S. Lancaster. 1983. "Parental investment: The hominin adaptation." In *How Humans Adapt: A Biocultural Odyssey,* ed. D. Ortner, 33–65. Washington, DC: Smithsonian Press.

Lancy, D. F. 2008. *The Anthropology of Childhood.* New York: Cambridge University Press.

Lassek, W. D., and S. J. C. Gaulin. 2008. "Waist-hip ratio and cognitive ability: Is gluteofemoral fat a privileged store of neurodevelopmental resources?" *Evolution and Human Behavior* 29: 26–34.

Laumann, E. O., J. H. Gagnon, R. T. Michael, and S. Michaels. 1994. *The Social Organization of Sexuality: Sexual Practices in the United States.* Chicago: University of Chicago Press.

Laumann, E. O., A. Nicolosi, D. B. Glasser, A. Paik, C. Gingell, E. Moreira, and T. Wang. 2005. "Sexual problems among women and men aged 40–80 y: Prevalence and correlates identified in the Global Study of Sexual Attitudes and Behaviors." *International Journal of Impotence Research* 17: 39–57.

Laumann, E. O., A. Paik, J. H. Kang, T. Wang, B. Levinson, E. D. Moreira Jr., A. Nicolosi, and C. Gingell. 2006. "A cross-national study of subjective sexual well-being among older women and men: Findings from the global study of sexual attitudes and behaviors." *Archives of Sexual Behavior* 35: 145–161.

Lee, R. B. 1992. "Work, sexuality, and aging among !Kung women." In *In Her Prime: New Views of Middle-Aged Women,* 2nd ed., ed. V. Kerns and J. K. Brown, 35–48. Urbana: University of Illinois Press.

Lehr, S. A., G. Willard, C. Merrill, P. B. Gray, and M. Banaji. 2012. "Moral judgment of homosexual behavior: Mate availability accounts for cultural variation." Paper presented at Society for Personality and Social Psychology conference.

Leitner, S., R. C. Marshall, B. Leisler, and C. K. Catchpole. 2006. "Male song quality, egg size, and offspring sex in captive canaries *(Serinus canaria).*" *Ethology* 112: 554–563.

LeMarre, A. K., L. Q. Paterson, and B. B. Gorzalka. 2003. "Breastfeeding and postpartum maternal sexual functioning: A review." *Canadian Journal of Human Sexuality* 12: 151–168.

Levay, S., and J. Baldwin. 2009. *Human Sexuality,* 3rd ed. Sunderland, MA: Sinauer.

Levin, R. J., and W. van Berlo. 2004. "Sexual arousal and orgasm in subjects who experience forced or non-consensual sexual stimulation: A review." *Journal of Clinical Forensic Medicine* 11:82–88.

Lévi-Strauss, C. 1949. *The Elementary Structures of Kinship.* Boston: Beacon.

Lieberman, D. E. 2011. *The Evolution of the Human Head.* Cambridge, MA: Belknap Press of Harvard University Press.

Lieberman, D., J. Tooby, and L. Cosmides. 2003. "Does morality have a biological basis? An empirical test of the factors governing moral sentiments relating to incest." *Proceedings of the Royal Society of London B* 270: 819–826.

Lindau, S. T., L. P. Schumm, E. O. Laumann, W. Levinson, C. A. O'Muircheartaign, and L. J. Waite. 2007. A study of sexuality and health among older adults in the United States. *New England Journal of Medicine* 357: 762–774.

Lippa, R. 2005. *Gender, Nature, and Nurture,* 2nd ed. Mahwah, NJ: Lawrence Erlbaum.

Little, A. C., T. K. Saxtan, S. C. Roberts, B. C. Jones, L. M. DeBruine, J. Vukovic, D. I. Perrett, D. R. Feinberg, and T. Chenore. 2010. "Women's preferences for masculinity in male faces are highest during reproductive age range and lower around puberty and post-menopause." *Psychoneuroendocrinology* 35: 912–920.

Lloyd, E. A. 2005. *The Case of the Female Orgasm.* Cambridge, MA: Harvard University Press.

Lock, M. 1993. *Encounters with Aging: Mythologies of Menopause in Japan and North America.* Berkeley: University of California Press.

Lonsdorf, E. V., L. E. Eberly, and A. E. Pusey. 2004. "Sex differences in learning in chimpanzees." *Nature* 428: 715–716.

Lottker, P., M. Huck, E. W. Heynmann, and M. Heistermann. 2004. "Endocrine correlates of reproductive status in breeding and nonbreeding wild female moustached tamarins." *International Journal of Primatology* 25: 919–937.

Lovejoy, C. O. 2009. "Reexamining human origins in light of *Ardipithecus.*" *Science* 326: 74e1–7.

Low, B. S. 2000. *Why Sex Matters.* Princeton, NJ: Princeton University Press.

Ma, H.-M., M.-L. Du, X.-P. Luo, et al. 2009. "Onset of breast and pubic hair development and menses in urban Chinese girls." *Pediatrics* 124: e269–e277.

Maccoby, E. 1998. *The Two Sexes: Growing up Apart, Coming Together.* Cambridge: Harvard University Press.

Maestripieri, D. 2007. *Macachiavellian Intelligence: How Rhesus Macaques and Humans Have Conquered the World.* Chicago: University of Chicago Press.

Mah, K., and Y. M. Binik. 2002. "Do all orgasms feel alike? Evaluating a two-dimensional model of the orgasm experience across gender and sexual context." *Journal of Sex Research* 39: 104–113.

Maier, T. 2009. *Masters of Sex: The Life and Times of William Masters and Virginia Johnson, the Couple Who Taught America How to Love.* New York: Basic Books.

Malinowski, B. 1929. *The Sexual Lives of Savages in North-Western Melanesia: An Ethnographic Account of Courtship, Marriage and Family Life among the Natives of Trobriand Islands, British New Guinea.* London: Routledge.

Mallants, C., and K. Casteels. 2008. "Practical approach to childhood masturbation: A review." *European Journal of Pediatrics* 167: 1111–1117.

Manson, J. H., S. Perry, and A. R. Parish. 1997. "Nonconceptive sexual behavior in bonobos and capuchins." *International Journal of Primatology* 18: 767–786.

Mantegazza, P. 1935. *The Sexual Relations of Mankind.* New York: Eugenics Publishing Company.

Margulis, S. W., S. Atsalis, A. Bellem, and N. Wielebrowski. 2007. "Assessment of reproductive behavior and hormonal cycles in geriatric western Lowland gorillas." *Zoo Biology* 26: 117–139.

Marlowe, F. 1998. "The nobility hypothesis: The human breast as an honest signal of residual reproductive value." *Human Nature* 9: 263–271.

Marlowe, F. 2000a. "Paternal investment and the human mating system." *Behavioural Processes* 51: 45–61.

———. 2000b. "The patriarch hypothesis: An alternative explanation of menopause." *Human Nature* 11: 27–42.

———. 2004. "Is human ovulation concealed? Evidence from conception beliefs in a hunter-gatherer society." *Archives of Sexual Behavior* 33: 427–432.

———. 2005. "Hunter-gatherers and human evolution." *Evolutionary Anthropology* 14: 54–67.

———. 2010. *The Hadza: Hunter-Gatherers of Tanzania.* Berkeley: University of California.

Marshall, D. S. 1971. "Sexual behavior on Mangaia." In *Human Sexual Behavior*, ed. D. S. Marshall and R. C. Suggs. New York: Basic Books.

Martin, R. 2007. "The evolution of human reproduction: A primatological perspective." *Yearbook of Physical Anthropology* 50: 59–84.

Massey, S. G., A. M. Merriwether, and J. R. Garcia. 2013. "Modern prejudice and same-sex parenting: Shifting judgements in positive and negative parenting situations. *Journal of GLBT Family Studies.*

Masters, W. H., and V. E. Johnson. 1966. *Human Sexual Response.* Boston: Little, Brown and Company.

Maston, G. A., and M. Ruvolo. 2002. "Chorionic gonadotropin has a recent origin within primates and an evolutionary history of selection." *Molecular Biology and Evolution* 19: 320–335.

Matzuk, M. M., and D. J. Lamb. 2008. "The biology of infertility: Research advances and clinical challenges." *Nature Medicine* 14: 1197–1213.

McCarthy, M. M., and A. P. Arnold. 2011. "Reframing sexual differentiation of the brain." *Nature Neuroscience* 14: 677–683.

McIntyre, M. H. 2006. "The use of digit ratios as markers for perinatal androgen action." *Reprodutive Biology and Endocrinology* 2: 10.

McIntyre, M. H., and C. Hooven. 2009. "Human sex differences in social relationships: Organizational and activational effects of androgens." In *Endocrinology of Social Relationships,* ed. P. T. Ellisonand P. B. Gray, 225–245. Cambridge, MA: Harvard University Press.

McLean, C. Y., et al. 2011. "Human-specific loss of regulatory DNA and the evolution of human-specific traits." *Nature* 471: 216–219.

McNeilly, A. S. 2006. "Suckling and the control of gonadotropin secretion." In *Knobil and Neill's Physiology of Reproduction,* 3rd ed., vol. 2, ed. J. D. Neill, 2511–2552. Amsterdam: Academic Press.

Meana, M. 2010. "Elucidating women's (hetero)sexual desire: Definitional challenges and content expansion." *Journal of Sex Research* 47: 104–122.

Meizner, I. 1987. "Sonographic observation of in utero fetal 'masturbation.' " *Journal of Ultrasound* 6: 111.

Melby, M. K., M. Lock, and P. Kaufert. 2005. "Culture and symptom reporting at menopause." *Human Reproduction Update* 11: 495–512.

Meredith, S. L. 2012. "Identifying proximate and ultimate causation in the development of primate sex-typed social behavior." In *Building Babies: Primate Development in Proximate and Ultimate Perspective,* ed. K. B. H. Clancy, K. Hinde, and J. N. Rutherford. New York: Springer.

Meston, C. M. 1997. "Aging and sexuality." *Western Journal of Medicine* 167: 285–290.

Meston, C. M., and D. M. Buss. 2007. *Why Women Have Sex: Understanding Sexual Motivations from Adventure to Revenge (and Everything in Between).* New York: Henry Holt.

Miletski, H. 2005. "Is zoophilia a sexual orientation? A study." In *Bestiality and Zoophilia,* ed. A. M. Beetz and A. L. Podberscek, 82–97. Lafayette, IN: Purdue University Press.

Miller, E. A. 2011. "The reproductive ecology of breastmilk immunity in Ariaal women of northern Kenya." Paper presented at the American Association of Physical Anthropology conference.

Miller, G. 2000. *The Mating Mind: How Sexual Choice Shapes the Evolution of Human Nature.* London: Heineman.

Miller, G., J. M. Tybur, and B. D. Jordan. 2007. "Ovulatory cycle effects on tip earnings by lap dancers: Economic evidence for human estrus?" *Evolution and Human Behavior* 28: 375–381.

Miller, L. 2006. *Beauty Up: Exploring Contemporary Japanese Body Aesthetics.* Berkeley: University of California Press.

Moalem, S. 2009. *How Sex Works.* New York: Harper Perennial.

Montagu, M. F. A. 1946. *Adolescent Sterility.* Springfield, IL: C. C. Thomas.

Morbeck, M. E., A. Galloway, and A. L. Zihlman. 1997. *The Evolving Female: A Life-History Perspective.* Princeton, NJ: Princeton University Press.

Morris, C. E., C. Reiber. 2011. "Frequency, intensity, and expression of post-relationship grief." *EvoS Journal* 3: 1–11.

Muehlenbein, M., ed. 2010. *Human Evolutionary Biology.* New York: Cambridge University Press.

Muller, M. N., M. E. Emery Thompson, and R. W. Wrangham. 2006. "Male chimpanzees prefer mating with old females." *Current Biology* 16: 2234–2238.

Muller, M. N., and R. W. Wrangham, eds. 2009. *Sexual Coercion in Primates and Humans.* Cambridge, MA: Harvard University Press.

Mundy, L. 2012. *The Richer Sex: How the New Majority of Female Breadwinners is Transforming Sex, Love and Family.* New York: Simon and Schuster.

Mundy, L. 2007. *Everything Conceivable: How Assisted Reproduction Is Changing Our World.* New York: Anchor.

Murdock, G. P. 1949. *Social Structure.* New York: Free Press.

———. 1967. *The Ethnographic Atlas.* Pittsburgh: University of Pittsburgh Press.

Murdock, G. P., and D. R. White. 1969. "Standard Cross-Cultural Sample." *Ethnology* 8: 329–369.

Nelson, E., C. Rolian, L. Cashmore, and S. Shultz. 2011. "Digit ratios predict polygyny in early apes, *Ardipithecus,* Neanderthals and early modern humans but not in *Australopithecus.*" *Proceedings of the Royal Society of London B* 278: 1556–1563.

Nelson, R. J. 2011. *An Introduction to Behavioral Endocrinology,* 4th ed. Sunderland, MA: Sinauer.

Nunn, C. L. 1999. "The evolution of exaggerated sexual swellings in primates and the graded-signal hypothesis." *Animal Behaviour* 58: 229–246.

Nunn, C. L., and S. M. Altizer. 2006. *Infectious Diseases in Primates: Behavior, Ecology and Evolution.* New York: Oxford University Press.

O'Connell, J. F., and J. Allen. 2007. "Pre-LGM Sahul (Pleistocene Australia–New Guinea) and the archaeology of early modern humans." In *Rethinking the Human Revolution,* ed. P. Mellars, K. Boyle, O. Bar-Yosef, and C. Stringer, 395–410. Cambridge: McDonald Institute for Archaeological Research.

Odling-Smee, J. F. 2003. *Niche Construction: The Neglected Process in Evolution.* Princeton, NJ: Princeton University Press.

Ogas, O., and S. Gaddam. 2011. *A Billion Wicked Thoughts.* New York: Dutton.

O'Hara, K., and J. O'Hara. 1999. "The effect of male circumcision on the sexual enjoyment of the female partner." *BJU International* 83 Suppl 1: 79–84.

Palombit, R. A., R. M. Seyfarth, and D. L. Cheney. 1997. "The adaptive value of 'friendships' to female baboons: Experimental and observational evidence." *Animal Behavior* 54: 599–614.

Parish, W. L., E. O. Laumann, and S. A. Mojola. 2007. "Sexual behavior in China: Trends and comparisons." *Population and Development Review* 33: 729–756.

Parzeller, M., C. Raschka, and H. Bratzke. 1999. "Sudden cardiovascular death during sexual intercourse: Results of a legal medicine autopsy study." *Zeitschrift für Kardiologie* 88: 44–48.

Pazol, K. 2003. "Mating among the Kakamega Forest blue monkeys *(Cercopithecus mitis)*: Does female sexual behavior function to manipulate paternity assessment?" *Behavior* 140: 473–499.

Pedersen, W. C., A. Putcha-Bhagavatula, and L. C. Miller. 2011. "Are men and women really that different? Examining some of sexual strategies theory (SST)'s key assumptions about sex-distinct mating mechanisms." *Sex Roles* 64: 629–643.

Perry, S., and J. H. Manson. 2008. *Manipulative Monkeys: The Capuchins of Lomas Barbudal.* Cambridge, MA: Harvard University Press.

Persky, H., H. I. Lief, D. Strauss, W. R. Miller, and C. P. O'Brien. 1978. "Plasma testosterone level and sexual behavior of couples." *Archives of Sexual Behavior* 7: 157–173.

Petraglia, F., P. Florio, and M. Torricelli. 2006. Placental endocrine function. In *Knobil and Neill's Physiology of Reproduction,* 3rd ed., vol. 2, ed. J. D. Neill, 2847–2898. Amsterdam: Academic Press.

Pfaff, D. W. 2010. *Man and Woman: An Inside Story.* New York: Oxford University Press.

Pinker, S. 2011. *The Better Angels of Our Nature: Why Violence Has Declined.* New York: Viking.

Piperata, B. A., and D. Guatelli-Steinberg. 2011. "Offsetting the costs of reproduction: The role of social support in human evolution." Paper presented at American Association of Physical Anthropology conference.

Pisani, E. 2008. *The Wisdom of Whores.* London: Granta.

Place, S. S., P. M. Todd, L. Penke, and J. B. Asendorpf. 2010. "Humans show mate copying after observing real mate choices." *Evolution and Human Behavior* 31: 320–325.

Platek, S. M., R. L. Burch, and G. G. Gallup. 2001. "The reproductive priming effect." *Social Behavior and Personality* 29: 245–248.

Platek, S. M., and D. Singh. 2010. "Optimal waist-to-hip ratios in women activate neural reward centers in men." *PLoS One* 5: e9042.

Plavcan, J. M. 2012. "Sexual size dimorphism, canine dimorphism, and male-male competition in primates: Where do humans fit in?" *Human Nature* 23: 45–67.

Poiani, A. 2010. *Animal Homosexuality.* New York: Cambridge University Press.

Popenoe, R. 2004. *Feeding Desire: Fatness, Beauty, and Sexuality among a Saharan People.* New York: Routledge.

Potts, M., and R. V. Short. 1999. *Ever since Adam and Eve: The Evolution of Human Sexuality.* New York: Cambridge University Press.

Power, M. L., and J. Schulkin. 2009. *The Evolution of Obesity.* Baltimore: Johns Hopkins University Press.

Priest, R. 2001. "Missionary positions: Christian, modernist, postmodernist." *Current Anthropology* 42: 29–68.

Proos, L. A., Y. Hofvander, and T. Tunevo. 1991. "Menarcheal and growth pattern in Indian girls adopted in Sweden: 1. Menarcheal age." *Acta Paediatrica Scandinavia* 80: 852–858.

Pusey, A., J. Williams, and J. Goodall. 1997. "The influence of dominance rank on the reproductive success of female chimpanzees." *Science* 277: 828–831.

Pusey, A., and M. Wolf. 1996. "Inbreeding avoidance in animals." *Trends in Ecology and Evolution* 11: 201–206.

Puts, D. A. 2010. "Beauty and the beast: Mechanisms of sexual selection in humans." *Evolution and Human Behavior* 31: 157–175.

Puts, D. A., K. Dawood, and L. L. M. Welling. 2012. "Why women have orgasms: an evolutionary analysis." *Archives of Sexual Behavior* 41: 1127–1143.

Puts, D. A., S. J. C. Gaulin, R. J. Sporter, and C. H. McBurney. 2004. "Sex hormones and finger length: What does 2D:4D indicate?" *Evolution and Human Behavior* 25: 182–199.

Puts, D. A., M. A. McDaniel, C. L. Jordan, and S. M. Breedlove. 2008. "Spatial ability and prenatal androgens: Meta-analyses of congenital adrenal hyperplasia and digit ratio (2D: 4D) studies." *Archives of Sexual Behavior* 37: 100–111.

Quinlan, R., and M. Quinlan. 2008. "Human lactation, pair-bonds and alloparents: A cross-cultural analysis." *Human Nature* 19: 87–102.

Quinn, A. E., A. Georges, S. D. Sarre, F. Guarino, T. Ezaz, and J. A. M. Graves. 2007. "Temperature reversal implies sex gene dosage in a reptile." *Science* 316: 411.

Ralls, K. 1976. "Mammals in which females are larger than males." *Quarterly Review of Biology* 51: 245–276.

Ravel, J., P. Gajer, Z. Abdo, G. M. Schneider, S. S. K. Koenig, S. L. McCulle, S. Karlebach, R. Gorle, G. Russell, C. O. Tacket, R. M. Brotman, C. C. Davis, K. Ault, L. Peralta, and L. J. Forney. 2011. "Vaginal microbiome of reproductive-age women." *Proceedings of the National Academy of Sciences* 108: 4680–4687.

Regalski, J. M., and S. J. C. Gaulin. 1993. "Whom are Mexican infants said to resemble? Monitoring and fostering paternal confidence in the Yucatan." *Ethology and Sociobiology* 14: 97–113.

Regan, P. C. 1999. "Hormonal correlates and causes of sexual desire: A review." *Canadian Journal of Human Sexuality* 8: 1–16.

Reiber, C., and J. R. Garcia. 2010. "Hooking Up: Gender differences, evolution, and pluralistic ignorance." *Evolutionary Psychology* 8: 390–404.

Reich, D., R. E. Green, M. Kirchner, J. Krause et al. 2010. "Genetic history of an archaic hominin group from Denisova Cave in Siberia." *Nature* 468: 1053–1060.

Reiches, M. W., P. T. Ellison, S. F. Lipson, K. C. Sharrock, E. Gardiner, and L. G. Duncan. 2009. "Pooled energy budgets and human life history." *American Journal of Human Biology* 21: 421–429.

Rhen, T., and A. Schroeder. 2010. "Molecular mechanisms of sex determination in reptiles." *Sexual Development* 4: 16–28.

Rice, W. R. 2000. "Dangerous liaisons." *Proceedings of the National Academy of Sciences* 97: 12953–12955.

———. 2002. "Experimental tests of the adaptive significance of sexual recombination." *Nature Reviews Genetics* 3: 241–251.

Richerson, P. J., and R. Boyd. 2005. *Not by Genes Alone.* Chicago: University of Chicago Press.

Ricklefs, R. E., and C. E. Finch. 1995. *Aging: A Natural History.* New York: Scientific American.

Riddle, J. M. 1992. *Contraception and Abortion from the Ancient World to the Renaissance.* Cambridge: Harvard University Press.

———. 1997. *Eve's Herbs: A History of Contraception and Abortion in the West.* Cambridge, MA: Harvard University Press.

Rizvi, K., J. P. Hampson, and J. N. Harvey. 2002. "Do lipid-lowering drugs cause erectile dysfunction? A systematic review." *Family Practice* 19: 95–98.

Roach, M. 2008. *Bonk: The Curious Coupling of Science and Sex.* New York: W. W. Norton.

———. 2010. *Packing for Mars: The Curious Science of Life in the Void.* New York: W. W. Norton.

Robertson, S. A. 2005. "Seminal plasma and male factor signaling in the female reproductive tract." *Cell Tissue Research* 322: 43–52.

Robillard, P.-Y., G. Dekker, G. Chaouat, J. Chaline, and T. C. Hulsey. 2008. "Possible role of eclampsia/preeclampsia in evolution of human reproduction." In *Evolutionary Medicine and Health: New Perspectives,* ed. W. R. Trevathan, E. O. Smith, and J. J. McKenna, 216–225. New York: Oxford University Press.

Robson, S. L., and B. Wood. 2008. "Hominin life history: Reconstruction and evolution." *Journal of Anatomy* 212: 455–458.

Rodrigues, S. M., L. R. Saslow, N. Garcia, O. P. John, and D. Keltner. 2009. "Oxytocin receptor genetic variation relates to empathy and stress reactivity in humans." *Proceedings of the National Academy of Sciences* 106: 21437–21441.

Rogoff, B. 2003. *The Cultural Nature of Human Development.* New York: Oxford University Press.

Rossouw, J. E., G. L. Anderson, R. L. Prentice, A. Z. Lacroix et al. 2002. "Risks and benefits of estrogen plus progestin in healthy postmenopausal women: Principal results from the Women's Health Initiative randomized controlled trial." *Journal of the American Medical Association* 288: 321–333.

Rosvall, K. A. 2011. "Intrasexual competition in females: Evidence for sexual selection?" *Behavioral Ecology* 22: 1131–1140.

Röttger-Rössler, B. "Voiced intimacies: Verbalized experiences of love and sexuality in an Indonesian society." In *Intimacies: Love and Sex Across Cultures,* ed. W. R. Jankowiak. New York: Columbia University Press.

Roughgarden, J. 2004. *Evolution's Rainbow: Diversity, Gender, and Sexuality in Nature and People.* Berkeley: University of California Press.

Rust, J., S. Golmbok, M. Hines, K. Johnston, and J. Golding. 2000. "The role of brothers and sisters in the gender development of preschool children." *Journal of Experimental Child Psychology* 77: 292–303.

Ryan, C., and C. Jetha. 2010. *Sex at Dawn: How We Mate, How We Stray, and What It Means for Modern Relationships.* New York: Harper Perennial.

Saad, G. 2011. *The Consuming Instinct: What Juicy Burgers, Ferraris, Pornography, and Gift Giving Reveal about Human Nature.* New York: Prometheus.

Saggar, A. K., and A. H. Bittles. 2008. "Consanguinity and child health." *Paediatrics and Child Health* 18: 244–249.

Scelza, B. A. 2011. "Female choice and extra-pair paternity in a traditional human population." *Biology Letters* 7: 889–891.

Schaefer, K., P. Mitteroecker, P. Gunz, M. Bernhard, and F. L. Bookstein. 2004. "Craniofacial sexual dimorphism patterns and allometry among extant hominids." *Annals of Anatomy* 186: 471–478.

Schapera, I. 1941. *Married Life in an African Tribe.* New York: Sheridan House.

Schiefenhovel, W., and A. Grabolle. 2005. "The role of maternal grandmothers in Trobriand adoption." In *Grandmotherhood,* ed. E. Voland, A. Chasiotis, and W. Schiefenhovel, 177–193. New Brunswick, NJ: Rutgers University Press.

Schlegel, A., and H. Barry III. 1991. *Adolescence: An Anthropological Inquiry.* New York: Free Press.

Schmitt, D. P., 2005. "Sociosexuality from Argentina to Zimbabwe: A 48-nation study of sex, culture, and strategies of human mating." *Behavioral and Brain Sciences* 28: 247–311.

Schmitt, D. P., and D. M. Buss. 2001. "Humane mate poaching: Tactics and temptations for infiltrating existing mateships." *Journal of Personality and Social Psychology* 80: 894–917.

Schmitt, D. P., and International Sexuality Description Project. 2003. "Universal sex differences in the desire for sexual variety: Tests from 52 nations, 6 continents, and 13 islands." *Journal of Personality and Social Psychology* 85: 85–104.

———. 2004. "Patterns and universals of mate poaching across 53 nations: The effects of sex, culture, and personality on romantically attracting another person's partner." *Journal of Personality and Social Psychology* 86: 560–584.

Schoentjes, E., D. Deboutte, and W. Friedrich. 1999. "Child sexual behavior inventory: A Dutch-speaking normative sample." *Pediatrics* 104: 885–893.

Schulz, K. M., H. A. Molenda-Figueira, and C. L. Sisk. 2009. "Back to the future: The organizational-activational hypothesis adapted to puberty and adolescence." *Hormones and Behavior* 55: 597–604.

Sear, R., and R. Mace. 2008. "Who keeps children alive? A review of the effects of kin and child survival." *Evolution and Human Behavior* 29: 1–18.

Segal, S. J. 2003. *Under the Banyan Tree.* New York: Oxford University Press.

Sell, A., L. Cosmides, J. Tooby, D. Sznycer, C. von Rueden, and M. Gurven. 2009. "Human adaptations for the visual assessment of strength and fighting ability from the body and face." *Proceedings of the Royal Society B* 276: 575–584.

Sellen, D. W., and R. Mace. 1997. "Fertility and mode of subsistence: A phylogenetic analysis." *Current Anthropology* 38: 878–889.

Servin, A., G. Bohlin, and L. Berlin. 1999. "Sex differences in 1-, 3-, and 5-year olds' toy-choice in a structured play-session." *Scandinavian Journal of Psychology* 40: 43–48.

Seymour, F. I., C. Duffy, and C. Koerner. 1935. "A case of authenticated fertility in a man of 94." *Journal of the American Medical Association* 105: 1423–1424.

Shepher, J. 1983. *Incest: A Biosocial View.* New York: Academic Press.

Shih, C. 2010. *Quest for Harmony: The Moso Traditions of Sexual Union and Family Life.* Stanford, CA: Stanford University Press.

Short, R. V. 1976. "The evolution of human reproduction." *Proceedings of the Royal Society of London B* 195: 3–24.

Sievert, L. L. 2006. *Menopause: A Biocultural Perspective.* New Brunswick, NJ: Rutgers University Press.

Simmons, Z. L., and J. R. Roney. 2011. "Variation in CAG repeat length of the androgen receptor gene predicts variables associated with intrasexual competitiveness in human males." *Hormones and Behavior* 60: 306–312.

Singapore Statistics. 2011. http://www.singstat.gov.sg, accessed October 6, 2011.

Singh, D. 1993. "Adaptive significance of female physical attractiveness: Role of waist-to-hip ratio." *Journal of Personality and Social Psychology* 65: 293–307.

Singh, D., B. J. Dixson, T. S. Jessop, B. Morgan, and A. F. Dixson. 2010. "Cross-cultural consensus for waist-to-hip ratio and women's attractiveness. *Evolution and Human Behavior* 31: 176–181.

Singh, N. K. 1997. *Divine Prostitution.* New Delhi: A. P. H. Publishing.

Smith, R. L., ed. 1984. *Sperm Competition and the Evolution of Animal Mating Systems.* New York: Academic Press.

Snowdon, C. T., and T. E. Ziegler. 2007. "Growing up cooperatively: Family processes and infant care in marmosets and tamarins." *Journal of Developmental Processes* 2: 40–66.

Soler, C., M. Núñez, R. Gutiérrez, J. Núñez, P. Medina, M. Sancho, J. Álvarez, and A. Núñez. 2003. "Facial attractiveness in men provides clues to semen quality." *Evolution and Human Behavior* 24: 199–207.

Solway, J. S. 1992. "Middle-aged women in Bakgalagadi society (Botswana)." In *In Her Prime: New Views of Middle-Aged Women,* 2nd ed., ed. V. Kerns and J. K. Brown, 49–60. Urbana: University of Illinois Press.

Spear, L. P. 2000. "The adolescent brain and age-related behavioral manifestations." *Neuroscience and Biobehavioral Reviews* 24: 417–463.

———. 2010. *The Behavioral Neuroscience of Adolescence.* New York: W. W. Norton.

Stanford, B. D., S. A. Synder, R. A. Trenholm, J. C. Holady, and B. J. Vanderford. 2010. "Estrogenic activity of U.S. drinking waters: A relative exposure comparison. *Journal of American Water Works Association* 102: 11.

Stanford, C. 1998. "The social behavior of chimpanzees and bonobos: Empirical evidence and shifting assumptions." *Current Anthropology* 39: 399–420.

Stanford, C., J. S. Allen, and S. C. Anton. 2011. *Biological Anthropology,* 3rd ed. New York: Prentice Hall.

Stearns, S. C. 1992. *The Evolution of Life Histories.* New York: Oxford University Press.

Sternberg, R. J. 1988. *The Triangle of Love: Intimacy, Passion, Commitment.* New York: Basic Books.

Stolting, K. N., and A. B. Wilson. 2007. "Male pregnancy in seahorses and pipefish: Beyond the mammalian model." *Bioessays* 29: 884–896.

Strassmann, B. I. 1992. "The function of menstrual taboos among the Dogon: Defense against cuckoldry?" *Human Nature* 3: 89–131.

Strassmann, B. I. 1996. "Energy economy in the evolution of menopause." *Evolutionary Anthropology* 5: 157–164.

Strassmann, B. I. 1997a. "The biology of menstruation in *Homo sapiens*: Total lifetime menses, fecundity, and nonsynchrony in a natural-fertility population." *Current Anthropology* 38: 123–129.

———. 1997b. "Polygyny as a risk factor for child mortality among the Dogon." *Current Anthropology* 38: 688–695.

Stribley, J. M., and C. S. Carter. 1999. "Developmental exposure to vasopressin increases aggression in adult prairie voles." *Proceedings of the National Academy of Sciences* 96: 12601–12604.

Strier, K. B. 2011. *Primate Behavioral Ecology,* 4th ed. Upper Saddle River, NJ: Prentice Hall.

Stumpf, R. 2011. "Chimpanzees and bonobos." In *Primates in Perspective,* 2nd ed., ed. C. J. Campbell, A. Fuentes, K. C. MacKinnon, S. K. Bearder, and R. Stumpf. New York: Oxford University Press.

Stumpf, R. M., and C. Boesch. 2005. "Does promiscuous mating preclude female choice? Female sexual strategies in chimpanzees *(Pan troglodytes verus)* of the

Tai National Park, Cote d'Ivore." *Behavioral Ecology and Sociobiology* 57: 511–524.

Suarez, S. S., and A. A. Pacey. 2006. "Sperm transport in the female reproductive tract." *Human Reproduction Update* 12: 23–37.

Suggs, D., and A. W. Miracle, eds. 1993. *Culture and Human Sexuality: A Reader.* Pacific Grove, CA: Brooks/Cole.

Suggs, R. 1966. *Marquesan Sexual Behavior.* Harcourt, Brace and World.

Swami, V., and A. Furnham. 2008. *The Psychology of Human Attraction.* New York: Routledge.

Swami, V., J. Jones, D. Einon, and A. Furnham. 2009. "Men's preferences for women's profile waist-to-hip ratio, breast size, and ethnic group in Britain and South Africa." *British Journal of Psychology* 100: 313–325.

Symons, D. 1979. *The Evolution of Human Sexuality.* New York: Oxford University Press.

Tanner, J. 1989. *Foetus into Man.* Cambridge, MA: Harvard University Press.

Tattersall, I. 2012. *Masters of the Planet: The Search for Our Human Origins.* New York: Macmillan.

Taylor, L. R. 2005. "Dangerous trade-offs: The behavioral ecology of child labor and prostitution in rural northern Thailand." *Current Anthropology* 46: 411–431.

Thornhill, R., and S. W. Gangestad. 1999. "Facial attractiveness." *Trends in Cognitive Science* 3: 452–460.

Townsend, J. M. 1993. "Sexuality and partner selection: Sex differences among college students. *Ethology and Sociobiology* 14 305–329.

———. 1998. *What Women Want, What Men Want.* New York: Oxford University Press.

Trevathan, W. 2010. *Ancient Bodies, Modern Lives: How Evolution Has Shaped Women's Health.* New York: Oxford University Press.

Trivers, R. 1972. "Parental investment and sexual selection." In *Sexual Selection and the Descent of Man,* ed. B. C. Campbell, 136–179. Chicago: Aldine.

Trivers, R. L. 1974. "Parent-offspring conflict." *American Zoologist* 14: 249–264.

Tsapelas, I., H. E. Fisher, and A. Aron. 2010. "Infidelity: When, where, why." In *The Dark Side of Close Relationships II*, eds. W. R. Cupach and B.H. Spitzberg (eds.), 175–196. New York: Routledge.

Tuljapurkar, S. D., C. O. Puleston, and M. D. Gurven. 2007. "Why men matter: Mating patterns drive evolution of human lifespan." *PLoS ONE* 2(8): e785.

Udry, J. R., and R. L. Cliquet. 1982. "A cross-cultural examination of the relationship between ages at menarche, marriage, and first birth." *Demography* 19: 53–63.

United Nations. 2008. *World Fertility Patterns* 2007. www.un.org.

Vallender, E. J., and B. T. Lahn. 2004. "How mammalian sex chromosomes acquired their peculiar gene content." *Bioessays* 26: 159–169.

Van Anders, S. M., L. Brotto, J. Farrell, and M. Yule. 2009. "Associations between physiological and subjective sexual responses, sexual desire, and salivary steroid hormones in healthy premenopausal women." *Journal of Sexual Medicine* 6: 739–751.

Van Anders, S. M., and P. B. Gray. 2007. "Hormones and human partnering." *Annual Review of Sex Research* 18: 60–93.

Van Anders, S. M., L. D. Hamilton, and N. V. Watson. 2007. "Multiple partners are associated with higher testosterone in North American men and women." *Hormones and Behavior* 51: 454–459.

Vasek, F. C. 1980. "Creosote bush: Long-lived clones in the Mojave Desert." *American Journal of Botany* 67: 246–255.

Vasey, P. L., and D. P. VanderLaan. 2007. "Birth order and male androphilia in Samoan *fa'afafine*." *Proceedings of the Royal Society of London B* 274: 1437–1442.

Vaupel, J. W. 2010. "Biodemography of human ageing." *Nature* 464: 536–542.

Vilain, E., and E. R. B. McCabe. 1998. "Mammalian sex determination: From gonads to brain." *Molecular Genetics and Metabolism* 65: 74–84.

Vitzthum, V. 2009. "The ecology and evolutionary endocrinology of reproduction in the human female." *Yearbook of Physical Anthropology* 49: 95–136.

Von Sydow, K. 1999. "Sexuality during pregnancy and after childbirth: A metacontent analysis of 59 studies." *Journal of Psychosomatic Research* 47: 27–49.

Voss, B. L. 2008. "Sexuality studies in archaeology." *Annual Review of Anthropology* 37: 317–336.

Walker, R. S., M. V. Flinn, and K. R. Hill. 2010. "Evolutionary history of partible paternity in lowland South America." *Proceedings of the National Academy of Sciences* 107: 19195–19200.

Walker, R. S., K. R. Hill, M. V. Flinn, and R. M. Ellsworth. 2011. "Evolutionary history of hunter-gatherer marriage practices." *PLoS ONE* 6: e19066.

Wallace, A. R. 1890. *The Malay Archipelago*. Singapore: Periplus.

Wallen, K. 2005. "Hormonal influences on sexually differentiated behavior in nonhuman primates." *Frontiers in Neuroendocrinology* 26: 7–26.

Wallen, K., and J. M. Hassett. 2009. "Sexual differentiation of behavior in monkeys: Role of prenatal hormones." *Journal of Neuroendocrinology* 21: 421–426.

Wallen, K., and E. A. Lloyd. 2011. "Female sexual arousal: Genital anatomy and orgasm in intercourse." *Hormones and Behavior* 59: 780–792.

Walter, A., and S. Buyske. 2003. "The Westermarck Effect and early childhood co-socialization: Sex differences in inbreeding avoidance." *British Journal of Developmental Psychology* 21: 353–365.

Walum, H., L. Westberg, S. Henningsson, J. M. Neiderhiser, D. Reiss, W. Igl, J. M. Ganiban, E. L. Spotts, N. L. Pedersen, E. Eriksson, and P. Lichtenstein. 2008. "Genetic variation in the vasopressin receptor 1a gene (AVPR1A) associates with pair-bonding behavior in humans." *Proceedings of the National Academy of Sciences* 105: 14153–14156.

Wardlow, H. 2008. "'She liked it best when she was on top': Intimacies and estrangements in Huli men's marital and extramarital relationships." In *Intimacies: Love and Sex across Cultures,* ed. W. Jankowiak, 194–223. New York: Columbia University Press.

Watts, D. P., and A. E. Pusey. 1993. "Behavior of juvenile and adolescent great apes." In *Juvenile Primates: Life History, Development, and Behavior,* ed. M. E. Pereira and L. Fairbanks, 148–167. New York: Oxford University Press.

Weatherford, J. 2004. *Genghis Khan and the Making of the Modern World.* New York: Crown.

Weisfeld, G. E. 1999. *Evolutionary Principles of Human Adolescence.* New York: Basic Books.

Weismantel, M. 2004. "Moche sex pots: Reproduction and temporality in ancient South America." *American Anthropologist* 106: 495–505.

Weitzer, R., ed. 2010. *Sex for Sale: Prostitution, Pornography, and the Sex Industry,* 2nd ed. New York: Routledge.

Wellings, K., M. Collumbien, E. Slaymaker, S. Singh, Z. Hodges, D. Patel, and N. Bajos. 2006. "Sexual behavior in context: A global perspective." *Lancet* 368: 1706–1728.

West-Eberhard, M. J. 2003. *Developmental Plasticity and Evolution.* Oxford: Oxford University Press.

Westermarck, E. 1921. *The History of Human Marriage.* London: Macmillan.

White, T. D., B. Asfaw, Y. Benene, Y. Haile-Selassie, C. O. Lovejoy, G. Suwa, G. Woldegabriel. 2009. "*Ardipithecus ramidus* and the paleobiology of early hominids." *Science* 326: 75–86.

Whiting, B. B., and C. P. Edwards. 1988. *Children of Different Worlds: The Formation of Social Behavior.* Cambridge, MA: Harvard University Press.

Whiting, J. W. M., and B. B. Whiting. 1975. Aloofness and intimacy of husbands and wives: A cross-cultural study. *Ethos* 3: 183–207.

Wich, S. A., R. W. Shumaker, L. Perkins, and H. De Vries. 2009. "Captive and wild orangutan (*Pongo* sp.) survivorship: A comparison and the influence of management." *American Journal of Primatology* 71: 680–686.

Wiegel, M., C. Meston, and R. Rosen. 2005. "The Female Sexual Function Index (FSFI): Cross-validation and development of clinical cutoff scores." *Journal of Sex and Marital Therapy* 31: 1–20.

Wiessner, P., and N. Pupu. 2012. "Toward peace: Foreign arms and indigenous institutions in a Papua New Guinea society." *Science* 337: 1651–1654.

Wilcox, A. J., D. D. Baird, D. B. Dunson, D. R. McConnaughey, J. S. Kesner, C. R. Weinberg. 2004. "On the frequency of intercourse around ovulation: Evidence for biological influences." *Human Reproduction* 19: 1539–1543.

Williams, G. C. 1957. "Pleiotropy, natural selection, and the evolution of senescence." *Evolution* 11: 398–411.

Wilson, C. G. 2008. "Male genital mutilation: An adaptation to sexual conflict." *Evolution and Human Behavior* 29: 149–164.

Wilson, E. O. 1998. *Consilience: The Unity of Knowledge.* New York: Random House.

Wilson, J. D., and C. Roehrborn. 1999. "Long-term consequences of castration in men: Lessons from the Skoptzy and the eunuchs of the Chinese and Ottoman courts." *Journal of Clinical Endocrinology and Metabolism* 84: 4324–4331.

Wilson, M., and M. Daly. 1985. "Competitiveness, risk-taking and violence: The young male syndrome." *Ethology and Sociobiology* 6: 59–73.

———. 1992. "The man who mistook his wife for a chattel." In *The Adapted Mind,* ed. J. H. Barkow, L. Cosmides, and J. Tooby, pp. 289–322. New York: Oxford University Press.

———. 1993. "An evolutionary psychological perspective on male sexual proprietariness and violence against wives." *Violence and Victims* 8: 271–294.

Wilson, R. A. 1966. *Feminine Forever.* New York: M. Evans.

Wingfield, J. C., R. E. Hegner, A. M. Dufty Jr., and G. F. Ball. 1990. "The 'challenge hypothesis': Theoretical implications for patterns of testosterone secretion, mating systems, and breeding strategies." *American Naturalist* 136: 829–846.

Winking, J., H. Kaplan, M. Gurven, and S. Rucas. 2007. "Why do men marry and why do they stray?" *Proceedings of the Royal Society B: Biological Sciences* 274: 1643–1649.

Winn, R. L., and N. Newton. 1982. "Sexuality in aging: A study of 106 cultures." *Archives of Sexual Behavior* 11: 283–298.

Wolf, A. P., and W. H. Durham, eds. 2004. *Inbreeding, Incest and the Incest Taboo.* Stanford, CA: Stanford University Press.

Wolk, L., N. B. Abdelli-Beruh, and D. Slavin. 2011. "Habitual use of vocal fry in young adult female speakers." *Journal of Voice* 36: 285–289.

Wood, B., and N. Lonergan. 2008. "The hominin fossil record: Taxa, grades and clades." *Journal of Anatomy* 212: 354–376.

Wood, J. W. 1994. *Dynamics of Human Reproduction.* New York: Aldine de Gruyter.

Woods, V., and B. Hare. 2011. "Bonobo but not chimpanzee infants use sociosexual contact with peers." *Primates* 52: 111–116.

World Health Organization. 2012. http://www.who.int/topics/hiv_aids/en/.

Wrangham, R. W. 1979. "On the evolution of ape social systems." *Social Science Information* 18: 335–368.

Wrangham, R. W., and D. Peterson D. 1996. *Demonic Males.* New York: Mariner.

Wynne-Edwards, K. E. 1987. "Evidence for obligate monogamy in the Djungarian hamster, *Phodopus campbelli*: Pup survival under different parenting conditions." *Behavioral Ecology and Sociobiology* 20: 427–437.

Yang, C. J., P. Gray, and H. G. Pope, Jr. 2005. "Male body image in Taiwan versus the West: Yanggang zhiqi meets the Adonis complex." *American Journal of Psychiatry* 162: 263–269.

Yapici, N., Y. Kim, C. Ribeiro, B. J. Dickson. 2008. "A receptor that mediates the post-mating switch in Drosophila reproductive behavior." *Nature* 451: 33–37.

Yovsi, R. D., and H. Keller. 2003. "Breastfeeding: An adaptive process." *Ethos* 31: 147–171.

Zahavi, A., and A. Zahavi. 1997. *The Handicap Principle.* New York: Oxford University Press.

Zaviacic, M., and B. Whipple. 1993. "Update on the female prostate and the phenomenon of female ejaculation." *Journal of Sex Research* 30: 148–151.

Zerjal, T., Y. Xue, G. Bertorelle, R. S. Wells et al. 2003. "The genetic legacy of the Mongols." *American Journal of Human Genetics* 72: 717–721.

Zhang, N., W. L. Parish, Y. Huang, S. Pan. 2012. "Sexual infidelity in China: Prevalence and gender-specific correlates." *Archives of Sexual Behavior* 41: 861–873.

Zheng, Z., and M. J. Cohn. 2011. "Developmental basis of sexually dimorphic digit ratios." *Proceedings of the National Academy of Sciences* 108: 16289–16294.

Ziegler, T. E. 2007. "Female sexual motivation during non-fertile periods: A primate phenomenon." *Hormones and Behavior* 51: 1–2.

Zuk, M. 2003. *Sexual Selections: What We Can and Can't Learn about Sex from Animals.* Berkeley: University of California Press.

———. 2007. *Riddled with Life: Friendly Worms, Ladybug Sex, and the Parasites Tthat Make Us Human.* New York: Houghton Mifflin Harcourt.

Index